Sensors for Next-Generation Electronic Systems and Technologies

The text covers fiber optic sensors for biosensing and photo-detection, graphene and CNT-based sensors for glucose, cholesterol, and dopamine detection, and implantable sensors for detecting physiological, bio-electrical, biochemical, and metabolic changes in a comprehensive manner. It further presents a chapter on sensors for military and aerospace applications. It will be useful for senior undergraduate, graduate students, and academic researchers in the fields of electrical engineering, electronics, and communication engineering.

The book

- Discusses implantable sensors for detecting physiological, bio-electrical, biochemical, and metabolic changes.
- Covers applications of sensors in diverse fields, including healthcare, industrial flow, consumer electronics, and military.
- Includes experimental studies, such as the detection of biomolecules using SPR sensors and electrochemical sensors for biomolecule detection.
- Presents artificial neural networks (ANN) based industrial flow sensor modeling.
- Highlights case studies on surface plasmon resonance sensors, MEMS-based fluidic sensors, and MEMS-based electrochemical gas sensors.

The text presents case studies on surface plasmon resonance sensors, MEMS-based fluidic sensors, and MEMS-based electrochemical gas sensors in a single volume. The text will be useful for senior undergraduate, graduate students, and academic researchers in the fields of electrical engineering, electronics, and communication engineering

Sensors for Next-Generation Electronic Systems and Technologies

Edited by

P. Uma Sathyakam

K. Venkata Lakshmi Narayana

CRC Press
Taylor & Francis Group
Boca Raton London New York

CRC Press is an imprint of the
Taylor & Francis Group, an **informa** business

Front cover image: Simakova Mariia/Shutterstock

First edition published 2023
by CRC Press
6000 Broken Sound Parkway NW, Suite 300, Boca Raton, FL 33487-2742

and by CRC Press
4 Park Square, Milton Park, Abingdon, Oxon, OX14 4RN

CRC Press is an imprint of Taylor & Francis Group, LLC

Library of Congress Cataloging-in-Publication Data
Names: Lakshmi Narayana, K. Venkata, editor. | Sathyakam, P. Uma, editor.
Title: Sensors for next-generation electronic systems and technologies /
edited by K. Venkata Lakshmi Narayana and P. Uma Sathyakam.
Description: First edition. | Boca Raton : CRC Press, [2023] | Includes
bibliographical references and index.
Identifiers: LCCN 2022053541 (print) | LCCN 2022053542 (ebook) | ISBN
9781032265155 (hbk) | ISBN 9781032265162 (pbk) | ISBN 9781003288633 (ebk)
Subjects: LCSH: Detectors.
Classification: LCC TK7872.D48 S453 2023 (print) | LCC TK7872.D48 (ebook)
| DDC 621.3815/36--dc23/eng/20230111
LC record available at https://lccn.loc.gov/2022053541
LC ebook record available at https://lccn.loc.gov/2022053542

ISBN: 978-1-032-26515-5 (hbk)
ISBN: 978-1-032-26516-2 (pbk)
ISBN: 978-1-003-28863-3 (ebk)

DOI: 10.1201/9781003288633

Typeset in Sabon
by SPi Technologies India Pvt Ltd (Straive)

Contents

Acknowledgements

The editors would like to thank all the authors who have contributed their chapters in a timely manner by committing their invaluable time and tireless efforts. Further, we would like to thank all the authors of research papers, books and monographs whose data has become the basis for writing this book. We assure that almost all the information presented here is cited properly and all the authors are given due recognition.

Next, we would like to thank our research scholars who have helped in arranging and retrieving data from the chapters at the final stages of editing.

We are thankful to our Head of the Department and the Dean of School of Electrical Engineering, Vellore Institute of Technology for giving time for our editing work in the busy academic schedule.

We would like to thank the CRC press editor Mr. Gauravjeet Singh Reen and Ms. Isha Ahuja for communicating and assisting us in a timely manner and providing inputs at different stages in the editing and publication of this book.

We are also indebted to our family members from whom we have taken a lot of time in order to complete this book successfully.

Preface

The main aspect of this book is to introduce new technologies in the field of sensors leveraging on the latest advancements in electronic materials and devices. This book definitely will be interesting to the readers and researchers who are fascinated to explore the future of sensors in various domains spanning from healthcare to space applications. In this modern world, sensors have impacted many areas of our daily life in the form of smart technology implementations. Even though there have been sensor applications explored in different domains, there is a need for a single platform, where the basic elements of modern day sensors with current technological advancement integrated with sensor development in different fields is covered. This markedly benefits beginners as well as domain experts in acquiring an overview of the advancement in sensor development and applications.

The book covers the various sensor technologies related to niche areas, like wearable technologies, healthcare monitoring, military and aerospace applications, nanomaterials like CNTs and Graphene, intelligent sensors, sensor arrays, sensors for flexible electronics, on-chip sensors for SoC, SoP, sensors for Lab-on-a-chip (LoC), magnetic sensors, smart sensors and, thermal sensors Hence, this book provides an interdisciplinary platform for researchers and promotes the research development in the above mentioned areas. Most of the reported findings in the areas of sensors are the motivation behind the IoT and telecommunication industry and special emphasis on signal processing and control automation that are helping to change the nature of human life. We are sure that the novel findings in the research chapters contained in this book will be found exciting and may inspire the research fraternity to work in the field of Sensors and Electronics. We foresee that this book is an initiative to integrate different domains of sensors into a single platform that can bring a great contribution to academic as well as industrial research.

About the editors

Dr. P. Uma Sathyakam is working as Associate Professor in the School of Electrical Engineering, Vellore Institute of Technology, Vellore, where he earned his MS (by Research) in Nanoelectronics and VLSI in 2011 and his PhD in Nanoelectronics in 2018. Prior to this, he earned his BSc. Electronics from University of Kerala, Thiruvananthapuram in 2005 and MSc. Applied Electronics from Bharathiar University, Coimbatore with distinction and Gold medal in 2007.

He served as a Research Fellow in a MHRD sponsored project on 'Development of online Lab in Microelectronics and VLSI' at VIT. He was also a Research Associate during his MS at VIT. He has more than 13 years of teaching and research experience.

His current research interests are nano-electrodes for supercapacitors, VLSI interconnects, carbon nanotube electronics, graphene, electronic materials and renewable energy materials. He has authored three books and 25 papers in peer reviewed journals and conferences. He is an Associate Editor of Nanotechnology for Environmental Engineering, a Springer Nature journal. He is also a peer reviewer for many journals. He has reviewed more than 75 papers for journals from IEEE, Elsevier, Springer, IET, Nature, Taylor & Francis, Trans Tech Publishers, and Wiley. He is a Senior Member of IEEE EDS/EPS.

Dr. K. Venkata Lakshmi Narayana is currently a Professor of the School of Electrical Engineering at Vellore Institute of Technology (VIT), Vellore, India. He received the BTech in Electronics and Instrumentation Engineering from the Nagarjuna University, Guntur, India in 2001, the MTech in Electrical Engineering from the Motilal Nehru National Institute of Technology (MNNIT), Allahabad, India in 2006, and the Ph.D. degree in Instrumentation engineering from the Andhra University, Visakhapatnam, in 2013. He achieved both his BTech and MTech with distinction. He is also the recipient of the Gold Medal in 2016 for his outstanding academic performance in MTech at MNNIT, Allahabad.

He has 18 years of experience in both teaching and research. He had completed one research project entitled "Design and Development of Smart Level and Pressure Transmitters for Ethiopian Chemical Industrial Processes," funded by the Ministry of Science of Technology, Ethiopia, as a principal investigator.

He rendered administrative services toward the constant maintenance of academic programs and instructional delivery as 'program chair' for Electronics and Instrumentation Engineering and the 'division chair' for the division of Industrial Automation and Instrumentation in the School of Electrical Engineering, VIT University, Vellore, India, from June 15, 2013 to January 4, 2016.

He is currently supervising PhD students (four PhDs have been awarded and one is on-going), has supervised 15 Dissertations at the P.G level, and supervised more than 30 undergraduate students for their final year projects.

As a published researcher, he has authored or co-authored over 50 research papers (five papers in IEEE Sensors Journals and six more in other SCI journals) and published in various peer reviewed journals, book chapters, and conferences of international repute. He has reviewed a large number of research papers in IEEE Access, IET Science, Measurement & Technology, Journal of Engineering Science and Technology (JESTEC), and Recent Advances in Electrical and Electronic Engineering.

His interests are in the area of sensors and signal conditioning, measurements, wireless sensor networks, optimization, process instrumentation, and virtual instrumentation.

Contributors

A.G.L.N. Aditya
Vellore Institute of Technology
Vellore, Tamil Nadu, India

Sajal Agarwal
Rajiv Gandhi Institute of Petroleum Technology
Jais, Amethi, Uttar Pradesh, India

S. Assa Aravindh
University of Oulu
Oulu, Finland

P. Chinnamuthu
National Institute of Technology Nagaland,
Dimapur, Nagaland, India

Bijit Choudhuri
National Institute of Technology Silchar
Silchar, Assam, India

Mohammadreza Faridafshin
Shahid Bahonar University of Kerman
Kerman, Iran

Komanapalli Gurumurthy
Vellore Institute of Technology
Amaravathi, Andhra Pradesh, India

Yasaman Heidari
Shahid Bahonar University of Kerman
Kerman, Iran

S. S. Jyothirmayee Aravind
Indian Institute of Technology Madras
Chennai, India

K. Karthikeyan
M. Kumarasamy College of Engineering
Karur, Tamil Nadu, India

Pratap Kollu
University of Hyderabad
Hyderabad, Telangana, India

Vaegae Naveen Kumar
Vellore Institute of Technology
Vellore, Tamil Nadu, India

Neeraj Kumar
Indian Institute of Science Bangalore
Bengaluru, Karnataka, India

Komanapalli Venkata Lakshmi Narayana
Vellore Institute of Technology
Vellore, Tamil Nadu, India

Alireza Nouroozi
Shahid Bahonar University of Kerman
Kerman, Iran

Mahshid Padash
Shahid Bahonar University of Kerman
Kerman, Iran

Yogendra Kumar Prajapati
Motilal Nehru National Institute of Information Technology Allahabad
Prayagraj, Uttar Pradesh, India

Parul Raturi
Omkaranand Sarswati Government Degree College
Devpryag, Uttarakhand, India

Elizabeth Rufus
Vellore Institute of Technology
Vellore, Tamil Nadu, India

L. Sujatha
Rajalakshmi Engineering College
Chennai, Tamil Nadu, India

Prakash Chander Thapliyal
CSIR-Central Building Research Institute
Roorkee, Uttarkhand, India

Meruga Udaya
University of Hyderabad
Hyderabad, Telangana, India

Chapter 1

Fabrication and study of fluidic MEMS device for toxic heavy metal ion sensing in water

K. Karthikeyan
M. Kumarasamy College of Engineering, Karur, India

L. Sujatha
Rajalakshmi Engineering College, Chennai, India

CONTENTS

1.1 INTRODUCTION

The toxic heavy metal ions, such as mercury, arsenic, lead, and cadmium, are extremely dangerous toward the health of humans and the environment. These ions cause a variety of ailments such as heart problems, neurological disorders, and other developmental illnesses. Water of unprocessed sewage, industrial garbage, pigments, gasoline, battery garbage, poor sanitation, garbage from medical refuse, and electronics garbage damage much of the

DOI: 10.1201/9781003288633-1

earth's surface and groundwater. It is critical to use low-cost, greater selectivity and more sensitive detection technologies to monitor the quantity of mercury ions in drinking water systems (Chao Wang and Yu, 2013, Ha Na Kim et al. 2012, Mohamed Shaban et al. 2014). Several approaches have been developed to detect Hg2+ in water. They include ICP-Atomic Emission Spectrometry (Xiaoping Zhu and Alexandratos, 2007), colorimetric (Ali A. Ensafi and Fouladgar, 2008), atomic absorption spectrometry (N. Pourreza and Ghanemi, 2009), surface-enhanced Raman spectroscopy (Daniel K. Sarfo et al. 2017), electrochemical methods (Aytug Gencoglu and Minerick, 2014, Gauta Gold Matlou et al. 2013, Razieh Salahandish et al. 2018, Sai Guruva Reddy Avuthu et al. 2016), fluorescence microscope (Dong-Nam Lee et al. 2009), and surface plasmon resonance spectroscopy (Shaopeng Wang et al. 2007). Traditional procedures include drawbacks such as expensive tools, highly specialized labor, lengthy execution times, and expensive chemicals. Instead, a sensor built with microfluidic technology provides a potential alternative to the traditional technique. It enables the implementation of comprehensive laboratory processes on a small device. Furthermore, a fluidic MEMS device provides a new technique for handling reagents and chemicals in the range of nano and micro liters.

Generally microfluidic devices are used for analyses of bio-samples, micro polymerase chain reaction (PCR) (Dolores Verdoya et al. 2012), identifying and differentiating tumor cells, and detecting and differentiating cancer cells (Rajapaksha et al. 2018). Detection of hormone substaffrdxnces (Nuno Miguel Matos Pires et al. 2013), detection of foodborne pathogens (Yuqian Jiang et al. 2016), fabrication of nanoparticles and biomaterials (Lung-Hsin Hung and Lee, 2007), and cell sorting are all examples of applications (C. Wyatt Shields et al. 2015). A zigzag microchannel based on microfluidics was reported for Pb2+ ions detection, using 11-mercaptoundecanoic acid and gold nanoparticles (Chunhui Fan et al. 2012). Similarly, The lead ion was detected using a Y-type micromixer with a fluorescent probe DANS4- Calix (Liyun Zhao et al. 2009), and deoxyribozyme immobilization on a microfluidic device to detect Pb2+ ions (Tulika S. Dalavoy et al. 2008).

A microfluidic device was used to detect cobalt ions by a chemiluminescence approach with the range of 5.610–11 mol/L (Xueye Chen et al. 2013). Hg2+ ions were detected using the chemiluminescence method and diodes with the range of detection is up to 100 nM from 66 nM (Heiko Schafer et al. 2008). Materials with strong photocatalytic activity, such as silver and gold nanoparticles, have been used for toxic metal ions detection (C. Joseph Kirubaharan et al. 2012, Jie Zhao et al. 2017, Yongchao Huang et al. 2018). According to several research papers, gold nanoparticles are used for toxic heavy metal ions detection. Generally, gold nanoparticles have good optical properties. A colorimetry sensor was reported as the lead ions detection using mercaptoundecanoic acid; the detection range is 2–50 M (Gulsu Sener et al. 2014, Jiehao Guan et al. 2015, Weinan Leng et al. 2015, Yang-Wei Lin et al. 2011). To detect mercury ions, a colorimetric sensor based on a digital microfluidic device was created (Anran Gao et al. 2011). A sensing probe of PTEC was used to detect $Hg2+$ ions, and color change was observed while detecting different concentrations of mercury ions, with the limit being 100 μM, from 6.25 μM using a 2μL sensing fluid. Using L-cysteine functionalized gold nano-rods, a colorimetry-based mercury ion detection method was created (Tiziana Placido et al. 2013).

A sensor was designed to detect three ppt using 3mL of sensing fluid. This volume is far larger than the proposed technique recommended by the World Health Organization. Numerous types of micromixers have been described as improved mixing processes; at the same time, building these structures is extremely difficult. Furthermore, pressure fell during fluid flow and total mixing was only accomplished at high flow rates (Arshad Afzal and Kim, 2014, Ravi L. Rungta et al. 2013, Shakhawat Hossain et al. 2010, Thomas P. Forbes and Kralj, 2012, Yan Du et al. 2010). The existing micromixer devices report that better mixing was attained at maximum flow rates; however, this flow rate will detach the connection of the device. The life span of the device is also shortened. Nevertheless, the metal ion detection employing gold nanoparticles remains hard in terms of sensitivity and selectivity. The gold nanoparticle (Au) with Rhodamine 6G has the fluorescence emission property, and its surface associated with the field of plasmonic was described. (Josiane P. Lafleur et al. 2013). Through their amino groups, amino acids may attach with the Au nanoparticle. These compounds create the complexes with heavy metal ions. In a micromixer device, the mercury ions were detected in water with the limit of 0.6 g/L, using rhodamine 6G – albumin serum with nanoparticles (gold) as the detection probe [Josiane P. Lafleur]. The micromixer described here has a relatively basic construction, resulting in reduced pressure drop and higher mixing efficiency at low flow rates. In this chapter, we discussed the HB tendencies micromixer device with the length of 95mm and fluid handling volume of roughly 2.8μL, while comparing the exciting, reported micromixer device, which was used to detect metal ions; the device length and consumption of sensing fluid volume is also less. This chapter discusses amino acid (L-Arginine) - rhodamine 6G and nanoparticles (gold) to detect $Hg2+$ by changing the fluorescence intensity of the complex. In addition, a low-cost fluorometric microfluidic system is developed to sense the $Hg2+$ ions in water using the herringbone tendencies micromixer device. The device is fabricated using MEMS technology, The prime mold of the device is fabricated using SU8 by photolithography technique and replica mold is fabricated using Poly Dimethyl Siloxane (PDMS) by soft lithography technique. As a sensing probe, a microliter quantity of gold particles, rhodamine 6G and L-Arginine was utilised for sensing of $Hg2+$ ions in water.

1.2 DESIGN OF MICROMIXER DEVICE

One of the most basic types for mixing two liquids in straight channel micromixer. The fluid flow in micro channels is laminar due to their tiny widths (about 100 or 200 m). As a result, mixing of fluids in micro channels occurs solely because of diffusion, and mixing takes more time throughout the channel. As a result, straight channel micromixers are typically two to four cm long. To increase mixing performance in micro channels, several passive and active mixing approaches are applied. In this study, we suggest a micro-mixing herringbone shape micromixer for heavy metal ions detection. Figure 1.1 depicts the HB bent micromixer as having one outlet and two inlets. The entrance and outflow ports are 3000 μm in length. The channel has 10 cm, with 24 bends of mixier device construction, occupying the device length is 2 cm and width is 200 μm.

1.2.1 Simulation of Herringbone (HB) structure micromixer device

The Navier-Stokes equation in Equation (1.1) and Equation (s) shows the continuity equation, may be used in a micromixer to define the Newtonian incompressible flow,

$$\rho.\left(\frac{\partial u}{\partial t}+(u.\nabla)u\right)=f-\nabla p+\upsilon\nabla^2 u \tag{1.1}$$

$$\nabla.u=0 \tag{1.2}$$

Where,

ρ - Density in kg/m^3,
f - Body force in N/m^3,
p - Pressure in N,

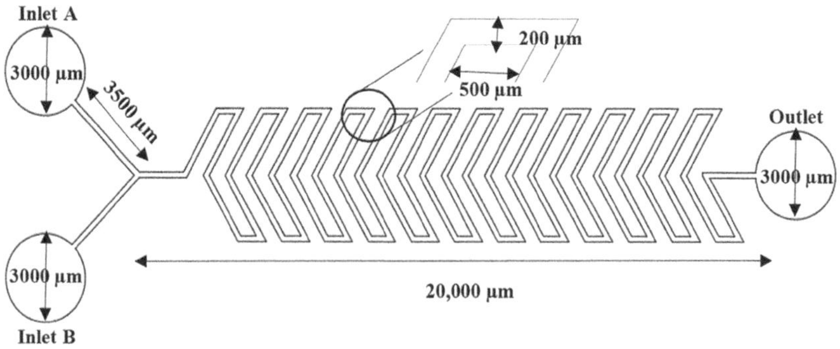

Figure 1.1 Y - herringbone bent micromixer.

u - Velocity vector in mm/s,
v - Dynamic viscosity of the fluid in kg·s/m,
t - Time in s

The diffusion-convection is illustrated in Equations (1.3), can represent species movement in the systems,

$$\frac{\partial c}{\partial t} + (u.\nabla)c = D\nabla^2 c \tag{1.3}$$

Where,
D - Diffusion co-efficient in m^2/s,
c - Concentration of the species in kg/m^3

The Reynolds number (Re) is defined as follows:

$$Re = \frac{uL_c}{v} \tag{1.4}$$

Where,
Re - momentum-viscous friction ratio,
Lc - hydraulic diameter of non-circular pipes.

To examine the level of mixing is calculate by the following formula,

$$\sigma^2 = \frac{1}{N}\sum_{i=1}^{N}(c_i - \bar{c})^2 \tag{1.5}$$

When c_i represents the normalized concentration, Total sampling points, N and $\bar{c}$ represents the predicted concentration at normalized level. The mixing efficiency is estimated using the following formula,

$$M = 1 - \sqrt{\frac{1}{N}\sum_{i=1}^{N}\left(\frac{c_i - \bar{c}}{\bar{c}}\right)} \tag{1.6}$$

M denotes the mixing efficiency. The effectiveness of mixing varies from 0 to 1 (zero to complete mixing). For the application process the mixing efficiency among 80% and 100% is acceptable. [Karthikeyan et al. 2018, Karthikeyan et al. 2017].

The simulations were carried out with the help of the COMSOL Software. The micromixer with Herringbone bent micromixer is illustrated in Figure 1.1. In this micromixer, inlet A fluid has the sensing fluid and inlet B has Hg2+ ions with water. The fluid properties are given as follows, 1000 kg/m^3 as density of the fluid, 0.001 Pa.s as viscosity of the fluid, velocity is $8.33*10^{-4}$m/s and $15.3*10^{-12}$ m^2s^{-1} as diffusivity. The fluid inflow velocities in both inlets are assumed to be similar and concentrations of the fluid in inlet A and inlet B are expected to be 1 and 0 mol.m^{-3}.

Figure 1.2 depicts the simulated result of HB bent micromixer with concentration distribution scale. Fluids move toward the outlet through the HB bends in a laminar flow, and the percentage of mixing improves due to the long HB bends microchannel.

The mixing concentration level is shown at different locations of the HB bends micromixer device in Figure 1.3. Fluid concentration levels are tested at several places, including 1500, 12200, 22400, 32600, and 42800 μm. The fluids do not mix completely at 1500, 12200, 22400, and whole diffusion occurs at 42800 μm.

Figure 1.4 depicts the concentration of mixing at different location of the device. The complete mixing with 0.5 mol/m3 happens at 8300 μm in the x-direction.

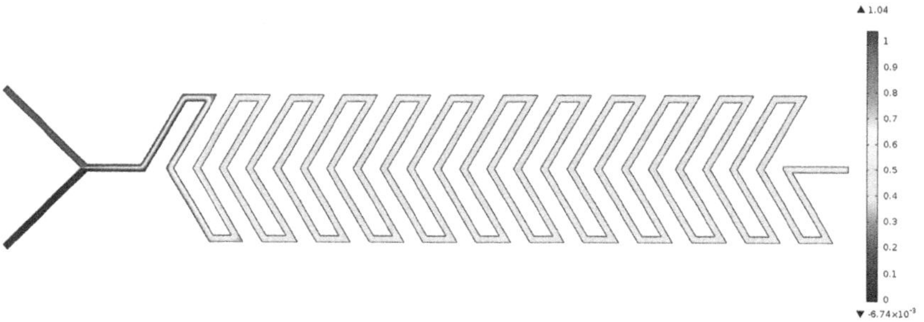

Figure 1.2 Simulated HB bent micromixer with concentration distribution scale.

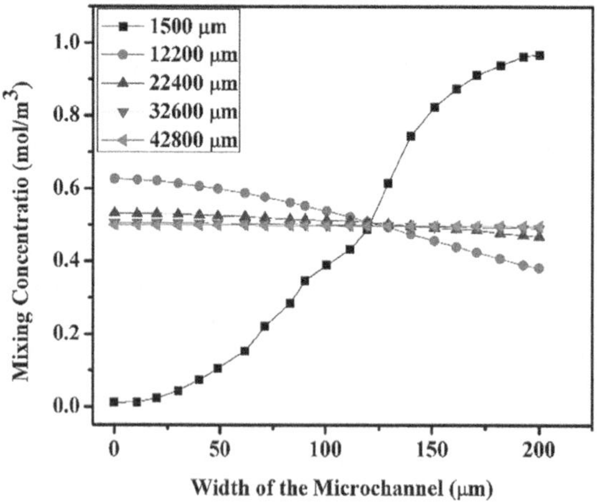

Figure 1.3 Mixing concentration at different location of the device.

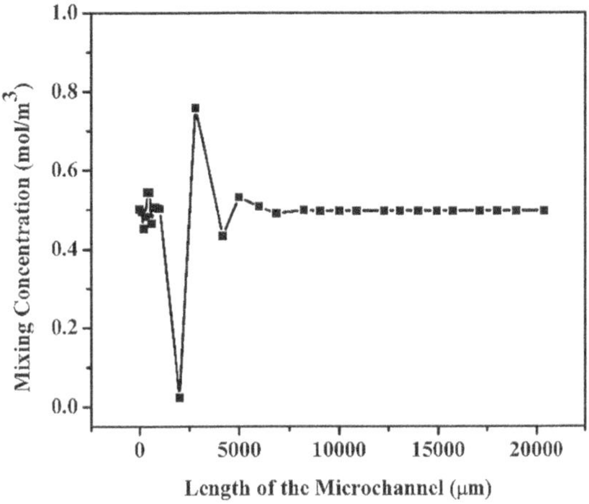

Figure 1.4 Mixing concentration level of the device.

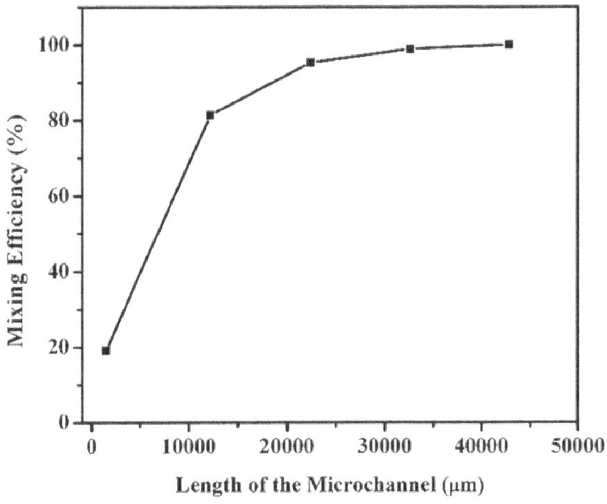

Figure 1.5 Mixing percentage of the HB bent micromixer.

Figure 1.5 depicts the mixing percentage of the device. The mixing efficiencies of the device with constant flow rate, 19.18 percent at 1500 μm, 78.96 percent at 12200 μm, 94.51 percent at 22400 μm, 98.63 percent at 32600 μm, and 100 percent at 42800 μm. The mixing is totally realized as a result of the HB bent micromixer in the channel. Figure 1.6 depicts the pressure level of the device at various lengths from inlet to outlet. The pressure is 66.94 Pa (maximum) at the input port. It decreases along the channel's length until it hits zero near the outlet port.

1.3 EXPERIMENTAL

1.3.1 Design of HB bent micromixer device

The proposed micromixer device with herringbone bends and its microchannel width is 200 μm, with two inlet ports such as inlet A, inlet B. The outlet port, has a 3 mm diameter of each port. The angle of the microchannel is 45 degrees with respect to x-axis, the inlets port length is 3.25 mm. The spacing between the neighboring channel is 500 μm. The mixing zone of the herringbone length of the microchannel is 95 mm, with a folding (8n's) and sensing zone of 5mm diameter. The overall chip length in the x-axis is 26 mm, while the chip width in the y-axis is 13.5 mm. Figure 1.7 depicts the design of a y-shaped herringbone micromixer.

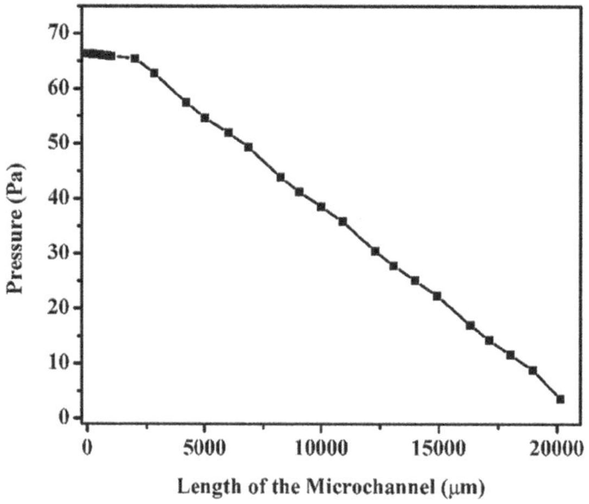

Figure 1.6 Pressure vs length of the microchannel.

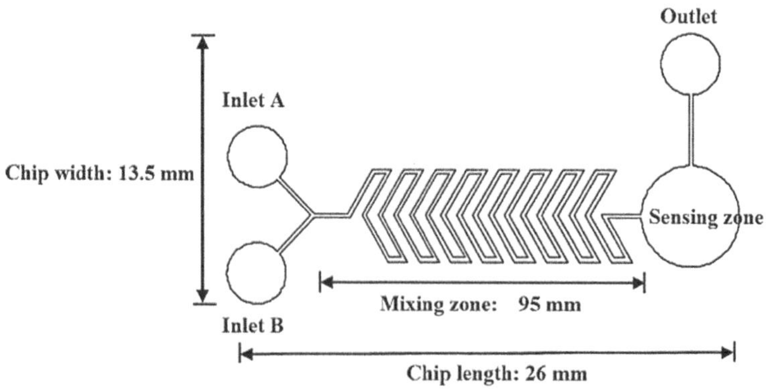

Figure 1.7 Specification of HB bent Micromixer.

1.3.2 Fabricated HB bent micromixer device

A stiff stamp (primary mold) bearing the pattern is utilized in soft lithography to make duplicate mold patterns (castings). The main mold in our studies was created using the UV lithography process using negative photoresist SU-8 negative. Photoresist was spin coated for 30 seconds at 2200 rpm on substrate and obtained a thickness of 100 µm. The spin-coated sample was pre-exposure baked at 65 degrees for 5 minutes Celsius and at 95 degrees Celsius at 15 minutes. The baked sample was under UV exposure for 12 seconds to a UV exposure system. The sample was post exposure baked at 95°C for 10 minutes. The sample was developed by an ultrasonicated bath with SU-8 developer solution aimed at 12 seconds.

The prepared sample was kept at 70°C for hard baking at one hour, with the required micropattern on the photoresist layer. In a petri dish, the constructed SU-8 primary mold structure was preserved. Following that, there was an application of a 10:1 ratio of silicone elastomer and curing agent as a PDMS pre-polymer (un-cured form of PDMS). During the mixing process of silicone elastomer and curing agent, air bubbles are formed. The vacuum desiccator was used to eliminate bubbles for one hour, which present on PDMS pre-polymer. Then the mixture solution was poured over the primary mold and cured at 70°C for three hours. The PDMS replica mold was scraped from the prime mold. This PDMS replica mold construction remains reverse pattern of SU8 mold of the device. This SU8 prime mold may be used to make the PDMS micro mold several times. The fabricated replica mold was bonded with glass substrate using oxygen plasma chamber. Using oxygen plasma treatment, the PDMS-based channel was eventually attached to the glass cover plate. Figure 1.8 depicts the constructed HB micromixer, which has two inlets and an output [K. Karthikeyan et al. 2018].

Figure 1.8 Fabricated HB bent micromixer device.

1.3.3 Preparation of gold nanofluids

The sodium citrate reduction process was used to make gold nanofluids (AuNFs). In a beaker, 50 ml of 1mM HAuCl4.3H2O was placed on a hot plate with magnetic stirring for 20–30 minutes at 80°C. After that, it was necessary to add 1% trisodium citrate (2 mL). The resultant substances were chilled to normal temperature, and gold nanofluids were stored in a dark place at 4°C (Remant Bahadur et al. 2014).

The Figure 1.9 shows the AuNFs surface morphology using a transmission electron microscope with high resolution - (a) gold nanoparticles with sphere-shaped (b) particle diameter (18 nm) (c) fringe spacing of 0.20 nm,

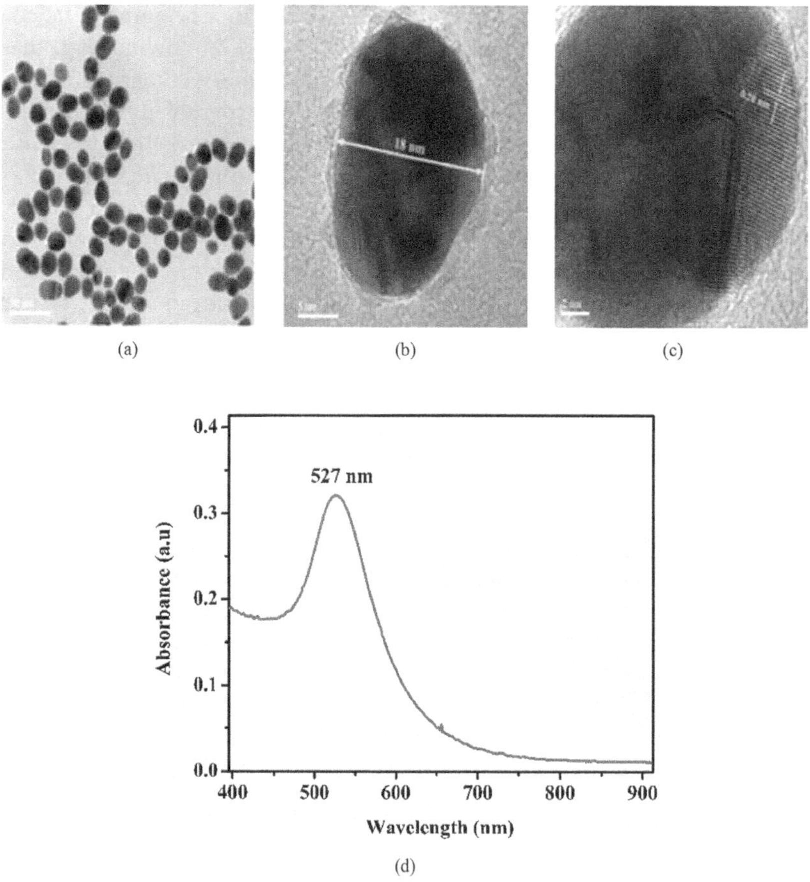

Figure 1.9 Gold nanoparticles HR-TEM images (a) Magnification at low scale (b) Magnification at high scale (c) Fringe spacing is 0.20 nm (d) UV-visible absorbance spectrum of gold nanofluids with the wavelength is 527 nm (See photograph of wine red colour gold nanofluids).

and (d) UV-visible absorbance spectra of gold nanofluids with a wavelength of 527 nm. The wine-red gold nanofluids are seen in the inset shot. L-arginine and Rhodamine 6G are used to functionalize the produced AuNFs. The function's method is as follows: In a vial containing AuNFs with Rhodamine 6G (0.5 mM) and L-arginine (0.2 mg) were mixed for 15 minutes at room temperature. As intake fluid A, the obtained sensing probe (L-arginine-AuNFs-Rhodamine 6G) will be employed.

1.4 SAMPLE FLUID PREPARATION

The prepared sample fluid is flow into the inlet B. This analyte inlet B holds a recognized amount of mercury ions. This Hg2+ ions values of 0, 2, 4, 6, 8, 10, 12, 14, and 16 nM in DI water.

1.4.1 Detection process of metal ions using microfluidic device

Fluids enter concurrently at ports A and B for mercury detection, and both are mixed as they travel through the micromixer. Both fluids were properly mixed in the mixing zone before being collected at the sensing zone, where the fluorescence inspection is being performed.

The fluid is discharged from the device via the outlet port. Through the sensing zone, the fluorescence response was observed at 350 nm. A portable spectrometer of fluorescence was used to measure the fluorescence response. The analysis of fluorescence demonstrates that the intensity changes with existence of mercury ions. The graphical working method of detection procedure depicted in Figure 1.10.

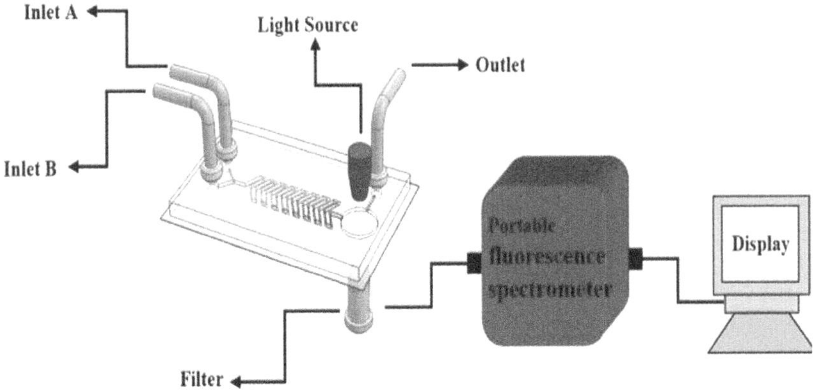

Figure 1.10 Mercury ion detection process.

1.5 RESULTS AND DISCUSSION

1.5.1 FTIR studies of sensing fluids

FTIR in the 400 cm^{-1}–4000 cm^{-1} region was used to explore the connections between the gold nanoparticles with Rhodamine 6G & L-Arginine. Figure 1.11 depicts the AuNPs FTIR spectrum (a). The peak points are attributed as the sodium citrate derivative O-H chemical bonds and C-H, C = O. The spectrum of gold nano particles is attained C-H bond appearances on 2942 cm^{-1}, At 3152 cm^{-1}, O-H bonds and at 1630 cm^{-1}, C = O bonds. Figure 1.11 depicts the FTIR spectra of L-Arginine with AuNPs (b). Water absorption of O-H stretching and molecules of COOH was observed in the peaks range at about 3853 cm^{-1} from 3638 cm^{-1}. The N-H stretching spectra was observed in AuNPs with the L-Arginine peak at 2355 cm^{-1}. N-H bending vibration in the primary amine group causes the peak distributed at 1546 cm^{-1}. Irregular distortion peak is found at 1450 cm^{-1} for vibration CH3 causes. The peaks at 1140 cm^{-1} suggest that the C-N stretching band is symmetrical.

L-arginine with AuNPs and COO-plane deformation are responsible for the band at 676 cm^{-1}. The NH3+ and COO- stretching of AuNPs is noticeably altered when L-arginine binds to their surface. This is probably caused by a change in the dipole moment. The FTIR spectra show a robust interaction between L-Arginine and AuNPs functional groups. Figure 1.11 depicts the Rhodamine 6G with L-Arginine and AuNPs FTIR spectra (c). Peaks at 693 and 661 cm^{-1} peaks are caused by plane bending of the C-C-C ring.

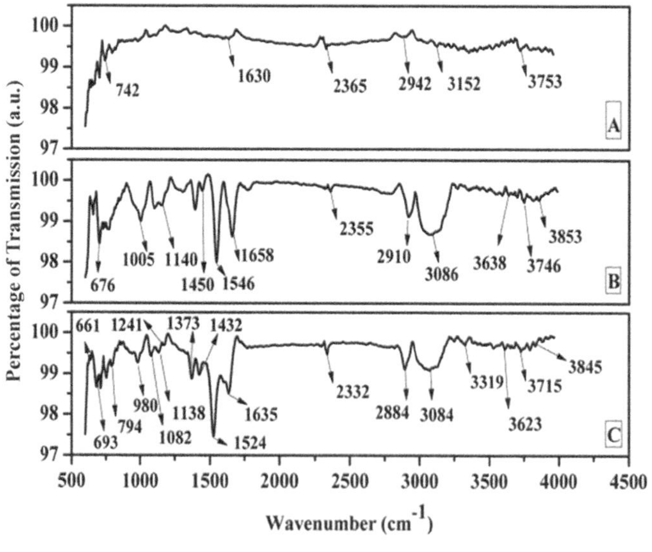

Figure 1.11 Fourier transform infrared spectroscopy spectrum of sensing fluids.

The C-H out plane bending peaks was observed at 794, 980 and 1082 cm^{-1}. The C-H in plane bending peaks at 1138 cm^{-1}. The peaks at 1432-1241 cm^{-1} are caused by the stretching of C-N and C-O-C. Aromatic C-C stretching is shown by peaks at 1635 cm^{-1} and 1524 cm^{-1}. The findings back up prior studies (A. Parvathy Rao and Venkateswara Rao, 2003 and A. L. Sunatkari et al. 2015).

1.5.2 Fluorescence studies

Figure 1.12 shows the fluorescence spectrum of sensing fluid for mercury ion detection at 0 nM to 16nM concentrations range. There is no peak found at 0nM of Hg2+ in the spectra. When the concentration of Hg2+ ion level gets increased from 2 nM to 16 nM, the fluorescence becomes more intense. The highest luminous intensity level 532 was attained at a concentration of 16 nM thanks to a remarkable variation in the intensity. The luminous amplification remains attributable to the construction of R-GN-L-Arg-Hg2+ complex. Figure 1.13 displays the error rate in fluorescence emission intensity produced by the creation of the R-GN-L-Arg-Hg2+ Vs combination at various concentration levels (0–16 nM).

Figure 1.13 shows the sensing selectivity of the mercury ions with various ions at a concentration of 16 nM and its velocity is 1μL/min. The selectivity study was done using sensing fluid R-GN-L-Arg. Following the passage of

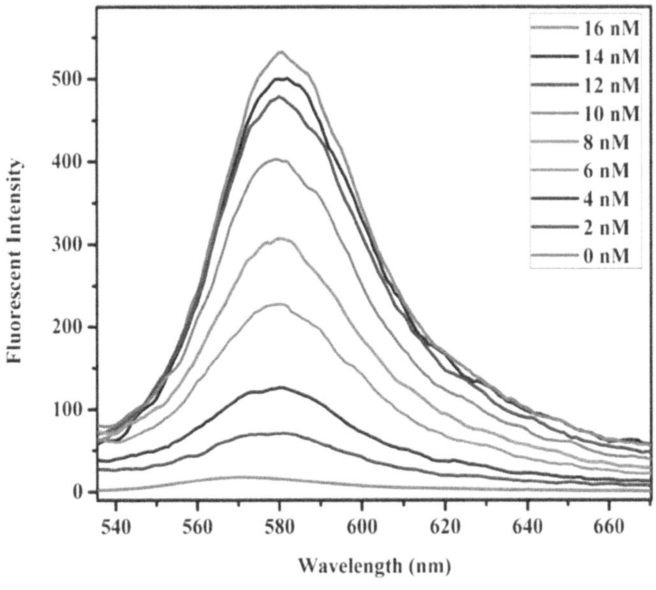

Figure 1.12 Fluorescence spectrum with different concentrations.

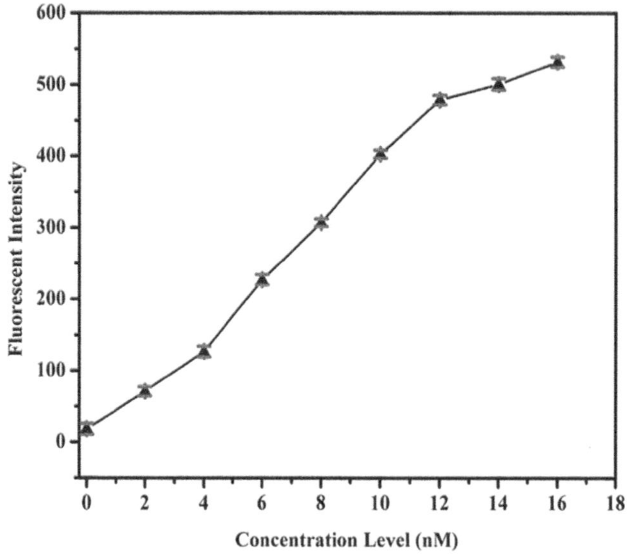

Figure 1.13 Fluorescent intensity of Hg^{2+} detection with error rate.

these fluids through the created micromixer unit, the intensity of the detected fluorescence emission was examined.

The variations in fluorescence intensity observed through the adding ions for instance Zn2+ Hg2+ Ba2+, Mg2+, Cu2+, Mn2+, Cr2+, Co2+, Pb2+, Al3+ and to R-GN-L-Arg solutions were measured. The R-GN-L-Arg-Hg2+ complex can detect metal ion concentrations of up to 16 nM Hg2+. The minimum detection limit of 2 nM and the linear detection limit is 12 nM. When increasing the concentration level of more than 12 nM, the intensity level is decreasing.

According to the findings of this investigation, fluorescence intensity of Pb2+ ions produce minor variations when related to other ions. However, the intensity change is relatively little when compared to Hg2+. Thus, when compared to other metal ions, the fluorescence emission intensity reported for Hg2+ has the best selectivity of R-GN-L-Arg. Figure 1.14 also depicts the error rates, with a 12 percent ambiguity in the fluorescence spectra with different metal ions.

At various flow rates of 1, 25, 50, 75, 100, 125, 150, 175 and 200 µL/min, Figure 1.15 demonstrations the intensity of fluorescence, 16nM Hg2+ ion mixer solution. The results demonstrate that the flow rate significantly affects the fluorescence signal. The analyte ion exchange is stronger as the flow rate is less and when it is mixed with the sensing solution, increasing detection, according to the fluorescence experiment. The fluorescence response reduces at larger flow rates because of inadequate mixing time, and the increased flow rate's high pressure may seriously harm the channel.

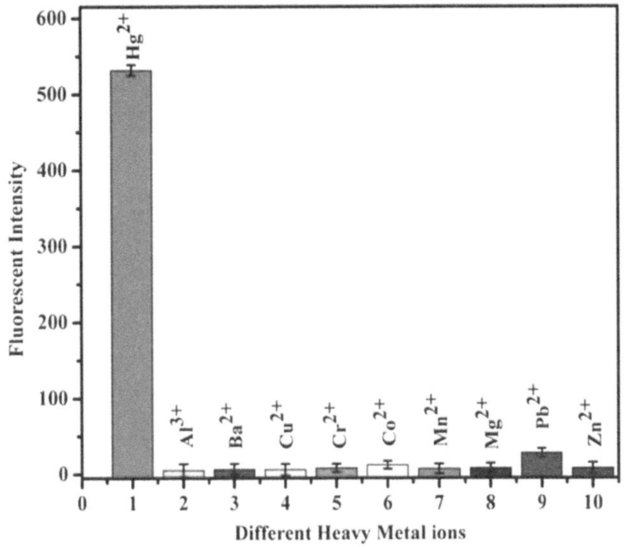

Figure 1.14 Error rates & selectivity.

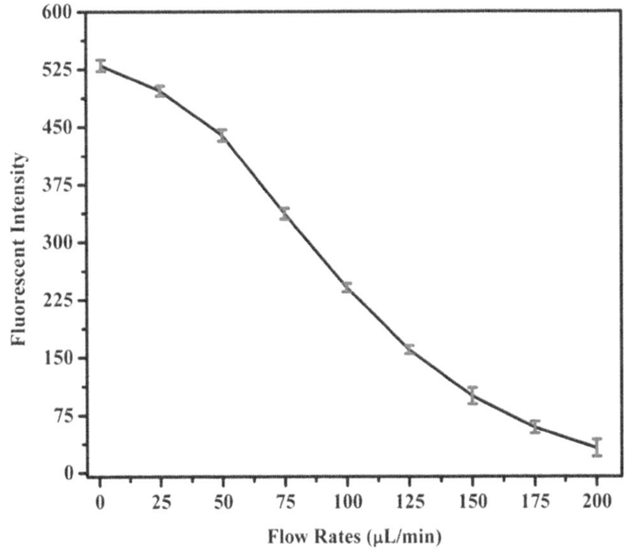

Figure 1.15 Error rates – sensitivity with different flow rate.

A sensing mechanism of metallophilic contacts of mercury ion is postulated in light of the experimental findings, and it causes increasing the intensity of R-GN-LAG. The suggested sensor's size and sensitivity are contrasted with those of previously reported Hg2+ sensors in Table 1.1.

Table 1.1 Evaluation of mercury ion detection

Method of Detection	*Sensing Materials*	*Metal ions Detection*	*Range of Detection*	*References*
Raman spectroscopy	Porous anodic alumina with gold nanoparticles	Hg^{2+}	1–20 ppb	Mohamed Shaban et al. 2014
Anodic stripping voltammetry with surface plasmon resonance	Gold electrode	Hg^{2+}	2 ppb	Shaopeng Wang et al. 2007
Cold vapor atomic absorption spectrometry	Modified agar powder with $SnCl_2$ and 2 -mercapto-benzimidazole	Hg^{2+}	0.040–2.40 $ngmL^{-1}$	N. Pourreza and Ghanemi, 2009
Surface enhanced Raman spectroscopy	Mercaptopropionic acid with Aminodibenzo-18-crown-6 coupled	Hg^{2+}	1×10^{-6} M to 1×10^{-11} M	Daniel K. Sarfo et al. 2017
Plasma Spectroscopy	Gold (III) chloride and Thiourea	Hg^{2+}	0.1–2.0mg L^{-1}	Xiaoping Zhu and Alexandratos, 2007
Spectrophotometric	Triacetyl cellulose with 1-[2-pyridylazo]-2-naphthol	Hg^{2+}	1.0–1000.0 μM	Ali A. Ensafi and Fouladgar, 2008
Square wave anodic stripping voltammetry	carbon nanotube with gold - dimethyl amino ethane thiol	Hg^{2+}	20–250 ppb	Gauta Gold Matlou et al. 2013
Screen Printed Phenanthroline based Flexible Electrochemical	Naphtho - dipyrido phenazine with phenanthroline	$Hg^{2}+$, Pb^{2+}	50 μM–1mM	Sai Guruva Reddy Avuthu et al. 2016
Electro-chemiluminescence	Trisphenanthroline ruthenium(II) complex	Hg^{2+}	60 nM	Heiko Schafer et al. 2008

Fluorescent spectra using using microfluidic device	2,6-pyridinedicarboxylic Acid and Mercaptopropionic acid - modified Au NPs	Hg^{2+}, Pb^{2+}, Cu^{2+}	9.6×10^{-8} to 6.4×10^{-6} M,	C. Joseph Kirubaharan et al. 2012
Colorimetric method	Gold nanoparticles with 11-mercaptoundecanoic acid	Hg^{2+}, Cd^{2+}, Fe^{3+}, Pb^{2+}, Al^{3+}, Cu^{2+}, Cr^{3+}	2–50 μM	Weinan Leng et al. 2015
Microfluidics based Colorimetric method	Oligonucleotide with triethylamino thiophene hydrochloride	Hg^{2+}	6.25–5000 μM	Yang-Wei Lin et al. 2011
microfluidics fluorescent spectrum	Bovine serum albumin - gold nanocluster	Hg^{2+}	15 μg L^{-1}–0.6 μg L^{-1}	Yan Du et al. 2010
Microfluidics based fluorescent spectrum	gold nanoparticles with Rhodamine 6G and L-Arginine	Hg^{2+}	2–12 nM	Proposed detection

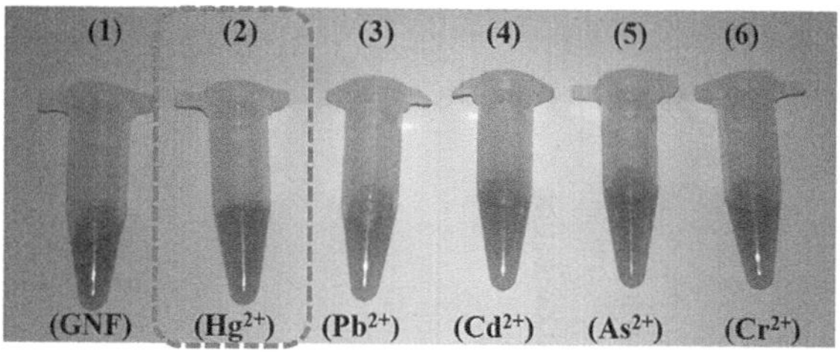

Figure 1.16 Selective Detection of 25 μM - Hg^{2+}.

1.6 COLORIMETRIC METHOD BASED HEAVY METAL IONS DETECTION

1.6.1 Colorimetric analysis in conventional tube

In the traditional approach for detecting mercury ions, 1 mL of sensing probe solution containing L-Arg-GNFs and 25 M concentration of Hg2+ is used. The colour modification produced by the surface plasmon resonance impact on AuNFs accumulation was used to detect Hg2+ using AuNFs. Figure 1.16 depicts the detection of Hg2+ selectively. In this procedure, a 25 M concentration of several harmful metal ions such as Pb2+, Cd2+, As2+, Hg2+, Cr2+ ions were used for selective detection of Hg2+. The color transitions from wine red in tube 1 to purple in tube 2. This procedure suggests that our sensing probe is extremely selective for Hg2+ detection.

1.6.2 Colorimetric analysis in microfluidic device

Because of its low toxicity, biocompatibility, and optical clarity, PDMS-based microfluidic devices are frequently employed in point-of-care and lab-on-chip investigations. A microfluidic chip with two inlets and an outlet of Y-type herringbone micro channels was created, and herringbone shaped channels were constructed to aid in solution mixing. Using a microfluidic syringe pump, 2 L of L-Arg-AuNFs solution and 2 L of 10 M concentrations of Hg2+ were pumped through each intake reservoir. Both inlet fluids flow together to the device Y junction and mix well in the channels. Because of herringbone bends and a lengthy channel with enough reaction time between L-Arg-AuNFs and Hg2+, mixing efficiency improves. The amine groups on the surface of L-Arg-AuNFs chelate and aggregate Hg2+. A black line appears in the fluid-fluid interaction layer during this process. The amount level of Hg2+ may be modified based on the intensity of the black layer, and 10μM of Hg2+ was effectively detected using an optical microscope. It was set up with a tiny amount (μL) of reagents for the detection procedure. Figure 1.17 depicts the detection of Hg2+ in a microfluidic

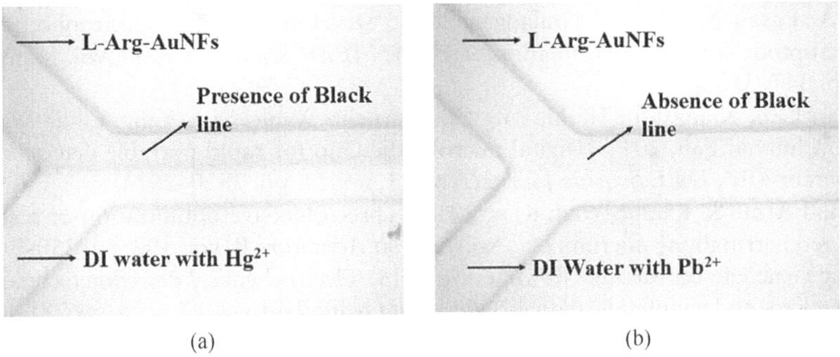

Figure 1.17 Detection of Hg2+ in microfluidic device (a) Presence of blackline (Hg2+) and (b) Absence of blackline.

device in the presence (a) and absence (b) of blackline (Hg2+). This discovery leads to the conclusion that using a high sensitivity colorimetry-based detection procedure on a microfluidic device with fluidic channels is not recommended.

1.7 CONCLUSION

For the effective design and construction of a microfluidic device with a herringbone structured micromixer for the detection of mercury ions in water. The sensing probe R-GN-L-Arg was created, reported, and utilized to find Hg2+. R-GN-L-Arg has a linear response for doses between 2 and 12 nM. When compared to other ions evaluated under the same circumstances, the device exhibits remarkable selectivity for Hg2+. The fluorometric-microfluidic system with the herringbone microchannel topology enabled highly sensitive fluorescence detection of Hg2+ ions. Similarly, a colorimetric sensor based on a microfluidic device is created to detect 10 M Hg2+ utilizing L-Arg-AuNFs. This microfluidic chip-based colorimetry technology outperforms the traditional tube quantitative method. When adopting a microfluidic chip-based detection procedure, less volume (L) of reagents are required, resulting in higher sensitivity. This allows for the detection of nanomolar concentrations of Hg2+, which may then be translated into a portable tiny device.

REFERENCES

A. Parvathy Rao & A. Venkateswara Rao, 2003. 'Luminiscent dye Rhodamine 6G doped monolithic and transparent TEOS silica xerogels and spectral properties', *Science and Technology of Advanced Materials*, vol. 4, pp. 121–129.

A. L. Sunatkari, S. S. Talwatkar, Y. S. Tamgadge, & G. G. Muley, 2015. 'Synthesis, characterization and optical properties of L-arginine stabilized gold nanocolloids', *Nanoscience and Nanotechnology*, vol. 5, no. 2, pp. 30–35.

Ali A. Ensafi & Masoud Fouladgar, 2008. 'Development of a spectrophotometric optode for the determination of Hg(II)', *IEEE Sensors Journal*, vol. 8, no. 4, pp. 347–353.

Anran Gao, Xiang Liu, Tie Li, Ping Zhou, Yuelin Wang, Qing Yang, Lihua Wang, & Chunhai Fan, 2011. 'Digital microfluidic chip for rapid portable detection of mercury(II)', *IEEE Sensors Journal*, vol. 11, no. 11, pp. 2820–2824.

Arshad Afzal & Kwang-Yong Kim, 2014. 'Three-objective optimization of a staggered herringbone micromixer', *Sensors and Actuators B*, vol. 192, pp. 350–360.

Aytug Gencoglu & Adrienne R. Minerick, 2014. 'Electrochemical detection techniques in micro- and nanofluidic devices', *Microfluid Nanofluid*, vol. 17, no. 5, pp. 781–807.

C. Joseph Kirubaharan, D. Kalpana, Yang Soo Lee, A. R. Kim, Don Jin Yoo, Kee Suk Nahm, & G. Gnana Kumar, 2012. 'Biomediated silver nanoparticles for the highly selective copper(II) ion sensor applications', *Industrial & Engineering Chemistry Research*, vol. 51, pp. 7441–7446.

C. Wyatt Shields IV, Catherine D. Reyes, & Gabriel P. López, 2015. 'Microfluidic cell sorting a review of the advances in the separation of cells from debulking to rare cell isolation', *Lab on a Chip*, vol. 15, pp. 1230–1249.

Chao Wang & Chenxu Yu, 2013. 'Detection of chemical pollutants in water using gold nanoparticles as sensors: a review agricultural and biosystems engineering', *Reviews in Analytical Chemistry*, vol. 32, no. 1, pp. 1–14.

Chunhui Fan, Shijiang He, Gang Liu, Lianhui Wang, & Shiping Song, 2012. 'A portable and power-free microfluidic device for rapid and sensitive lead (Pb2+) detection', *Sensors*, pp. 9467–9475.

Daniel K. Sarfo, Arumugam Sivanesan, Emad L. Izake & Godwin A, 2017. 'Ayoko Rapid detection of mercury contamination in water by surface enhanced Raman spectroscopy'. *RSC Advances*, vol. 7, pp. 21567–21575.

Dolores Verdoya, Ziortza Barrenetxeaa, Javier Berganzoa, Maria Agirregabiriab, Jesús M. Ruano-Lópezb, José M. Marimónc & Garbiñe Olabarríaa, 2012. 'A novel Real-Time micro PCR based Point-of-Care device for Salmonella detection in human clinical samples', *Biosensors and Bioelectronics*, vol. 32, pp. 259–265.

Dong-Nam Lee, Gun-Joong Kim & Hae-Jo Kim, 2009. 'A fluorescent coumarinyl-alkyne probe for the selective detection of mercury (II) ion in water', *Tetrahedron Letters*, vol. 50, pp. 4766–4768.

Gauta Gold Matlou, Duduzile Nkosi, Kriveshini Pillay & Omotayo Arotiba, 2013. 'Electrochemical detection of Hg(II) in water using self-assembled single walled carbon nanotube-poly (m-amino benzene sulfonic acid) on gold electrode', *Sensing and Bio-Sensing Research*, vol. 10, pp. 27–33.

Gulsu Sener, Lokman Uzun & Adil Denizli, 2014. 'Colorimetric sensor array based on gold nanoparticles and amino acids for identification of toxic metal ions in water', *ACS Applied Materials & Interfaces*, vol. 6, pp. 18395–18400.

Ha Na Kim, Wen Xiu Ren, Jong Seung Kim & Juyoung Yoon, 2012. 'Fluorescent and colorimetric sensors for detection of lead cadmium, and mercury ions', *Chemical Society Reviews*, vol. 41, pp. 3210–3244.

Heiko Schafer, Hengwei Lin, Michael Schmittel & Markus Bohm, 2008. 'A labchip for highly selective and sensitive Electro chemiluminescence detection of Hg2+ ions in aqueous solution employing integrated amorphous thin film diodes', *Microsystem Technologies*, vol. 14, pp. 589–599.

Jie Zhao, Son C. Nguyen, Rong Ye, Baihua Ye, Horst Weller, Gábor A. Somorjai, A. Paul Alivisatos & F. Dean Toste, 2017. 'A comparison of photocatalytic activities of gold nanoparticles following plasmonic and interband excitation and a strategy for harnessing interband hot carriers for solution phase photocatalysis', *ACS Central Science*, vol. 3, no. 5, pp. 482–488.

Jiehao Guan, Yi-Cheng Wang, & Sundaram Gunasekaran, 2015. 'Using L-arginine functionalized gold nanorods for visible detection of mercury(II) ion', *Journal of Food Science*, vol. 80, no. 4, pp. 1–6.

Josiane P. Lafleur, Silja Senkbeil, Thomas G. Jensen, & Jorg P. Kutter, 2013. 'Gold nanoparticle-based optical microfluidic sensors for analysis of environmental pollutants', *Lab on a Chip*, vol. 12, pp. 4651–4656.

K. Karthikeyan & L. Sujatha, 2018a. Fluorometric sensor for mercury ion detection in a fluidic MEMS device, *IEEE Sensors Journal*, vol. 18, no. 13, July, pp. 5225–5231.

K. Karthikeyan & L. Sujatha, 2018b. Detection of mercury ion in a microfluidic device using colorimetric method, *International Journal of Advances in Science, Engineering and Technology (IJASEAT)*, vol. 6, no. 2, pp. 67–70.

K. Karthikeyan & L. Sujatha, n.d.-a. Study of permissible flow rate and mixing efficiency of the micromixer devices, *International Journal of Chemical Reactor Engineering*, 20180047, ISSN (Online) 1542–6580.

K. Karthikeyan & L. Sujatha, n.d.-b. Design and fabrication of microfluidic device for mercury ions detection in Water, *IEEE Explore*, DOI:10.1109/ICNETS2.2017.8067934.

K. Karthikeyan, L. Sujatha, & N. M. Sudharsan, 2017. Numerical modeling and parametric optimization of micromixer for low diffusivity fluids, *International Journal of Chemical Reactor Engineering-DE GRUYTER*, vol. 16, no. 3, May, 20160231.

K. Karthikeyan, L. Sujatha, R. Sundar, & S. K. Sharma, 2018. Dimension tolerances in fabrication of polymer microfluidic devices, *Journal of Semiconductor Technology and Science-IEIE*, vol. 18, no. 2, April, pp. 262–269.

Liyun Zhao, Ting Wu, Jean-Pierre Lefevre, Isabelle Leray, & Jacques A. Delaire, 2009. 'Fluorimetric lead detection in a microfluidic device', *Lab on a Chip*, vol. 9, pp. 2818–2823.

Lung-Hsin Hung & Abraham Phillip Lee, 2007. 'Microfluidic devices for the synthesis of nanoparticles and biomaterials', *Journal of Medical and Biological Engineering*, vol. 27, no. 1, pp. 1–6.

Mohamed Shaban, Asmaa Gamal Abdel Hady, & Mohamed Serry, 2014. 'A new sensor for heavy metals detection in aqueous media', *IEEE Sensors Journal*, vol. 14, no. 2, pp. 436–441.

N. Pourreza & K. Ghanemi, 2009. 'Determination of mercury in water and fish samples by cold vapor atomic absorption spectrometry after solid phase extraction on agar modified with 2-mercaptobenzimidazole', *Journal of Hazardous Materials*, vol. 161, pp. 982–987.

Nuno Miguel Matos Pires, Tao Dong, Ulrik Hanke, & Nils Hoivik, 2013. 'Integrated optical microfluidic biosensor using a polycarbazole photodetector for point-of-care detection of hormonal compounds', *Journal of Biomedical Optics*, vol. 18, no. 9, p. 97001.

Rajapaksha W. R. L. Gajasinghe, Michelle Jones, Tan A. Ince, & Onur Tigli, 2018. 'Label and immobilization free detection and differentiation of tumor cells', *IEEE Sensors Journal*, vol. 18, no. 9, pp. 1–8.

Ravi L. Rungta, Hyun B. Choi, Paulo J. C. Lin, Rebecca W. Y. Ko, Donovan Ashby, Jay Nair, Muthiah Manoharan, Pieter R. Cullis, & Brian A. MacVicar, 2013. 'Lipid nanoparticle delivery of siRNA to silence neuronal gene expression in the brain', *Molecular Therapy Nucleic Acids*, vol. 2, p. 136.

Razieh Salahandish, Ali Ghaffarinejad, Seyed Morteza Naghib, Keivan Majidzadeh-A, & Amir Sanati-Nezhad, 2018. 'A novel graphene-grafted gold nanoparticles composite for highly sensitive electrochemical biosensing', *IEEE Sensors Journal*, vol. 18, no. 6, pp. 2513–2519.

Sai Guruva Reddy Avuthu, Jared Thomas Wabeke, Binu Baby Narakathu, Dinesh Maddipatla, Jaliya Samarakoon Arachchilage, Sherine O. Obare, & Massood Z. Atashbar, 2016. 'A screen printed phenanthroline-based flexible electrochemical sensor for selective detection of toxic heavy metal ions', *IEEE Sensors Journal*, vol. 16, no. 24, pp. 8678–8684.

Shakhawat Hossain, Afzal Husain, & Kwang-Yong Kim, 2010. 'Shape optimization of a micromixer with staggered-herringbone grooves patterned on opposite walls', *Chemical Engineering Journal*, vol. 162, pp. 730–737.

Shaopeng Wang, Erica S. Forzani, & Nongjian Taoe, 2007. 'Detection of heavy metal ions in water by high-resolution surface plasmon resonance spectroscopy combined with anodic stripping voltammetry', *Analytical Chemistry*, vol. 7, no. 9, pp. 4427–4432.

Thomas P. Forbes & Jason G. Kralj, 2012. 'Engineering and analysis of surface interactions in a microfluidic herringbone micromixer', *Lab on a Chip*, vol. 12, pp. 2634–2637.

Tiziana Placido, Roberto Comparelli, Marinella Striccoli, Angela Agostiano, Arben Merkoçi, & Maria Lucia Curri, 2013. 'Assembly of gold nanorods for highly, sensitive detection of Mercury ions', *IEEE Sensors Journal*, vol. 13, no. 8, pp. 2834–2841.

Tulika S. Dalavoy, Daryl P. Wernette, Maojun Gong, Jonathan V. Sweedler, Yi Lu, Bruce R. Flachsbart, Mark A. Shannon, Paul W. Bohnd, & Donald M. Cropek, 2008. 'Immobilization of DNAzyme catalytic beacons on PMMA for Pb2+ detection', *Lab on a Chip*, pp. 786–793.

Weinan Leng, Paramjeet Pati, & Peter J. Vikesland, 2015. 'Room temperature seed mediated growth of gold nanoparticles. mechanistic investigations and life cycle assessment' *Environmental Science: Nano*, vol. 2, pp. 440–453.

Xiaoping Zhu & Spiro D. Alexandratos, 2007. 'Determination of trace levels of mercury in aqueous solutions by inductively coupled plasma atomic emission spectrometry elimination of the memory effect', *Microchemical Journal*, vol. 86, pp. 37–41.

Xueye Chen, Chong Liu, Zheng Xu, Yuzhen Pan, Junshan Liu, & Liqun Du, 2013. 'An effective PDMS microfluidic chip for chemiluminescence detection of cobalt (II) in water', *Microsystem Technologies*, vol. 19, pp. 99–103.

Yan Du, Zhiyi Zhang, ChaeHo Yim, Min Lin, & Xudong Cao, 2010.'A simplified design of the staggered herringbone micromixer for practical applications', *BioMicrofluidics*, vol. 4, pp. 024105.

Yang-Wei Lin, Chih-Ching Huangb, & Huan-Tsung Chang, 2011. 'Gold nanoparticle probes for the detection of mercury, lead and copper ions', *Analyst*, vol. 136, pp. 863–871.
Yongchao Huang, Kunshan Li, Ying Lin, Yexiang Tong, & Hong Liu, 2018. 'Enhanced efficiency of electron–hole separation in Bi2O2CO3 for photocatalysis via acid treatment', ChemCatChem., vol. 10, pp. 1–7.
Yuqian Jiang, Shan Zoub, & Xudong Cao, 2016. 'Rapid and ultra-sensitive detection of foodborne pathogens by using miniaturized microfluidic devices a review', *Analytical Methods*, pp. 6668–6681.

Chapter 2

The review of micro-electromechanical systems-based biosensor

A cellular base perspective

A.G.L.N. Aditya and Elizabeth Rufus
Vellore Institute of Technology, Vellore, India

CONTENTS

2.1 INTRODUCTION

The quest for detection, diagnosis, and care in the field of medical science plays a crucial role, as has been proven in recent pandemic situations. The need and urge for diagnosis are fulfilled mainly by clinic-pathological exercises, conventional systems, imaging techniques, and biosensors (Hunter, 1955). But dynamic environments and the quest for knowledge have

DOI: 10.1201/9781003288633-2

increased the challenges in this field. To overcome challenges like sensitivity, rapid detection, larger sample volumes, feasibility, and power consumption of conventional systems, the incorporation of micro-electro-mechanical systems (MEMS) has brought profound applications in the diversified fields of medical and bio-systems in the detection of various agents. But the prominence of early detection and diagnosis is always credited in the field of medicine. The earlier the detection of the disease, the higher the chances of cure are in most scenarios. But several diagnostic techniques that are traditionally used depend upon clinical-pathological tests that are undergone in the laboratories after the onset of disease symptoms. But detection and diagnosis of the disease at early onset can facilitate the chances of cure. Techniques and technology in this aspect have progressed very little, which has created a bottleneck in the diagnosis of the disease state.

As the cell is the basic functional unit of any biological entity, cellular functions of the organisms can be utilized to understand the molecular behavior that is established with single cell analysis (Hamano et al., 2021; Nomura, 2020). These studies are not only useful in diagnosis, but also in estimating ecological and biogeochemical models (Cermak et al., 2016). But the paucity of content about mammalian cell growth across the evaluation process is always a quest with a technical hindrance. Most of the preceding works in the literature have been carried out on studies of population of cells with size homeostasis (Burg, 2015; Son et al., 2012). In such experiments, the cell cycle is related to the average size of the population, which yields minimal resolution, and the authenticity cell cycle and growth relation cannot be established. Further, these artifacts generated during the synchronization process because deviating cell behaviors add to further downgraded resolution. These daunting challenges emphasize more of the need for pursuing single-cell research (Mozharov et al., 2021; Son et al., 2012).

To contemplate single-cell studies, label-free detectors have gained prominence for developing throughput devices. Techniques like fluorescent and radioactive labeling have displayed higher sensitivity, but the factors of sample preparation and characterization made them less interesting (Burg, 2015; Burg & Manalis, 2003). Other redundant techniques, like quartz crystal microbalance (QCM) and surface Plasmon resonance (SPR), have a bottleneck with sensitivity, though the sample preparation limitation is overcome with larger sample volumes. So, these techniques are limiting the essence of biological detection with better resolutions, which is an appealing factor essential for single-cell studies. Emerging technologies depending upon a micro/nano fabrication process have potential and are viable methods to address the challenges. These technologies are optics-based sensors (SPR), microdevices (ISFET) and micro-mechanical sensors (Burg, 2003; Jorgensen, 2004). Though these techniques are poised equally with pros and cons, the resolution has common limitations and few or less scale-down prerogatives.

The MEMS-based resonant sensor is the paradigm technique to measure the bio-physical properties of cells. The principle of the resonant sensor is approached with frequency shift, while the shift in resonant frequency with and without load is equated to load added onto the device. Resonant MEMS sensors have higher sensitivity and better resolution when compared with their counterparts. The most advantageous parameter is that these devices can adopt scale-down approaches from MEMS to NEMS. These attributes facilitate their use in chemical and bio-sensing (Park, 2010). Previous work includes the detection of the single virus particle and mass of bacteria (Akin & Gupta, 2004; Davila et al., 2007; Gupta, 2004; Park, 2012), B. Anthrax spores in fluid (Davila et al., 2007), E.coli detection (Gfeller et al., 2005), the growth cycle of mammalian cells (Park, 2010) and suspended microchannel resonators to observe growth cycle and drug-induced studies and pedestal proof mass to observe apoptosis of H29T cells (Park, 2011).

This chapter emphasises discussion on various kinds of Biosensors that incorporate MEMS-based approaches, with their principle of operation and challenges involved, to use in ambient environments for potential applications. Recent advances in MEMS-based biosensors are brought forward in later sections, concentrating more on cell-based studies. These studies demand surface plasmon resonance and resonant beam mass sensors, which have gained larger prominence in the field of Biosensors. The understanding of basic structures, like cantilevers and beams used as biosensors, has been discussed, with mechanisms for various applications also reported.

2.2 THE PROLOGUE OF BIOLOGICAL CELLS

The main method to control the cell cycle is to concentrate the critical regulatory proteins, which is defined mainly by cell volume and by expression levels. Furthermore, the cell volume is interlinked to mass and energy requirements, which control the cell division and its survival. Changes to the rate of mass and volume accumulation are correlated to cell cycle position and can be used to measure the variations in cell density as well. So overall, all these parameters could be helpful in the detection of Homeostasis and cell-related diseases at early stages. Further, this system could provide a potentially viable solution for drug-based studies in cellular platforms, making a more effective and concise drug development process and, with recent advances, can further progression in omics.

2.2.1 The prologue of biological cell

Cell- anything smaller than a cell is considered to be non-living (referred to as a basic building block). The architecture of the cell can be explained by an outer membrane, dense structure nucleus, and chromosomes (source of genetic material DNA/RNA). The basic categories of cells fall under two

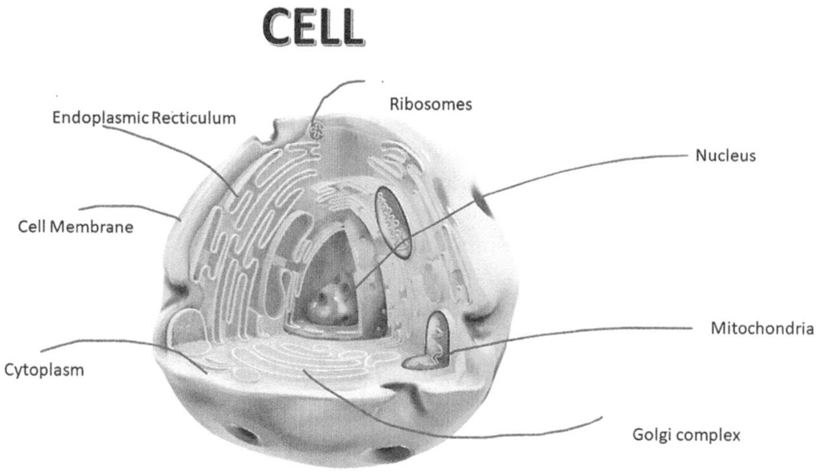

Figure 2.1 Internal structure and components of the cell.

groups, prokaryotic and eukaryotic, differentiated by cell wall membranes. type of cell consists of semi-fluidic membrane referred to as cytoplasm (a significant contributor to the volume of the cell) where cellular–chemical functionalities are sourced.

Besides this other major constituents of the cell are Organelles which are categorized as endoplasmic reticulum, mitochondria, Golgi complex, lysosomes, microbodies, and Vacuoles, all of which are very much absent in prokaryotic cells. So overall cells differ in shape, size, mass and volume, which can be considered biomarkers for various abnormal homeostatic processes (as shown in Figure 2.1).

2.2.2 Cell cycle and division

Cell division incorporates the multiplication of the cells with the growth phase of individual cells and tissues (proliferation) and then incorporates apoptosis (natural death of cell). During the first phase, proliferation is the process that involves DNA replication and cell growth. Though the cytoplasmic increase is a continuous process, DNA replication (synthesis) is specific to a particular stage of the cell cycle. And in the second phase of apoptosis, the cell undergoes natural/artificial death after a repeated number of proliferations.

Though most of the million cells divide within a 24 hour span (doubling rate), in actuality the rate of cell division varies among cell types and organisms (For example, the yeast doubling rate is 90 minutes (Bergman, n.d.). Division of a cell can be expressed in the majority of cases with two phases, interphase and M Phase (Shown in Figure 2.2).

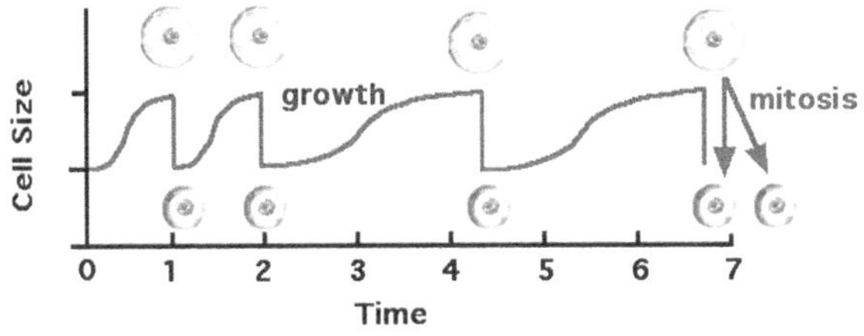

Figure 2.2 Cell division and cycle with the increase in the cytoplasmic increase *(cell growth-wiki).*

2.2.3 Mathematical modeling of Single cell

In a group of cells, division and growth are not uniform throughout the populations of cells, though the medium is constant throughout time. Few of the cells are indeed forced into proliferation, while the remaining undergo apoptosis. During the proliferation phase of cells, the mean deviation of the cells that induce proliferation also varies, depending on the conditions. So, assessment of individual cell growth patterns not only determines the disease conditions but also predicts the behavioral condition of the population and vice versa. But in the latter case, the generalized behavioral pattern of the population of cells cannot be attributed to the individual cells, due to uncertainty and deviation of cell patterns. But in the former case, properties of the individual cells depict or can be attributed to the behavior of the entire population (in healthy conditions). Under disease conditions, the cell cycle deviation from its checkpoints is evidence that can be further assessed and can be used in different fields of omics.

Most of the studies have emphasized developing models for population-based studies that determine the cell count and time for division (Ahmadian et al., 2020; Bell & Anderson, 1967; Bergman, n.d.; Charlebois, 2019; Crivelli et al., 2012; Cross et al., 2002; Rejaniak, 2005; Singhvi et al., 2014). A few other studies have concentrated on relative cell volume to age-base studies. But cell models with the division of cells and checkpoints are always expressed in terms of Cyclic Kinases and trigger points (Trombetta et.al., 2014).

It is understood that a cell should grow in critical size in terms of mass and volume for replication (DNA replication and Division) (Bell & Anderson, 1967). At the time of birth of the cell, typically the mass of the cell is less than that of the critical size, the starting point of the cell cycle, which can be expressed as the size-dependent phase, where the cell reaches to critical size ready for replication. The second phase is a time dependent phase where size is the independent parameter. There are various studies about cell-based studies that explain the yeast and eukaryotic cell as well, with different balanced conditions and regulation points (Pickering, 2019; Soifer, 2014).

2.2.4 Growth model

To initiate the growth model of the cell, the external parameter or trigger point is so essential to ambient temperatures and environments. In most cases, the trigger point is Insulin (IGF- insulin growth factors). With a series of kinase stimulations, the ribosomal triggers the S6 protein that is the base cause for the protein synthesis that makes the cells grow in size/growth phase (Shown in Figure 2.3). The basic assumption of this growth model is the proportionality of the protein synthesis to [S6] and total ribosomal content R. Later, these proteins are degraded at the same rate.

$$\frac{dm}{dt} = k_1[S6]R - k_2 m \tag{2.1}$$

in the above equation m-biomass of the cell (total), R- ribosomal content, k_1 & k_2 are the rate constants. From Figure 2.3, each activity of protein synthesis is proportional to protein concentration from the outer membrane of the cell to the inner, which is dependent upon the cell surface area A. So the concentration of [S6] can be expressed as

$$\frac{d[S6]}{dt} = k_3 A/V - k_4[S6] \tag{2.2}$$

Where V is volume and k_3 is the rate constant of [S6] concentration. From this equation, we can say that the rate of distribution of [S6] concentration is directly proportional over the surface area A to the volume V ratio. For steady-state concentration, means $\frac{d[S6]}{dt} = 0$, the above equation can be expressed as

$$\frac{k_3 A}{k_4 V} = [S6] \tag{2.3}$$

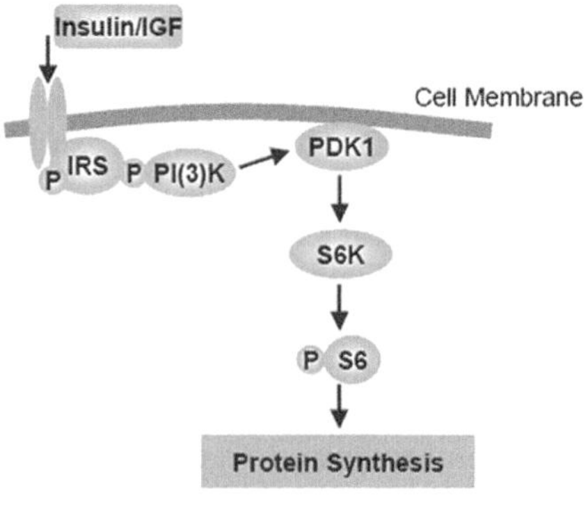

Figure 2.3 Signaling pathways of the growth model of cell.

In this state, the growth rate is fixed, which means the [S6] concentration is constant. From this, we express the rate of change of mass at the initial condition as

$$\frac{dm}{dt} = k_1 \frac{k_3 A}{k_4 V} R - k_2 m \tag{2.4}$$

In Equation 2.4 ($k_2 m$) is protein degradation, which is very minimal during the cell cycle process from which rate of change of mass at initial condition is dependent upon surface A and volume V. But during this phase, the volume doesn't change, and ribosomes are also intact, so R/V = R'(constant). So, cell growth at this phase is dependent upon the surface area of the cell. The larger the cell surface area, the rate is higher where it is correlated with the linear model (Equations 2.5 & 2.6).

$$\frac{dm}{dt} = \frac{k_1 k_3 R'}{k_4} A \tag{2.5}$$

$$\frac{dm}{dt} = KA \tag{2.6}$$

Where K = constant, A – surface area of cell $\frac{dm}{dt}$ = growth rate.

Take into account the cell population each of which contains the potential to grow and divide or perhaps to decease without doing so. Making predictions about the evolution of the population should be possible if one is aware of the characteristics of the individual cells and the mechanisms driving their growth and division. On the contrary, if sufficiently accurate measurement dynamics of a population over time are made, it will be feasible to infer information on the characteristics of the single cells, identifying things like their lifespan and growth rates. The relationship between the cell growth phase and the cell division phase is known to be tenuous and susceptible to disruption by the environment. On the opposing side, there must be some feedback mechanism that persists between them for maintaining stability (Bell & Anderson, 1967).

$$N(V) = N\left(\frac{V_O{}^2}{V}\right) \tag{2.7}$$

Where V_0 is the volume of the cell at its conception and V is the volume at a time before the division.

Despite the fact that cell volume is a desirable statistic because of its simplicity and accuracy of measurement, it may be less closely associated with the cell's fundamental synthesis rates than dry mass or total protein.

Consequently, the findings of certain researchers (Terasima & Tolmach, 1963; Bell & Anderson, 1967) give a suggestion that significant imbibition of water can occur during the premeiotic period, causing the cell volume to more than double the average birth volume for a brief period of the life cycle. The authors indicated although they have not discovered any observations that support an extreme effect of this kind, it is still possible that variable levels of hydration could have a major impact on the apparent rate law that was seen. Overall, researchers have argued that it is difficult to differentiate between linear and exponential growth, based on the calculation of volume or mass as a time-dependent; the largest difference between these two possibilities is just 6% (Bell & Anderson, 1967) as depicted in Figure 2.4. The majority of research on cell proliferation, growth, and death employs proteins having stoichiometric relationships (Chen et al., 2004). These stoichiometric reactions are formally expressed by the CDK-involved kinetic equation. However, these are uncontrollable and unpredictable internal patterns that may result in mismatches that necessitate continuously improved formulations.

Various modeling tools, such as deterministic models' Boolean networks and stochastic models, are used in population-based studies to investigate the roles of distinct gene and protein relationships in the robust progression of the cell cycle. The majority of these models take a deterministic approach. In this method, the time-dependent change of biochemical reaction of each molecular species is characterized by a nonlinear ordinary differential equation (ODE) within which the concentrations of the substance are treated as a continuous variable that evolves predefined with time. However, this approach has failed to provide an accurate assessment of cell-to-cell divergence studies that employ standard deviation models that contain population-wide averages (Ahmadian et al., 2020). Thus, the stochastic simulation algorithm (SSA) introduced by Gillespie (Ahmadian et al., 2020), is the most effective way of implementing fluctuating molecular interactions, which are

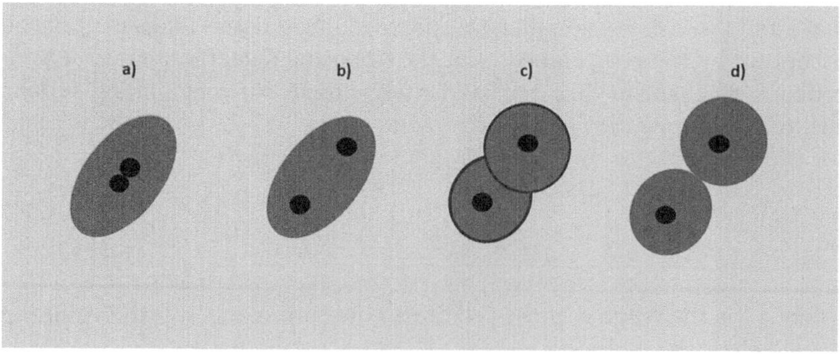

Figure 2.4 Major phases of the cell division cycle (a) nucleus division; (b) cell growth; (c) contractile ring formation; (d) Division and origination of two daughter cells.

described by the mass action kinematics summarized in the preceding equation 2.8. Despite the fact that these studies have increasingly featured non-linearity in mass kinematics reactions, cell cycle and growth models continue to be the focus of research.

$$\frac{dx}{dt} = \frac{\left(k_3' + k_3''A\right)\left(1 - x\right)}{J_3 + 1 - x} - \frac{K_4 myx}{J_4 + x} \tag{2.8}$$

$$\frac{dy}{dt} = k_1 - \left(k_2' + k_2''x\right)y \tag{2.9}$$

$$\frac{dm}{dt} = \mu m\left(1 - \frac{m}{m_*}\right) \tag{2.10}$$

2.2.5 Cancer prognosis

Cancer is commonly regarded as a disorder of the cell cycle. As a result, it is unsurprising that dysregulation of cell growth is among the most common modifications during tumor growth. The progression of the cell cycle is a highly structured, tightly controlled process including many checkpoints that evaluate extracellular growth factor receptors, cell size, and DNA stability. Essential deleterious regulators are cyclin-dependent kinase inhibitors (CKIs), that can function as barriers to impede cell cycle progression in response to regulatory signals. Cyclin-dependent kinases (CDKs) and their cyclin counterparts are significant beneficial regulators or accelerators that accelerate cell cycle progression. Cancer is the result of the aberrant expression or stimulation of favorable regulators and the functional repression of negative regulators. It is believed to imply that the profound acquisition of mutations and gene regulation abnormalities in the expression of various genes with highly diverse functions is required for tumorigenesis, the correlation between age and cancer incidence has long been interpreted as indicating that tumorigenesis is a long-term process. An important subset of these genes is engaged in stages of the cell cycle, which are places of control that guarantee the correct order of cell cycle activities and incorporate DNA synthesis with cell-cycle advancement.

In two of these phases, the cells carry out the fundamental activities of cell replication, such as the production of a single, accurate copy of their genetic material (synthetic or S-phase) and the segregation of all cell functions between both identical daughter cells (G1 phase) (mitosis or M phase). The remaining two stages of the cell (G1 and G2) are transitional phases where the cells prepare for their completion of the S and M phases, correspondingly. When cells quit proliferating as a result of antimitogenic signals or a lack of mitogenic activity, they leave the cell cycle and reach a non-proliferating, quiescent state called G0. In contrast, the cell cycle can be halted at the G1 or

G2 checkpoints, which evaluate cell size, extracellular growth impulses, and DNA stability (Park & Lee, 2003).

Cell cycle regulators are commonly altered in human malignancies, highlighting the significance of maintaining cell cycle commitment in the control of chronic cancer, which is revealed by molecular analysis.

2.3 TECHNIQUES INVOLVED IN DETECTION OF BIOPHYSICAL PROPERTIES OF THE CELL

2.3.1 Coulter devices

Cell separation enables the detection of cells, and the issues associated with cell separation include the localization of cell types and the extraction of single cells enabling biophysical characterizations. These devices, formerly known as resistive pulse sensors, are highly developed instruments used to determine the size and concentration of suspended living cells and nano particles in a solution. The solutions containing particles are permitted to pass via a microchannel dividing into separate chambers, or micro-channels, in Coulter devices. When a particle moves throughout a channel, the electrical conductivity of the channel is altered due to resistance change. These changes in resistance can be detected as pulses of current/voltage, which can be linked to particle bio-physical characteristics - size, shape, mobility, surface charge density, and concentration.

Coulter counters are comprised of two chambers separated by an insulating membrane and a single channel (shown in Figure 2.5, Henriquez et al.,

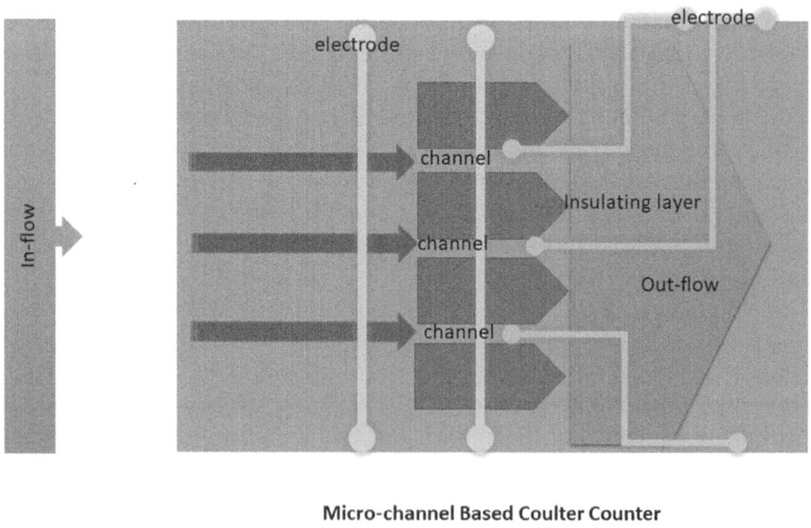

Figure 2.5 Schematic of Microchannel (Multi-analyte) Coulter Counter.

2004). An ionic current is driven across the channel by electrodes submerged in a liquid electrolyte present within every chamber. If, in contrast to the solution, there are particles of a size comparable to the channel's diameter, then these particles may enter the channel and lower the magnitude of the ionic current. A Coulter counter's output is a graph of current versus time (Henriquez et al., 2004) containing an array of current pulses. W. Coulter patented the Coulter counter in 1953, which has been routinely used in medical laboratories to determine biological cell densities over the past fifty years. Due to their ease of use, high sensitivity, and dependability, Coulter-type machines have been employed in a wide variety of applications, including the study of blood cells (Zimmerman, 1993; Zimmermann et al., 1974), the observation and counting of colloidal particles (DeBlois & Bean, 1970; Saleh & Sohn, 2001), pollen (Zheng, 2013), metal ions, and viruses (DeBlois & Wesley, 1977). Several groups have recently demonstrated the use of Coulter counter devices with a single nanoscale channel for the detection of latex microparticles (Henriquez et al., 2004; Saleh & Sohn, 2001), single molecules (Bayley & Martin, 2000), DNA (Kasianowicz et al., 1996), and antibody–antigen binding (Saleh & Sohn, 2001). Pathogenic bioparticles, including food poisoning bacteria, can be recovered in an electrolyte solution, using normal extraction methods (Zhao, 2016; Zhao et al., 2004). However, an issue posed by such gadgets is decreased sensitivity. Traditional Coulter counters with a single channel have a throughput that is proportional to the cube of the channel's diameter. If the device is designed to count submicron or nanometer-sized particles, the detector's diameter must be made proportionally smaller such that the inclusion of a particle has a significant and detectable effect on the channel resistance; otherwise, the device's sensitivity is reduced and, with a single-channel device that can deal with a very small volume, restricting the system's throughput in the presence of degraded sensitivity because at the conjunction with this bioactive particle that is typically present in extremely dilute amounts, detection with background noise is quite challenging. Hence, it would take an impractically long time to detect a significant volume (>100 ml) of the bioparticle solution without processing by using a mono-channel detector (Wu, 2006). To prevent such volumetric difficulties, techniques such as the multichannel Coulter counter are used (Zhe et al., 2007).

Currently available technologies enable the creation of smaller, more precisely defined channels than was before conceivable. Second, the interest in characterizing nanoscale things is at an all-time high. Third, there is a substantial commercial motivation to design portable (compact, lightweight, and low energy consumption), sensitive and selective label-free biosensors.

2.3.2 Fluorescent-based techniques

Fluorescence microscopy is the most widely used method of cellular examination for cell biology at the single cell level. Flow cytometry is utilized to analyze specifically a large number of cells just at the individual cell level, whereas fluorescence microscopy is employed to observe cells continuously in real time over duration (Lu & Rosenzweig, 2000). However, miniature fluorescence sensors for single cell detection have recently been designed. The sensor needs to be around 100 times smaller than the examined cell to achieve full non-invasive intracellular analysis (Lu & Rosenzweig, 2000).

One must surpass the conceptual resolution limit of light microscopy, referred to as the diffraction limit, in order to perform opto-chemical sensing in nanoscale level. The fluorescence excitation wavelength, l, determines the diffraction limit, which is equal to l/2. Realizing that, while light can't be focused to a point lower than l/2 but can be apertured to generate a spot of these tiny dimensions forms the basis for the development of fluorescence sensors having resolution above the diffraction limit. This idea was first applied to the creation of near-field optical microscopy (NSOM). Thus, near-field opto-chemical sensors are another name for submicron fluorescence sensors. By integrating the detecting reagent inside a micropipette that has a tapered tip of submicron dimensions by using a micropipette puller, Lewis et al (Aaron Lewis and Klony Lieberman, 1991) created the first submicron fluorescence sensor in 1991. Kopelman et al. (Tan et al., 1992) created the first nanometer based optical fibre sensor. The majority of studies concentrate on the capacity to detect minimal concentrations of fluorescence from the sensor as well as the miniaturization of sensors utilizing micromachining techniques. Recent improvements in digital fluorescence imaging equipment have substantially aided the field of fluorescent nano sensors (Rizzuto et al., 1998). For instance, it is now possible to identify individual fluorescent molecules and track the kinetics of their diffusion in aqueous solution at ambient temperature (Xu & Yeung, 1997). These developments make it possible to see signal variations in fluorescent nanodevices. Fluorescent nano-sensors' analytical capabilities enable precise investigation of biological systems.

The biocompatibility of fluorescent nanodevices is a significant challenge that needs to be addressed when using this approach. Although the method is minimally invasive, the prolonged presence of a nanoparticle or fibre optic sensor tip in a cell may have harmful effects that cause cancer or cell death. Therefore, it is crucial to research how sensing materials affect cells and choose sensing materials that have a negligible impact on the survivability and functionality of cells (Lu & Rosenzweig, 2000).

2.3.3 Flow cytometry

In a standard flow cytometer, discrete particles move through an illuminating zone at a rate of about 1,000 cells per second (although considerably higher speeds are feasible with specialist instruments) and appropriate sensors determine the amplitude of a pulse that represents the amount of light scattered. The amplitude of such pulses is electronically sorted as "bins" or "channels," allowing the presentation of histograms relating the number of cells with a particular quantitative attribute to the quantity of channels. Despite the fact that many applications of flow cytometry, including such microbial identification, demand only qualitative results. The angle dependence of light scattering delivers further data about the nature of the scattering particle, but more crucially, fluorophores can be added to the liquid culture in order to better see the scattering particles that are stains that bind to (react) specific molecules like DNA, RNA, or protein; fluorogenic substrates that reveal distributions in lipase activity; markers that change their property as a result of PH change.

2.3.4 Mass spectrometry

Because of its outstanding sensitivity, selectivity, accuracy, and high throughput screening capabilities, mass spectrometry (MS), one of the most potent analytical instruments, is progressively being used in a wide range of scientific domains. It can be used in a variety of fields, such as forensics (Ojanperä et al., 2012; Wood et al., 2006), sports co-doped (Ojanperä et al., 2012; Peters et al., 2010), food certification and quality (Gan et al., 2014; Vaclavik et al., 2012), and biological and pharmaceutical studies (Hopfgartner et al., 2012). By analyzing the mass-to-charge ratio (m/z) readings of a charged particles and maldi-tof mass spectrometric fragments, Mass Spectrometry can reveal structural information about the analyte. There are both high- and low-resolution MS instruments available on the market today, and each one has three essential components: an ionized source, a mass analyzer, and a sensor (El-Aneed et al., 2009)(8). Multiple instruments should be employed concurrently to address difficult research problems because MS instruments come in a wide variety of configurations and have variable capacities. The initial step is the transformation of target analytes first from liquid or solid phases into gas-phase ionized species, regardless of the MS apparatus or the application. Since the aforementioned process denatures the cellular characteristics, this is the main obstacle and challenge in detecting the bio-physical characteristics of individual cells (Awad et al., 2015).

2.3.5 Surface plasmon resonance

In the 1980s, the first investigations on the application of surface plasmons for sensing (Lin et al., 2008) and probing operations on the interfaces of metal layers (Gordon & Ernst, 1980) appeared. The first SPR- biosensors were developed and used to study biomolecular interactions in the early 1990s (Löfås et al., 1991). Traditional SPR biosensors allow for the real-time measurements of interactions between biological items immobilized upon that surface of a metal supporting plasmon and their equivalents in a liquid sample. Surface plasmon resonance imaging (SPR imaging) and microscopy are techniques that aim to enable localization of these interactions by imaging of the metal surface (Špačková et al., 2016). The collective oscillations of free conductive electrons, also called electron gas, in the metal close to the metal-dielectric interface are known as surface plasmons. Surface plasmons have a confined electromagnetic field that decays into both the metal and the dielectric at the interface (Špačková et al., 2016). Because of this quality, surface plasmons are useful for examining processes on metal surfaces (Lin et al., 2008; Maier, 2009). There are numerous distinct surface plasmon modes on various metal-dielectric (nano)structures (Špačková et al., 2021). The surface plasmon mode that is most frequently used in SPR imaging and microscopy is the propagating surface plasmon (PSP), supported by a continuous metallic sheet. Despite the fact that SPR imagery and microscopy are frequently employed to identify a variety of biological organisms, this work focused only on reporting cells. Argoul's team reported on their investigation of adherent cells with the help of a high-resolution SPR microscope (Streppa et al., 2016). They examined the adhesion and motility of C2C12 mouse myoblast cells grown on an SPR chip. The development of the adhering cells' shapes was monitored, and the dynamics of filopodia and lamellipodia indentations were seen locally. Additionally, they watched as C2C12 mouse myoblast cells moved, adhered, detached, and attached to a gold substrate. Tu et al. looked at the way that individual cells adhere (Lin et al., 2008; Špačková et al., 2016). On a nanoporous array system integrated with microfluidic devices, including single cell trapping devices, they used SPR microscopy. The cultivation of human tumor cells (HeLa) and mouse embryonic stem cells (C3H10) was observed, and it was shown that the single-cell attachment process adheres to the logistic delayed growth model (with different parameters for the different cells). Tao's team investigated the movement of mitochondria along primary rat hippocampal neurons (Špačková et al., 2016).

2.4 MICRO ELECTRO MECHANICAL SYSTEMS BASED MASS SENSORS

Using microfluidic systems with sample sizes as small as nanoliters, biomolecule detection is made possible by microfabricated transducers. By boosting analytical analysis, with its combination of microfluidic sample processing into lab-on-a-chip devices can tremendously leverage experimental work in systems biology and pharmaceutical research that significantly lowering reagents cost. Strong and compact detection equipment with real-time functionalities for point-of-use applications can also be made possible by microdevices (Burg, 2006).

Changes in a physical characteristic, such as charge, refractive index, contact stress, or mass at a solid-liquid interface, are commonly measured by sensors that detect unlabeled biomolecules. Immobilized receptor preferentially attaches the analytes of interest to offer specificity by changing the surface's characteristics and producing a signal. This method does not necessitate a monoclonal antibodies or fluorescent labeling of the molecules, in contrast to traditional protein microarrays (Burg, 2006). As a result, label-free detection techniques enable investigations where labeling might impede the binding process or when real-time observations of the detection test are important. Additionally, label-free approaches are appropriate in many situations in which a fluorescent plate assay cannot be employed because no monoclonal antibody is needed for detection.

The two most popular techniques for label-free interacting protein analysis at the present moment are surface plasmon resonance (SPR) (Burg & Manalis, 2003; Son et al., 2012) and quartz crystal microbalance (QCM) (Fritz et al., 2002; Skládal, 2003), with usages spanning from basic systems biology research to pharmaceutical development and quality control. Commercially available laboratory instruments based on SPR and QCM exist, although these two concepts are difficult to miniaturize and produce in large quantities. Electronic, optical, and mechanical transducers are the three main types of microfabricated sensing devices for biochemistry, each of which has unique benefits and drawbacks. Despite the fact that charge screening makes electronic field-effect devices highly sensitive to the binding of charged molecules of any molecular weight, these sensors require low ionic strength solutions and tightly bound ligands to the surface (Cui, 2001). Integrated optical sensors must precisely align external optical components to the microfabricated part in order to detect the refractive index in the emission spectrum of planar waveguides, which typically restricts the depth of the sensitive layer to less than 100 nm. (Lukashenko, 2018; Thundat, 1995) Recently, numerous research projects have focused on micromechanical interface stress sensors, in particular due to their ease of use and high sensitivity in some tests ((B., 2004) Lange, 2002; White & Wenzel, 1988) Sensitivity to various target molecules can vary significantly because surface stress is dependent on a number of physical characteristics of the interface, including charge, hydrophobicity, and hydrophobic interactions between adsorbed molecules.

Assay development is further complicated by the requirement for distinct surface functioning on the device's top and bottom sides. MEMS resonators with a mass measurement provided by their natural frequency make up a different class of micromechanical transducers; chemical sensing in gaseous conditions has proved incredibly successful with resonant mass sensors (Wang et al., 1998; Weinberg, 2003). However, the poor-quality factor and highly effective mass caused by viscous drag in liquids reduce the sensitivity and resolution of resonant sensors. Although some designs have the ability to significantly reduce these restrictions, their sensitivity is still less sensitive than resonators based on air or vacuum (O'Sullivan, 2019; O'Sullivan & Guilbault, 1999). Several organizations have used the "dip and dry" method to solve this issue, which involves measuring the resonant frequencies in air both prior to and after the device has been confined to the sample (B., 2004)

The resonating element must be kept clean at all costs to guarantee reliability. Furthermore, real-time data for the study of binding kinetics are not possible with the approach. Mass boosting labels can be employed in some experiments to improve the S/N ratio in fluid at the expense of more time and effort spent on sample processing and experimentation (Wang et al., 1998).

Due to their miniaturization capabilities and batch production capabilities, MEMS-based resonant mass sensors are being utilized increasingly frequently. Additionally, these tools make it possible to work with nanoliter- and picolitre-sized samples that are extremely small. Resonant mass sensors have the ability to overcome these difficulties for the detection of biophysical features of the cell that are complicated in nature as compared to traditional devices such as SPR and Mass spectrometry (Burg, 2006).

During the division and proliferation of cells in multicellular organisms over the last 50 years, there has been a tremendous amount of interest in figuring out how these processes are controlled by both single cells and cell groups. This search has included direct measurements of fluctuations in human cell mass against growth rate (Cooper, 2006; Killander & Zetterberg, 1965; Mitchison, 2003). These measurements have the ability to shed light on the intrinsic mechanism that coordinates cellular proliferation and cell growth (Brooks & Shields, 1985; Paul Jorgensen et al., 2002) and reveal whether the growth rate is constant regardless of cell size and cell cycle (T P Burg & Manalis, 2003) or whether it is proportionate to cell size (Gupta et al., 2004). The linearly increasing model is predicated on the idea that the amount of DNA needed to start the transcription process, or "gene dosage," limits the pace of biosynthesis (Cooper, 2006). Contrarily, the exponentially growing model is predicated on the idea that the amount of cytoplasm and ribosome machinery within a cell determines how much that cell mass increases. As a result, a cell's capacity to produce more mass and accelerate growth increases as it gets bigger (or heavier). Theoretically, whereas exponential growth necessitates a size-dependent mechanism for size homeostasis, linear growth can preserve cell-size homeostasis without one (Bryan, 2010; Tzur, 2011). In recent years, significant progress has been made in this direction thanks to interferometric measures of dry cell mass (Godin, 2010), population measurements of buoyant mass of suspended cells (Burg & Manalis, 2003; (Thomas P. Burg et al., 2009), or volume estimates of softly synchronized subgroups of suspended mammalian cells (Tzur, 2009).

2.4.1 Technique of mass sensing using resonant mass sensor

According to theory, a cantilever's resonance frequency, f, sensitively fluctuates with mass loading (m). k is the cantilever's spring constant, m^* is its effective mass, and α is a mathematical constant (Datar et al., 2009). A cantilever's surface area can be increased through nanopatterning, which can increase adsorbed mass and detection sensitivity. In order to increase the adsorbed mass, Lee et al. (2000), presented a cantilever containing nanofabricated perforations (Datar et al., 2009). When used under solution medium, mass detection employing using cantilever resonance frequency does have very poor mass resolution even if it is ideally suited for measuring mass in vacuum and air. As a result, resonance frequency fluctuations are rarely employed for the extremely sensitive sensing of attached mass in liquid environments. Therefore, when biomarkers need to be found in bodily fluids like serum, the surface stress changes are the method of choice. The SMR, a hollow cantilever idea, was recently developed by Burg et al. (2003) and is capable of identifying biological interactions in liquids with seemingly unbelievable sensitivity.

$$f = \frac{1}{2\pi}\sqrt{\frac{k}{m^* + \propto \Delta m}}$$

2.4.2 Cantilever mechanics

Surface free energy is reduced as a result of molecular adsorption. A differentially surface stress is created between the two sides of a cantilever beam even though the adsorption of molecules is primarily restricted to one side. The Shuttleworth equation can be used to link surface stress, measured in g, and surface free energy, measured in E. where the ratio of surface area to the total area is known as the surface stress. The strain component is sometimes overlooked since the cantilever's bending is so modest in comparison to its length. But there is lively debate over this topic in the literature (Datar et al., 2009). Cantilever bending is caused by the variable surface stress caused by molecule adsorption. The cantilever deflection, h, and the differential in surface stress, g, between the chemically changed and untreated surfaces are related by Stoney's equation, where v is the material's Poisson's ratio, E is the cantilever material's Young's (elastic) modulus, and t and L are the cantilever's thickness and length, respectively (Datar et al., 2009). The surface stress can alternatively be conceptualized as a shift in the surface energy density or surface tension. It is obvious that a cantilever's sensitivity to measuring surface stresses increases with length (Datar et al., 2009).

$$g = \gamma + \frac{d\gamma}{d\varepsilon}$$

$$\Delta g = \frac{E\Delta h}{4(1-\upsilon)}\left(\frac{t}{L}\right)2$$

2.5 MASS SENSORS REPORTED IN LITERATURE

Though the traditional cantilever structure is widely used for the mass sensing at different regimes of micro to nano scale, dimensions have the challenges of the stiction, surface to volume ratio, and functionalization of surfaces for labeling. Along with these issues, rectangular cantilevers suffer from sensitivity and quality factor issues. In the literature, most reviewed and reported resonant mass sensors are pedestal mass sensors and suspended micro resonating channels (SMR).

2.5.1 Pedestal mass sensor

Due to the cantilever beam structure's basic geometry and suitability for extreme downsizing, higher mass sensitivity can be attained. However, as is widely known, the mass sensitivity of these cantilevers beam resonant mass sensors are spatially non-uniform (Park, 2012). When the added mass is positioned at the cantilever's free end, the mass sensitivity is at its highest, and it drops to zero as the added mass approaches the fixed end. If there will be a lot of target entities coupled and they are considerably smaller than the sensor (Park & Bashir, 2009), one may assume a uniform mass distribution and apply an average mass sensitivity that is simple to determine analytically. But if only a few or a single target entity need Ansys to be connected to the sensor, one cannot assume that the target mass will be distributed uniformly. Instead, one must either limit the attachment site to the cantilever's end or adjust the extracted mass using the mass distribution from optical pictures of cantilevers (Burg & Manalis, 2003). These methods, however, lessen the mass sensor's actual sensitivity and diminish its usefulness.

The sensors consist of four partially layered springs and a platform in the middle on which the object would be attached. The topmost silicon layer has a thickness of 2 micron. On the platform areas and the springs, a small layer of metal is coated to enable magnetic actuation by electric current. A XeF_2 etcher is used to etch the silicon substrate underneath the spring and cantilever, releasing the device.

The square of the vibration amplitude at a given position determines the mass sensitivity of a resonating mass sensor there (Park & Bashir, 2009). A homogeneous mass sensitivity can be attained by maintaining a constant vibration amplitude at the target attachment site. The flexural vibrational bending on the beam spring is partially converted into torsional bending at the folded region of the beam spring because of the special structure of the sensor, which causes the platform to vibrate with a consistent amplitude. The platform's variation in vibration amplitude is just 2.0% of the maximum value. As a result, there is a 4.1% discrepancy between the mass sensitivity's minimum and highest value. The HT29 adherent cells are calibrated with the help of this pedestal mass sensor. The HT29 cells proliferate on median 3.25% of their mass per hour, which results in a mass doubling time of around 2214 log 2 log 1.032 hours, according to our examination of the growth rates of individual cells. We expanded our research to include pedestals that caught numerous adhering cells in addition to just single adherent cells (Park, 2010) Figure 2.6.

2.5.2 Suspended micro-resonating channel (SMR)

Target molecules are passed in a vibrating suspended microchannel during SMR detection, where they are acquired by receptor molecules connected to the inner channel walls. The method by which a frequency signal is generated from biomolecules that have bound to the interior wall surfaces of a suspended micro-fluidic channel. The total effective mass and the spring constant work together to generate the physical resonant frequency in the suspended fluid channel (Burg et al., 2006) The liquid inside the channel, the mass of any deposited substance on the channel walls, and the constant mass of the channel walls make up the effective mass. The mass deposited to the channel walls contributes significantly if the surface area to volume proportion of the channel is high. The bulk density of proteins in aqueous solutions is roughly (Burg & Manalis, 2003; Burg et al., 2007; Cermak et al., 2016; Datar et al., 2009; Son et al., 2012), which is larger than that of the densities of the pure solution, making them detectable by this approach. Proteins that bind to the receptors are drained from the solution and aggregate on the sidewalls as they are constantly replenished by the passage of sample through the device, if indeed the channel walls are derivatized with bio-receptors. Therefore, even if a target molecule's concentration in liquid is too low to be determined by a change in bulk density, it is still possible to detect its existence.

A particle's mass as it travels through the SMR channel is determined by the temporary resonant frequency shift. Multiple successive measurements of individual particles are added up to determine the propagation of particles in solution. The SMR determines the difference in mass of a substance with regard to that of the displaced fluid, Md = V (rh_{op}–rh_{of}), where V is the particle's volume and p and f are the densities of the particle and the carrier fluid, respectively. The frequency response of the cantilever is defined by the total mass, that of the fluid and of a suspended particle. V = 4/3piR^3 yields the radius of a particle R that is believed to be spherical. When the particle density exceeds that of the surrounding solution, the differential mass M d is positive; if the nanoparticle is buoyant in the solution, it is negative. A mixture of 708.6 nm polystyrene (p =1.05 g/cm^3) and 99.5 nm gold (p 19.3 g/cm^3) nanoparticles was fed through the resonator at a concentration of 108 particles/ ml to demonstrate the impact of liquid density on the measured differential mass. It demonstrates that the average frequency shift for the mixture is between 25 and 3 mHz, indicating that both the gold as well as the polystyrene particles exhibit positive differential masses in PBS solution of 1.0 g/cm^3(Burg, 2006; Burg et al., 2007). Although the masses of

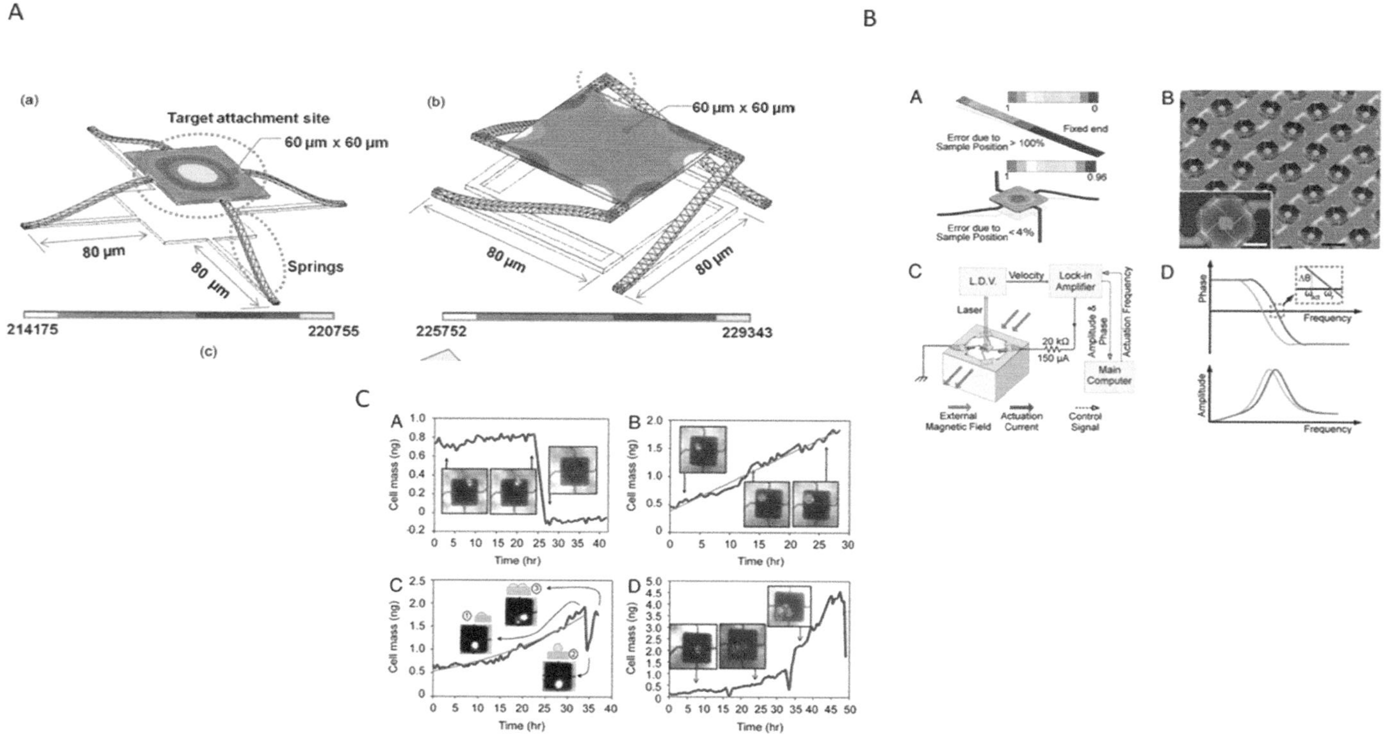

Figure 2.6 Pedestal Mass Sensor for adherent cell detection using uniform mass sensitivity. (a) Design and modelling of pedestal mass sensor (Park, 2012); (b) Fabrication and characterization of pedestal mass sensor; (c) Characterisation of adherent cells on sensor for mass sensing applications – explaining about the proliferation, dead cells and division (Park, 2011).

the polystyrene 195.6 fg and gold 9.8 fg particles are very different from one another, their divergent masses with respect to the transport fluid PBS are both close to 9 fg. (Burg, 2006) Figure 2.7.

2.6 FRACTAL MEMS STRUCTURES

The scalability and change of the micro cantilever's dimensions, structures, and materials determine its sensitivity. The sensitivity for a specified dimension can be changed by the materials (Kirstein et al., 2005) and structural modifications that demonstrate a greater shift-resonant frequency per absorbed unit of mass. Using fractal structures, which have a large surface area for increased selectivity in addition to sensitivity, is one way to increase sensitivity. This study takes into account the design of cantilevers employing fractal structures to boost sensitivity while dimensional stabilization for mass measurements utilizing cantilever resonant frequency phase. Mandelbrot (Mendelbrot, 1977) (first used the term "fractal" in 1981. He defined it as a structure that is self-similar, has repeating patterns, and has reduced dimensions from pattern to pattern, which is the central concept and how it is implemented from one order (level) to another across the structure. Heldge Von-Koch and Georg Cantor's modeling and other mathematical representation techniques were used in 1904 to illustrate the central concept and the design of the object (Zhang, 2014). All of them support the statement "Continuous everywhere but differentiable nowhere" (Zhang, 2013) that fractals are. Since their introduction, these fractals have been applied to geometry, designs, pattern recognition, city planning, etc. Additional fractal geometry research is being applied to the respiratory and circulatory systems of mammals (Zhang, 2014). Inductors are made using fractal geometry designs in electric systems because of their high surface-to-volume ratios (Chen & Cheng, 2002). These results show an improvement in sensitivity after fractal porous structure on the surface of cantilevers. The fractal geometry cantilever will be introduced in this paper to improve the sensitivity and selectivity of cantilevers without compromising their dimensional scalability (Aditya & Ru, 2017).

2.6.1 Fractal tree geometry realization

The effectiveness of all living things is best demonstrated by nature, which is organized using one of the fractal methods. The tree is one of the best illustrations of fractal geometry since it starts with a surface-level trunk, divides into branches, and then into twigs. Each of these components is identical to the trunk in shape, but they have smaller dimensions. The branches get smaller and more intricate as the order gets higher, but they still resemble the first order. For maximal strength to support its full mass, it splits the root system in a similar manner, even below the surface, as seen in Figure 2.8

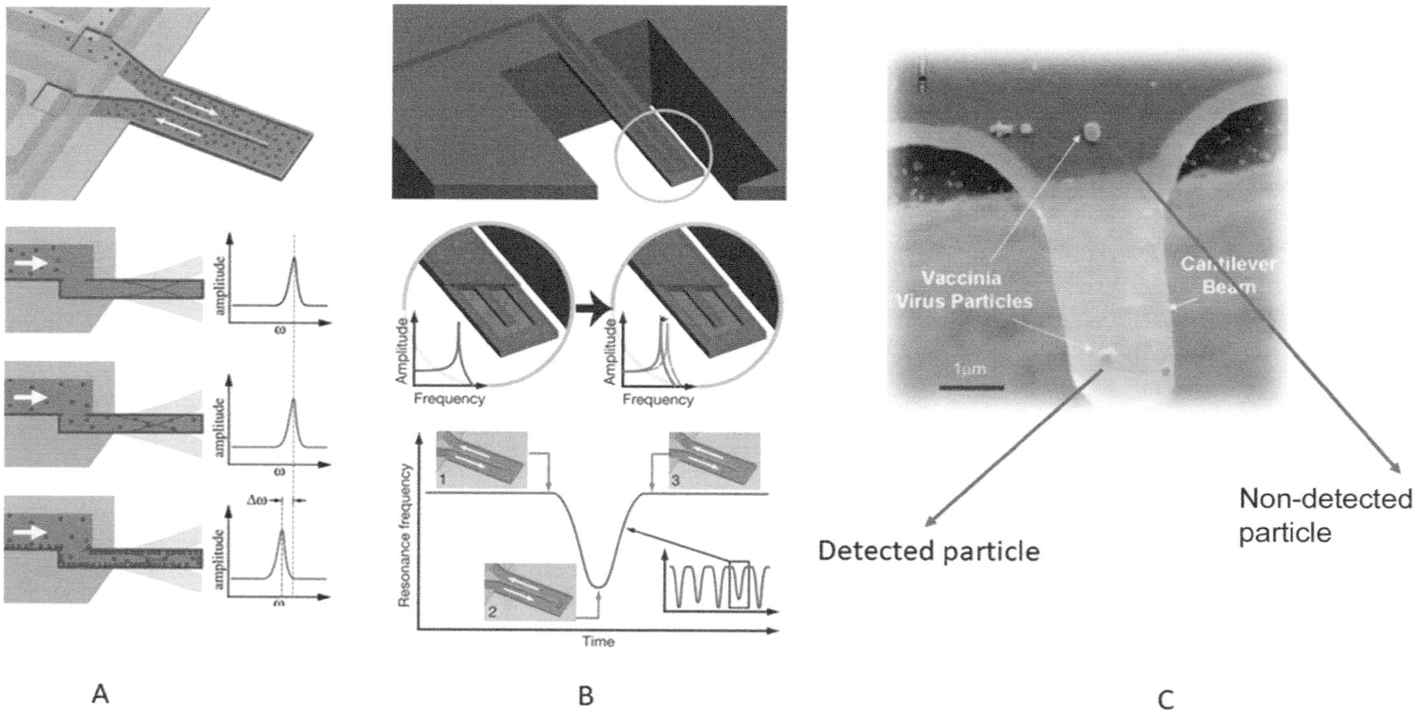

Figure 2.7 (a) Suspended Micro resonating channel; (b) Fabricated suspended resonating channels for label free detection of single channels; (c) challenging with traditional cantilevers were sensitivity is compromised when attached at the fixed end (T P Burg, 2006; Gagino, 2020; Gupta et al., 2004; Lee, 2011).

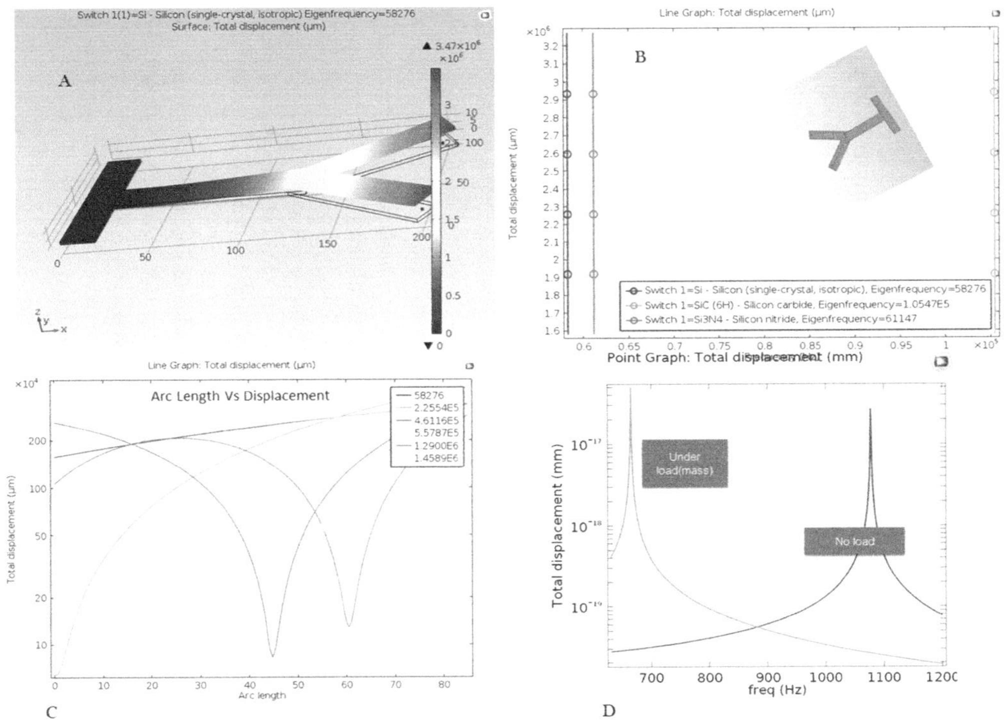

Figure 2.8 Fractal geometry realized for mass sensing application using gold nano particle. (a) Principle mode fractal cantilevers K = I stage; (b) Material sweep analysis of the fractal structure; (c) Deflection of the fractal cantilever in different Eigen frequency modes; (d) Amplitude vs Frequency shift from no load and load conditions

In nature, dimensions may follow symmetry in structure but not in size, however when designing fractal structures, higher order fractals require mass distribution symmetry (Aditya & Ru, 2017).

$$V = n_k M_k$$

$$V = n_k L_k \, M_k$$

The degree or order of fractals at which the trees is differentiable is depicted in Figure 2.8(c). The tree begins at the base as a trunk of length for iteration K = 0. Then, for the purpose of increasing the order of K, branches can be divided into scaled copies that are identical to one another. The scalability at each level in that specific sequence of the tree can be achieved with nk *L or nk *B. Scaling may be nk *L /B for an iteration level of K = i. Each tree level in this realization has a pattern with a 30° angle with respect to plain. The 100-micron level zero (K = 0) initiator, also known as the fractal cantilever structure, is thought to have a 20-micron breadth. Given that the cantilever's scalability is initially assumed to be around 0.9 due to the intricate geometrical structure of the fractal cantilever tree, the length and width are scaled to 90 micrometers and 18 micrometers with a reference angle of angle of 30 ° for K = 1 and 81 micron and 16.2 micron, respectively, for K = 2 when there are 4 elements.

REFERENCES

Aditya, A. L. G. N., & Rufus, E. (2017). *Modelling and simulation of fractal cantilever's MEMS based for mass sensing applications*. Conference: 2017 International Conference on Nextgen Electronic Technologies: Silicon to Software (ICNETS2), 154–157.

Ahmadian, M., Tyson, J. J., & Peccoud, J. (2020). Open a hybrid stochastic model of the budding yeast cell cycle. *Npj Systems Biology and Applications*, 1–10. https://doi.org/10.1038/s41540-020-0126-z

Akin, D., & Gupta, A. (2004). Detection of bacterial cells and antibodies using surface micromachined thin silicon cantilever resonators. *Other Nanotechnology Publications*, 22. https://doi.org/10.1116/1.1824047

Awad, H., Khamis, M. M., & El-Aneed, A. (2015). Mass spectrometry, review of the basics: Ionization. *Applied Spectroscopy Reviews*, 50(2), pp. 158–175. doi: 10.1080/05704928.2014.954046

Bayley, H., & Martin, C. R. (2000). Resistive-pulse sensing - from microbes to molecules. *Chemical Reviews*, *100*(7), 2575–2594. https://doi.org/10.1021/cr980099g

Bell, G. I., & Anderson, E. C. (1967). Cell growth and division I. A mathematical model with applications to cell volume distrlutions. *Biophysical Journal*, 7(4), 329–351. https://doi.org/10.1016/S0006-3495(67)86592-5

Bergman, L. W. (n.d.). Growth and maintenance of yeast. *Methods in Molecular Biology*, *177*, 9–14.

Brooks, R. F., & Shields, R. (1985). Cell growth, cell division and cell size homeostasis in Swiss 3T3 cells. *Experimental Cell Research*, *156*(1), 1–6. https://doi.org/10.1016/0014-4827(85)90255-1

Bryan, A. K., Goranov, A., Amon, A., & Manalis, S. R. (2010). Measurement of mass, density, and volume during the cell cycle of yeast. *Proceedings of the National Academy of Sciences of the United States of America*. https://doi.org/10.1073/pnas.0901851107

Burg, T. P. (2003). Suspended microchannel resonators for biomolecular detection. *Applied Physics Letters*. https://doi.org/10.1063/1.1611625

Burg, T. P. (2006). Vacuum-packaged suspended microchannel resonant mass sensor for biomolecular detection. *Journal of Microelectromechanical Systems*. https://doi.org/10.1109/JMEMS.2006.883568

Burg, T. P. (2015). Devices with embedded channels. *Advanced Micro and Nanosystems*. https://doi.org/10.1002/9783527676330.ch11

Burg, T. P., Godin, M., Knudsen, S. M., Shen, W., Carlson, G., Foster, J. S., Babcock, K., & Manalis, S. R. (2007). Weighing of biomolecules, single cells and single nanoparticles in fluid. *Nature*, *446*(7139), 1066–1069. https://doi.org/10.1038/nature05741

Burg, T. P., & Manalis, S. R. (2003). Suspended microchannel resonators for biomolecular detection. *Applied Physics Letters*, *83*(13), 2698–2700. https://doi.org/10.1063/1.1611625

Burg, T. P., Mirza, A. R., Milovic, N., Tsau, C. H., Popescu, G. A., Foster, J. S., & Manalis, S. R. (2006). Vacuum-packaged suspended microchannel resonant mass sensor for biomolecular detection. *Journal of Microelectromechanical Systems*, *15*(6), 1466–1476. https://doi.org/10.1109/JMEMS.2006.883568

Burg, T. P., Sader, J. E., & Manalis, S. R. (2009). Nonmonotonic energy dissipation in microfluidic resonators. *Physical Review Letters* https://doi.org/10.1103/PhysRevLett.102.228103

Cermak, N., Becker, J. W., Knudsen, S. M., Chisholm, S. W., Manalis, S. R., & Polz, M. F. (2016). Direct single-cell biomass estimates for marine bacteria via Archimedes' principle. *The ISME Journal: Multidisciplinary Journal of Microbial Ecology*, *11*, 825–828. https://doi.org/10.1038/ismej.2016.161

Charlebois, D. A. (2019). Modeling cell population dynamics. *In Silico Biology*, *13*, 21–39. https://doi.org/10.3233/ISB-180470

Chen, K. C., Calzone, L., Csikasz-Nagy, A., Cross, F. R., Novak, B., & Tyson, J. J. (2004). *Integrative Analysis of Cell Cycle Control in Budding Yeast 15*(August), 3841–3862. https://doi.org/10.1091/mbc.E03

Chen, Y., & Cheng, P. (2002). Heat transfer and pressure drop in fractal tree-like microchannel nets. *International Journal of Heat and Mass Transfer*, *45*(13), 2643–2648. https://doi.org/10.1016/S0017-9310(02)00013-3

Cooper, S. (2006). Distinguishing between linear and exponential cell growth during the division cycle: Single-cell studies, cell-culture studies, and the object of cell-cycle research. *Theoretical Biology and Medical Modelling*, *3*, 1–15. https://doi.org/10.1186/1742-4682-3-10

Crivelli, J. J., Földes, J., Kim, P. S., Wares, J. R., Crivelli, J. J., Földes, J., Kim, P. S., & Wares, J. R. (2012). A mathematical model for cell cycle-specific cancer virotherapy. *Journal of Biological Dynamics*, 3758. https://doi.org/10.1080/17513758.2011.613486

Cross, F. R., Archambault, V., & Miller, M. (2002). Testing a mathematical model of the yeast cell cycle. *Molecular Biology of the Cell*, *13*(January), 52–70. https://doi.org/10.1091/mbc.01

Cui, Y. (2001). Nanowire nanosensors for highly sensitive and selective detection of biological and chemical species. *Science*. https://doi.org/10.1126/science.1062711

Datar, R., Kim, S., Jeon, S., Hesketh, P., Manalis, S., Boisen, A., & Thundat, T. (2009). Cantilever sensors: Nanomechanical tools for diagnostics. *MRS Bulletin*, *34*(6), 449–454. https://doi.org/10.1557/mrs2009.121

Davila, A. P., Jang, J., Gupta, A. K., Walter, T., Aronson, A., & Bashir, R. (2007). Microresonator mass sensors for detection of Bacillus anthracis Sterne spores in air and water. *Biosensors and Bioelectronics*, *22*(12), 3028–3035. https://doi.org/10.1016/j.bios.2007.01.012

DeBlois, R. W., & Bean, C. P. (1970). Counting and sizing of submicron particles by the resistive pulse technique. *Review of Scientific Instruments*, *41*(7), 909–916. https://doi.org/10.1063/1.1684724

DeBlois, R. W., & Wesley, R. K. (1977). Sizes and concentrations of several type C oncornaviruses and bacteriophage T2 by the resistive-pulse technique. *Journal of Virology*, *23*(2), 227–233. https://doi.org/10.1128/jvi.23.2.227-233.1977

El-Aneed, A., Cohen, A., & Banoub, J. (2009). Mass spectrometry, review of the basics: Electrospray, MALDI, and commonly used mass analyzers. *Applied Spectroscopy Reviews*, *44*(3), 210–230. https://doi.org/10.1080/05704920902717872

Fritz, J., Cooper, E. B., Gaudet, S., Sorger, P. K., & Manalis, S. R. (2002). Electronic detection of DNA by its intrinsic molecular charge. *Proceedings of the National Academy of Sciences of the United States of America*, *99*(22), 14142–14146. https://doi.org/10.1073/pnas.232276699

Gan, H. H., Soukoulis, C., & Fisk, I. (2014). Atmospheric pressure chemical ionisation mass spectrometry analysis linked with chemometrics for food classification - A case study: Geographical provenance and cultivar classification of monovarietal clarified apple juices. *Food Chemistry*, *146*, 149–156. https://doi.org/10.1016/j.foodchem.2013.09.024

Gfeller, K. Y., Nugaeva, N., & Hegner, M. (2005). Micromechanical oscillators as rapid biosensor for the detection of active growth of Escherichia coli. *Biosensors and Bioelectronics*, *21*, 528–533. https://doi.org/10.1016/j.bios.2004.11.018

Godin, M., Tabard-Cossa, V., Miyahara, Y., Monga, T., Williams, P. J., Beaulieu, L. Y., Bruce Lennox, R., & Grutter, P. (2010). Cantilever-based sensing: The origin of surface stress and optimization strategies. *Nanotechnology*. https://doi.org/10.1088/0957-4484/21/7/075501

Gordon, J. G., & Ernst, S. (1980). Surface plasmons as a probe of the electrochemical interface. *Surface Science*, *101*(1–3), 499–506. https://doi.org/10.1016/0039-6028(80)90644-5

Gupta, A. (2004). Single virus particle mass detection using microresonators with nanoscale thickness. *Applied Physics Letters*. https://doi.org/10.1063/1.1667011

Gupta, A., Akin, D., & Bashir, R. (2004). Single virus particle mass detection using microresonators with nanoscale thickness. *Applied Physics Letters*, *1976*(January), 9–12. https://doi.org/10.1063/1.1667011

Hamano, M., Nomura, S., Iida, M., Komuro, I., & Yamanishi, Y. (2021). Prediction of single-cell mechanisms for disease progression in hypertrophic remodelling by a trans-omics approach. *Scientific Reports*, 1–17. https://doi.org/10.1038/s41598-021-86821-y

Henriquez, R. R., Ito, T., Sun, L., & Crooks, R. M. (2004). The resurgence of Coulter counting for analyzing nanoscale objects. *Analyst*, *129*(6), 478–482. https://doi.org/10.1039/b404251b

Hopfgartner, G., Tonoli, D., & Varesio, E. (2012). High-resolution mass spectrometry for integrated qualitative and quantitative analysis of pharmaceuticals in biological matrices. *Analytical and Bioanalytical Chemistry*, *402*(8), 2587–2596. https://doi.org/10.1007/s00216-011-5641-8

Hunter, D. (1955). The art of diagnosis. *The Central African Journal of Medicine*, *1*(5), 207–215. https://doi.org/10.1056/nejm198205273062104

Ilic, B., Yang, Y., & Craighead, H. G. (2004). Virus detection using nanoelectromechanical devices. *Applied Physics Letters*. https://doi.org/10.1063/1.1794378

Jorgensen, P. (2004). How cells coordinate growth and division. *Current Biology*. https://doi.org/10.1016/j.cub.2004.11.027

Jorgensen, P., Nishikawa, J., Breitkreutz, B.-J., & Tyers, M. (2002). Systematic identification of pathways that couple cell growth and division in yeast. *Science (New York, N.Y.)*, *297*, 395–400. https://doi.org/10.1126/science.1070850

Kasianowicz, J. J., Brandin, E., Branton, D., & Deamer, D. W. (1996). Characterization of individual polynucleotide molecules using a membrane channel. *Proceedings of the National Academy of Sciences of the United States of America*, *93*(24), 13770–13773. https://doi.org/10.1073/pnas.93.24.13770

Killander, D., & Zetterberg, A. (1965). Quantitative cytochemical studies on interphase growth. I. Determination of DNA, RNA and mass content of age determined mouse fibroblasts in vitro and of intercellular variation in generation time. *Experimental Cell Research*, *38*(2), 272–284. https://doi.org/10.1016/0014-4827(65)90403-9

Kirstein, K. U., Li, Y., Zimmermann, M., Vancura, C., Volden, T., Song, W. H., Lichtenberg, J., & Hierlemannn, A. (2005). *Cantilever-based biosensors in CMOS technology. Proceedings-Design, Automation and Test in Europe, DATE '05, II*, 1340–1341. https://doi.org/10.1109/DATE.2005.90

Lange, D. (2002). Complementary metal oxide semiconductor cantilever arrays on a single chip: Mass-sensitive detection of volatile organic compounds. *Analytical Chemistry*. https://doi.org/10.1021/ac011269j

Lewis, Aaron & Lieberman, Klony (1991). 'The Optical Near Field And Analytical Chemistry', 63(11), p. 12.

Lin, K., Lu, Y., Chen, J., Zheng, R., Wang, P., & Ming, H. (2008). Surface plasmon resonance hydrogen sensor based on metallic grating with high sensitivity. *Optics Express*, *16*(23), 18599. https://doi.org/10.1364/oe.16.018599

Löfås, S., Malmqvist, M., Rönnberg, I., Stenberg, E., Liedberg, B., & Lundström, I. (1991). Bioanalysis with surface plasmon resonance. *Sensors and Actuators: B. Chemical*, *5*(1–4), 79–84. https://doi.org/10.1016/0925-4005(91)80224-8

Lu, J., & Rosenzweig, Z. (2000). Nanoscale fluorescent sensors for intracellular analysis. *Fresenius' Journal of Analytical Chemistry*, *366*(6–7), 569–575. https://doi.org/10.1007/s002160051552

Lukashenko, S. Y., et al. (2018). Resonant mass detector based on carbon nanowhiskers with traps for nanoobjects weighing. *Physica Status Solidi (A)*. https://doi.org/10.1002/pssa.201800046

Maier, T., & Güell, M., & Serrano, L. (2009). Correlation of mRNA and protein in complex biological samples. *FEBS Letters*. https://doi.org/10.1016/j.febslet.2009.10.036

Mitchison, J. M. (2003). Growth during the cell cycle. *International Review of Cytology*. https://doi.org/10.1016/S0074-7696(03)01004-0

Mozharov, A., Berdnikov, Y., Solomonov, N., Novikova, K., Nadoyan, I., Shkoldin, V., Golubok, A., Kislov, D., Shalin, A., Petrov, M., & Mukhin, I. (2021). Nanomass sensing via node shift tracing in vibrations of coupled nanowires enhanced by fano resonances. *ACS Applied Nano Materials*, *4*(11), 11989–11996. https://doi.org/10.1021/ACSANM.1C02558

Nomura, S. (2020). Single-cell genomics to understand disease pathogenesis. *Journal of Human Genetics*. https://doi.org/10.1038/s10038-020-00844-3

null. (2014). Preparation of single-cell RNA-Seq libraries for next generation sequencing. *Current Protocols in Molecular Biology/Edited by Frederick M Ausubel [et Al]*. https://doi.org/null

O'Sullivan. (2019). Developments in transduction, connectivity and ai/machine learning for point-of-care testing. *Sensors*. https://doi.org/10.3390/s19081917

O'Sullivan, C. K., & Guilbault, G. G. (1999). Commercial quartz crystal microbalances - Theory and applications. *Biosensors and Bioelectronics*, *14*(8–9), 663–670. https://doi.org/10.1016/S0956-5663(99)00040-8

Ojanperä, I., Kolmonen, M., & Pelander, A. (2012). Current use of high-resolution mass spectrometry in drug screening relevant to clinical and forensic toxicology and doping control. *Analytical and Bioanalytical Chemistry*, *403*(5), 1203–1220. https://doi.org/10.1007/s00216-012-5726-z

Park, K., & Bashir, R. (2009). MEMS-based resonant sensor with uniform mass sensitivity. *TRANSDUCERS 2009 – 15th International Conference on Solid-State Sensors, Actuators and Microsystems*, 1956–1958. https://doi.org/10.1109/SENSOR.2009.5285673

Park, M., & Lee, S. (2003). Cell cycle and cancer. *Journal of Biochemistry and Molecular Biology*, *36*(1), 60–65.

Park, K., Kim, N., Morisette, D. T.; Aluru, N. R., & Bashir, R. (2012). Resonant MEMS mass sensors for measurement of microdroplet evaporation. *Journal of Microelectromechanical Systems*. https://doi.org/10.1109/jmems.2012.2189359

Park, K., Millet, L. J., & Kim, N., et al. (2010). *Measurement of adherent cell mass and growth*. *Proceedings of the National Academy of Sciences of the United States of America*. https://doi.org/10.1073/pnas.1011365107

Park, K., Millet, L., Kim, N., Li, H., Hsia, K. J., Aluru, N. R., & Bashir, R. (2011). MEMS mass sensors with uniform sensitivity for monitoring cellular apoptosis. *2011 16th International Solid-State Sensors, Actuators and Microsystems Conference*. https://doi.org/10.1109/transducers.2011.5969307

Peters, R. J. B., Stolker, A. A. M., Mol, J. G. J., Lommen, A., Lyris, E., Angelis, Y., Vonaparti, A., Stamou, M., Georgakopoulos, C., & Nielen, M. W. F. (2010). Screening in veterinary drug analysis and sports doping control based on full-scan, accurate-mass spectrometry. *TrAC – Trends in Analytical Chemistry*, *29*(11), 1250–1268. https://doi.org/10.1016/j.trac.2010.07.012

Pickering, M. (2019). Fission yeast cells grow approximately exponentially. *Cell Cycle*. https://doi.org/10.1080/15384101.2019.1595874

Rejaniak, K. A. (2005). A single-cell approach in modeling the dynamics of tumor microregions Katarzyna A. Rejniak. *Mathematical Biosciences and Engineering*, *2*(3), 643–655. http://www.mbejournal.org/

Rizzuto, R., Carrington, W., & Tuft, R. A. (1998). Digital imaging microscopy of living cells. *Trends in Cell Biology*, *8*(7), 288–292. https://doi.org/10.1016/S0962-8924(98)01301-4

Saleh, O. A., & Sohn, L. L. (2001). Quantitative sensing of nanoscale colloids using a microchip Coulter counter. *Review of Scientific Instruments*, 72(12), 4449–4451. https://doi.org/10.1063/1.1419224

Singhvi, R., Kumar, A., Lopez, G. P., Stephanopoulos, G. N., Daniel, I., Wang, C., Whitesides, G. M., Ingber, D. E., Singhvi, R., Kumar, A., Lopez, G. P., Stephanopoulos, G. N., Wang, D. C., Whitesides, G. M., & Ingber, D. E. (2014). *Engineering Cell Shape and Function*. 264(5159), 696–698.

Skládal, P. (2003). Piezoelectric quartz crystal sensors applied for bioanalytical assays and characterization of affinity interactions. *Journal of the Brazilian Chemical Society*, *14*(4), 491–502. https://doi.org/10.1590/S0103-50532003000400002

Soifer, I., & Barkai, N. (2014). Systematic identification of cell size regulators in budding yeast. *Molecular Systems Biology*. https://doi.org/10.15252/msb.20145345

Son, S., Tzur, A., Weng, Y., Jorgensen, P., Kim, J., Kirschner, M. W., & Manalis, S. R. (2012). Direct observation of mammalian cell growth and size regulation. *Nature Methods*, August, 1–4. https://doi.org/10.1038/nmeth.2133

Špačková, B., Šípová-Jungová, H., Käll, M., Fritzsche, J., & Langhammer, C. (2021). Nanoplasmonic–Nanofluidic single-molecule biosensors for ultrasmall sample volumes. *ACS Sensors*. https://doi.org/10.1021/acssensors.0c01774

Špačková, B., Wrobel, P., Bocková, M., & Homola, J. (2016). Optical biosensors based on plasmonic nanostructures: A review. *Proceedings of the IEEE*, *104*(12), 2380–2408. https://doi.org/10.1109/JPROC.2016.2624340

Streppa, L., Berguiga, L., Boyer Provera, E., Ratti, F., Goillot, E., Martinez Torres, C., Schaeffer, L., Elezgaray, J., Arneodo, A., & Argoul, F. (2016). Tracking in real time the crawling dynamics of adherent living cells with a high resolution surface plasmon microscope. *Plasmonics in Biology and Medicine XIII*, *9724*, 97240G. https://doi.org/10.1117/12.2211331

Tan, W., Shi, Z. Y., & Kopelman, R. (1992). Development of submicron chemical fiber optic sensors. *Analytical Chemistry*, *64*(23), 2985–2990. https://doi.org/10.1021/ac00047a019

Terasima, T., & Tolmach, L. J. (1963). Growth and nucleic acid synthesis in synchronously dividing populations of HeLa cells. *Experimental cell research*, *30*(2), 344–362.

Thundat, T. (1995). Detection of mercury vapor using resonating microcantilevers. *Applied Physics Letters*. https://doi.org/10.1063/1.113896

Tzur, A. (2009). Cell growth and size homeostasis in proliferating animal cells. *Science*. https://doi.org/10.1126/science.1174294

Tzur, A. (2011). Optimizing optical flow cytometry for cell volume-based sorting and analysis. *PLoS ONE*. https://doi.org/10.1371/journal.pone.0016053

Vaclavik, L., Schreiber, A., Lacina, O., Cajka, T., & Hajslova, J. (2012). Liquid chromatography-mass spectrometry-based metabolomics for authenticity assessment of fruit juices. *Metabolomics*, *8*(5), 793–803. https://doi.org/10.1007/s11306-011-0371-7

Wang, A. W., Kiwan, R., White, R. M., & Ceriani, R. L. (1998). A silicon-based ultrasonic immunoassay for detection of breast cancer antigens. *Sensors and Actuators, B: Chemical*, *49*(1–2), 13–21. https://doi.org/10.1016/S0925-4005(98)00127-0

Weinberg, M. S. (2003). Fluid damping in resonant flexural plate wave device. *Journal of Microelectromechanical Systems*. https://doi.org/10.1109/JMEMS.2003.818452

White, R. M., & Wenzel, S. W. (1988). Fluid loading of a Lamb-wave sensor. *Applied Physics Letters*, *52*(20), 1653–1655. https://doi.org/10.1063/1.99047

Wood, M., Laloup, M., Samyn, N., del Mar Ramirez Fernandez, M., de Bruijn, E. A., Maes, R. A. A., & De Boeck, G. (2006). Recent applications of liquid chromatography-mass spectrometry in forensic science. *Journal of Chromatography A*, *1130*(1 SPEC. ISS.), 3–15. https://doi.org/10.1016/j.chroma.2006.04.084

Wu, J. (2006). Biased AC electro-osmosis for on-chip bioparticle processing. *IEEE Transactions on Nanotechnology*, *5*(2), 84–88. https://doi.org/10.1109/TNANO.2006.869645

Xu, X. H., & Yeung, E. S. (1997). Direct measurement of single-molecule diffusion and photodecomposition in free solution. *Science*, *275*(5303), 1106–1109. https://doi.org/10.1126/science.275.5303.1106

Zhang, N.-H., Meng, W.-L., & Tan, Z.-Q. (2013). A multi-scale model for the analysis of the inhomogeneity of elastic properties of DNA biofilm on microcantilevers. *Biomaterials*. https://doi.org/10.1016/j.biomaterials.2012.11.023

Zhang, P., & Wang, S. (2014). Designing fractal nanostructured biointerfaces for biomedical applications. *ChemPhysChem*. https://doi.org/10.1002/cphc.201301230

Zhao, Y., Gaur, G., Retterer, S. T., Laibinis, P. E., & Weiss, S. M. (2016). Flow-through porous silicon membranes for real-time label-free biosensing. *Analytical Chemistry*. https://doi.org/10.1021/acs.analchem.6b02521

Zhao, X., Hilliard, L. R., Mechery, S. J., Wang, Y., Bagwe, R. P., Jin, S., & Tan, W. (2004). A rapid bioassay for single bacterial cell quantitation using bioconjugated nanoparticles. *Proceedings of the National Academy of Sciences of the United States of America*, *101*(42), 15027–15032. https://doi.org/10.1073/pnas.0404806101

Zhe, J., Jagtiani, A., Dutta, P., Hu, J., & Carletta, J. (2007). A micromachined high throughput Coulter counter for bioparticle detection and counting. *Journal of Micromechanics and Microengineering*, *17*(2), 304–313. https://doi.org/10.1088/0960-1317/17/2/017

Zheng, Y., Nguyen, J., Weia, Y., & Sun, Y. (2013). Recent advances in microfluidic techniques for single-cell biophysical characterization. *Lab on a Chip*. https://doi.org/10.1039/c3lc50355k

Zimmerman, S. B. (1993). Macromolecular crowding effects on macromolecular interactions: Some implications for genome structure and function. *Biochimica et Biophysica Acta*. https://doi.org/10.1016/0167-4781(93)90142-Z

Zimmermann, U., Pilwat, G., & Riemann, F. (1974). Dielectric breakdown of cell membranes. *Biophysical Journal*, *14*(11), 881–899. https://doi.org/10.1016/S0006-3495(74)85956-4

Chapter 3

MEMS-based electrochemical gas sensor

Yasaman Heidari, Mahshid Padash and Mohammadreza Faridafshin
Shahid Bahonar University of Kerman, Kerman, Iran

CONTENTS

3.1 INTRODUCTION

There are various types of gas molecules (natural and artificial) that form the atmosphere around us. Some of them are a necessity to breathe, while others are produced from the industrial processes. Due to their toxicity, which is dangerous to human health, detection and monitoring of gases are crucial. Gas sensors are devices that have been utilized for detecting gas molecules in an environment by converting gas volume fractions into electrical signals. During the 1970s the first gas sensors, like semiconductor gas sensors, solid electrolyte oxygen sensors, and humidity sensors for basic use, were commercialized [1]. The importance of using, and wide applications of, gas sensor technology in modern society for improving the quality of life is the impetus for many studies that have been done to design and

DOI: 10.1201/9781003288633-3

advance before-mentioned sensors and any new gas sensors, while guaranteeing energy saving, safety, health, amenity, etc.

A gas sensor is a transducer that can only detect the presence of some specific gases. The filter guarantees that the sensor only responds to the intended analyte. We should consider that the response can be encouraged to emerge in different guises, according to sundry factors [2].

The basic construction of a gas sensor (or indeed any chemical sensor) is shown in Figure 3.1. In general, a gas sensor is like a transducer that identifies gas molecules and converts them into measurable electrical signals. The filter allows the sensor to determine the specific gas (sometimes called the measurand or analyte) and not other gases. In fact, it either operates as a

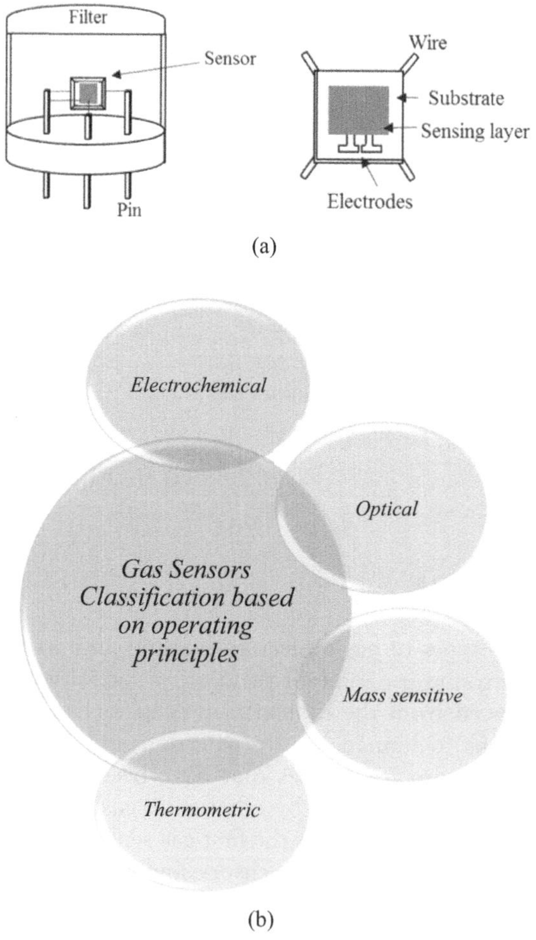

Figure 3.1 (a) Schematic construction of a gas sensor (b) classification of gas sensors.

filter that permits only the flow of certain gas molecules into the capture zone or it is a more practical instrument that identifies the target gas based on its optical absorption wavelength properties. Note that the obtained result may be a single response or a time-dependent signal or electronically or otherwise processed.

Due to properties such as low price, small size, or energy-efficiency, gas sensors are widely employed in the detection of flammable and toxic gases, respiratory analysis in medical diagnostics, food safety monitoring, and other industrial fields. Different devices have been used to detect gases, such as fire alarms that sense and detect smoke or heat. Gas sensors can detect the leak and make a sound alarm; they can detect flammable and combustion gases and control the content of the pollution. Also, gas sensors can be utilized in the food industry for monitoring the quality of food [3]. Some major gas contaminants and their side effects are listed in Table 3.1.

With the growing demand for enhancement in gas sensors of higher sensitivity, selectivity, and stability, meticulous efforts are in progress to find new materials with functional properties. The application of gas sensors in different fields is listed in Table 3.2.

Table 3.1 Example of pollutant gas

Gases	*Source*	*Side-Effect*
Carbon dioxide	Industrial products, power generation, motor vehicle	Oxygen deficiency, and make asphyxiation
Carbon dioxide	Defective combustion of natural gas, coal, etc., exhaust smoke of motor vehicles	Cause headache, dizzying, vomiting, and weakness
Methane	Main component of natural gas changes the earth's weather that is produced by anaerobic decomposition, biomass burning	Temperature and climate
Nitrogen oxide	Product of the reaction between O_2 and N_2, in combustion processes, open fireplace, burning of fossil fuels	Causes headaches, breathing problems, eye irritation
Sulphur dioxides	Product of combustion of oil and volcanic eruptions	Eye and nose irritation bronchitis, shortness of breath during activity and exercise
Volatile organic compounds (VOCs)	Incomplete fuel burning, petroleum refining, and factories	Results in some diseases such as cancer and the death of plants
Ammonia	Agriculture process and fertilizer factories	Burning and irritation of the eyes, nose, coughing

Sources: https://www.epa.gov/air-trends, https://www.britannica.com/science/air-pollution, and http://en.wikipedia.org/wiki/Air_pollution

Table 3.2 Some applications of gas sensors in different fields

Application	*Detected gas*
Environment	Atmospheric pollutants, including CO_2, volatile organic compounds, NO_x, Ammonia, SO_2, and CO
Agriculture	Measure and control the humidity and CO_2, concentration for optimal crop growth, and methane emissions from agriculture
Medical diagnosis	Biomarker detection exhaled, human breath, including inorganic particles such as NO, NH_3, or CO, and volatile organic molecules such as acetone and isoprene
Food Safety Monitoring	Expelled gases from rotten food, ethyl acetate, a volatile component produced in the initial bacterial putrefaction of meat

Sources: Ramsden (2009) Nanotechnology Perceptions, Korotecenkov (2013) springer science.

3.2 GAS SENSORS CLASSIFICATION

Gas sensors are classified into categories based on the material used, transduction mechanism, etc. Considering operating principles, they are classified into four types: 1) electrochemical gas sensor, 2) Thermometric gas sensor, 3) optical gas sensor, and 4) mass sensitive gas sensor [4].

3.2.1 Mass-sensitive gas sensors

These sensors are based on high-frequency mechanical vibrations. Interaction between modified surfaces to the chemical species results in frequency shift that is measured for the concentration of mass loading on the electrode's surface [5]. The methods, such as deflecting a micromechanical structure, migratory acoustic wave, and evaluating the frequency specifications of a resonating structure are three methods to monitor mass change [4]. Mass-sensitive gas sensors utilize a piezoelectric material to produce the acoustic wave on the surface or throughout the structure's bulk. The two most suitable types of mass-sensitive gas sensors have been employed, such as quartz crystal micro-balance (QCM), and surface acoustic waves (SAW). QCM contains piezoelectric quartz crystal between two electrodes, and the sensing metal layer is on the surface of the electrode that senses the target gas. Generally, electrodes are made of noble metal to avoid oxidation. Quartz, in comparison to other piezoelectric crystals, is more desirable due to having a linear relationship between mass loading and frequency response and its mechanical and chemical structure stability [6].

SAW sensors include two metal thin-film transducers on the piezoelectric crystal substrate and the gas-sensitive layer (generally polymer) that is located on top of the substrate. They are favored owing to their low expense, sensitivity, and ability to employ a vast variety of gases. Temperature and humidity are two important factors that should be controlled. Any variation in the

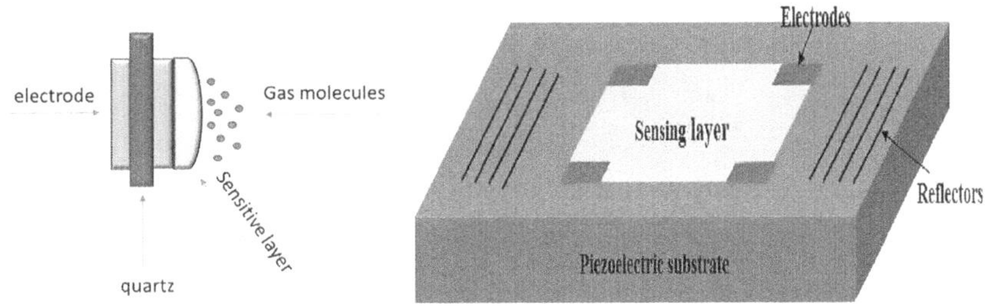

Figure 3.2 Schematic design operation of mass-sensitive gas sensor.

mentioned factors results in a frequency shift during interaction with gas species [4, 7]. The figures for Mass-sensitive gas sensors are depicted in Figure 3.2.

3.2.2 Optical gas sensors

When gas molecules interact with the material layer, the receptor layer measures and transforms any variation in light into analytical signals, and monitors them at optical wavelength [8]. The operation is based on sensing the intrinsic optical property of the gas molecule or optical feature of the indicator. Generally, the molecule has intrinsic optical properties that are utilized in infrared and ultraviolet spectroscopes, and the indicator-mediated molecule is used for luminescence and absorption spectroscopy [4]. A group of sensors in optical gas sensors is fabricated using an optical waveguide, especially optical fiber [9]. The regular design of the optical gas sensors is shown in Figure 3.3.

3.2.3 Thermometric sensor

The thermometric gas sensor is known as calorimetric gas sensing. Generally, this sensor is used for sensing combustible gases. The operation process is

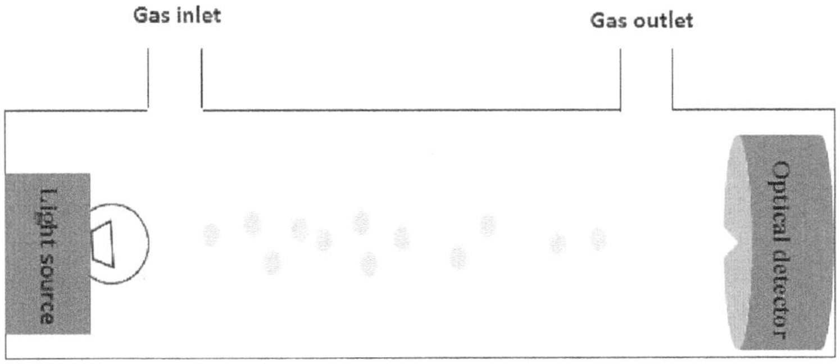

Figure 3.3 Schematic representation of the optical gas sensor.

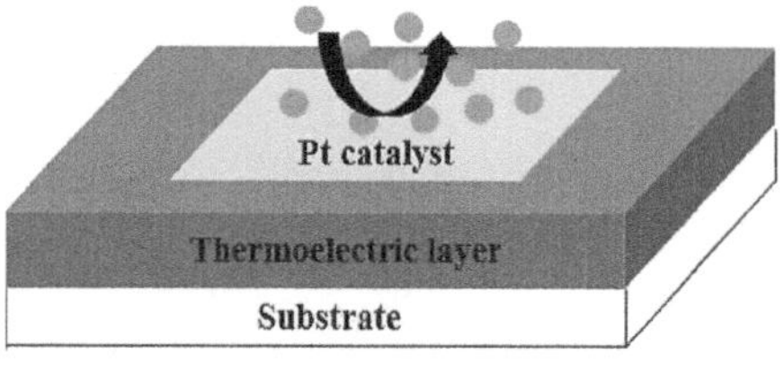

Figure 3.4 Schematic diagram of The thermometric gas sensor.

based on the radical interaction of the analyte and transduces and converts any changes in temperature and resistance into a measurable electrical signal. Radical interaction needs a high temperature that is provided by the heat of combustion. It is ordered according to operation in catalytic combustion: thermal conductivity and adsorption/desorption heat. Among them, some advantages such as stable, accurate, and long operating life, caused the catalytic layer to be a widely used alarm sensor for the detection of flammable gases in hazard-prone conditions. By passing the current through platinum (Pt) as the underlying coil, the catalytic layer gets warm and results in the oxidation of combustible gas (Figure 3.4). The generated heat increases the temperature of the catalyst and changes electrical resistance, which is related to the gas concentration [4, 10].

3.2.4 Electrochemical sensors

Electrochemical sensors are utilized for the detection of ambient gas due to selectivity, low LOD, sensitivity to fluctuation, ease of miniaturization and fabrication, quick response, and the capability to be portable. The obtained information is based on changing electrical signals by either reduction or oxidation of analytes on the surface of electrodes. The concentration of gas molecules as analytes affects the results of electrochemical sensors [11, 12].

Electrochemical sensors are ordered based on modifying the fabrication and surface of electrodes with different techniques, like amperometry, potentiometry, and conductometry, that improve the detection of organic species [4]. A schematic shape of electrochemical sensor operation is presented in Figure 3.5.

Amperometric electrochemical gas sensors impose various voltages between electrodes, the response is obtained by measuring resistance. Oxidation or reduction of the analyte on the surface of the electrode generates a measurable current. The current is affected by the concentration of the diffused analyte. There is a linear relation between sensitivity and concentration of gas molecules. Recently, researchers use the pulsed-potential method to overcome its low sensitivity [13, 14].

Voltammetric electrochemical gas sensors apply to alter potential between work and reference electrodes, and the analyte will be oxidized or reduced. The resulting current of the analyte oxidation-reduction is measured and is

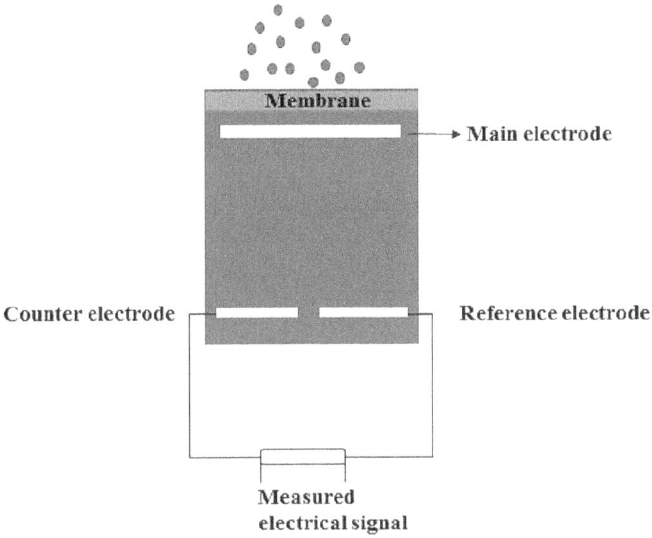

Figure 3.5 Schematic design of the electrochemical gas sensor.

ploted vs. applied potential named as voltammogram. Depending on the applied potential, the rudction and oxidation reaction occurs. According to the rudction and oxidation reaction, the resulted current is called cathodic and anodic current, respectively [15].

Potential measurement in potentiometric electrochemical gas sensors rely upon the interaction between the electrode and the molecules that changes the potential between electrodes. In general, the measurement of electrical potential can be done in the absence of current [13].

Some advantages and disadvantages of any type of sensor are listed in Table 3.3.

Table 3.3 Some advantages and disadvantages of sensor types

Sensor type	*Advantages*	*Disadvantages*
Mass-sensitive gas sensors	Acceptable precision and sensitivity, fast response, inexpensive, small	Low signal-to-noise ratio temperature-sensitive
Optical gas sensors	High sensitivity and selectivity, flexible capability in the detection of different gases, portability	Expensive to operate, low resistance to corrosion, complex system
Thermometric gas sensor	Fast response, high specificity, inexpensive	High-temperature operation, only utilized for combustion gas
Electrochemical gas sensors	Low power consumption, easy operation, fast response	Low selectivity, limited sensor coatings

Source: Holden Li (2018), published by MDPI. Yunusa (2014) published by IFSA.

Table 3.4 Application of electrochemical gas sensor

Electrochemical application	*Detected Gases*	*Reference*
GO-covered ZnO MWs	H_2	[23]
Zinc oxide NWs	NO_2	[24]
ZnO	C_2H_2	[25]
WO_3 NPs embedded in chitosan membrane	H_2S	[26]
RGO-CGN	O_2	[27]
Pd (NCs)- NiO nanosheets	H_2	[28]
Pt–WO_3 NWs	H_2	[29]
WO_3 film	H_2S	[30]
PANI/cellulose/WO_3	acetone	[31]

Furthermore, an electrochemical gas sensor is designed and rearranged depending on the material used on the surface of electrodes to ameliorate sensitivity and selectivity. Some examples of electrochemical gas sensor applications are shown in Table 3.4.

For modification of sensor and increasing efficiency of signal, many materials like graphene oxide, metal oxide semiconductors, and composite materials, are used.

3.3 FABRICATION MATERIALS

3.3.1 Metal oxide semiconductors

The first practical utilization of metal oxide semiconductors (MOS) as Taguchi gas sensors goes back several decades. In recent years, owing to some profits, like particular structure, big surface-to-volume ratio, physical and chemical structure stability, simplicity, and low-pricing, they are applied as sensors, for monitoring gases in various fields, such as agriculture, food safety, industry, and medicine [16].

For years, several n-type and p-type of MOS, have been used as gas sensors (TiO2, WO3, CuO, etc.). Among all of them, SnO2 is widely used due to specific physical properties such as stability. The surface of MOS material turns any changes in the concentration and conductance of gases into an electrical signal. Generally, due to the diffusion of gas molecules on the surface of MOS, interaction, and changes in conductivity take place. Although the MOS-type gas sensors have lots of advantages like high sensitivity and adsorption of gas molecules, low LOD, fast response time, and the ability to detect gas species, they still need the temperature to catalyze the gas reaction by increasing gas absorption [17, 18].

3.3.2 Graphene

GO is a practical porous carbon nanomaterial that is used as a sensing layer in electrochemical gas sensors. For decades, GO, due to some advantages such as thermal and physical stability, doped with other materials and 2D structures. Another development of nanocarbon material is carbon nanotubes. Single-walled carbon nanotubes (SWCNTs), and multi-walled carbon nanotubes (MWCNTs) with specific morphology, high sensitivity, low-priced, high porosity, and large surface area that increases adsorption of analytes, and change its conductivity [19, 20]. Because of the accepting functional group and modifications such as doping metal on the surface or sidewall of CNT, they can enhance the adsorption of gas species [21].

3.3.3 Composite materials (CM)

CM such as polypyrrole (PPy), polythiophene (PTH), and polyaniline (PANI), can be easily modified and used as sensing layers, due to intrinsic conducting and specific properties such as high sensitivities and short response time, chemical and physical stability, and long-chain structure. Some chemical and electrochemical oxidation processes utilize conducting polymer as powder or coating on the electrode's surface as an active layer for sensing gas species. Polymerization needs an initiator to oxidize monomers and produce a polymer chain [22].

3.4 MEMS ELECTROCHEMICAL GAS SENSOR

Term MEMS (micro-electromechanical system) is applied to every miniaturized device fabricated and derived from microelectronics. It is highly utilized in different applications due to its high performance at small size and low-cost, portable, fast response, ability to enhance the surface-to-volume ratio, and decrease the operating temperature of the metal oxide gas sensors.

As described, the interaction between molecules and the active layer generates electrical signals. So, the surface area has a crucial role in sensitivity enhancement. Increasing the surface area of electrodes needs to amplify the size of the sensor. To maintain the high surface area where reducing to a small size of devices needs to increase the ratio of the surface area to volume, MEMS technology provides increased surface area while maintaining a small size [7, 32].

Electrochemical gas sensors are divided into two kinds, liquid electrolytes, and solid electrolytes. Solid electrolytes, in comparison to liquid electrolytes, can be used at higher temperatures. Metal oxide semiconductors are one of the categorized electrochemical gas sensors which can be easily fabricated in small sizes. Any conductance change in the metal oxide active layer is measured with

MOS sensors. Recently, MOS has evolved due to having advantages like being inexpensive, having excellent efficiency, low consumption, high sensitivity, and capability of matching with silicon microelectronics.

3.5 STRUCTURE

MOS active layer is coated on alumina or silica substrate where electrodes are located on the top, and the embedded microheater generates the temperature up to 450 °C. By increasing temperature, oxygen species adsorb on the MOS active layer and capture the electron conduction band that changes the resistance of the MOS layer. The concentration of molecules affects the resistance of the sensor. As known, MOS is classified, depending on the response to reducing or oxidizing gases, into two kinds, either n-type (such as SnO_2, In_2O_3, WO_3, ZnO) or p-type (such as NiO, CuO, Co_3O_4). The thickness of the receptor is an effective parameter for the sensitivity of the MOS. Considering n-type MOS, adsorption of oxygen and turn into ion molecules (O2-), (O-), (O2-), electron-depletion occurs at the surface. Therefore, any interaction with reducing or oxidizing gas affected the depletion process and sequentially decreases or increases resistance [33–35].

3.6 FABRICATION OF MEMS GAS SENSORS

Usually, the MEMS technique component of the gas sensor consists of the sensitive layer, substrate, electrodes, and microheater [32].

One of the disadvantages of the gas sensors, especially MOS, is operating at high temperatures to increase sensitivity, which leads to an increase in power consumption. In the MEMS technique, to decrease power consumption, we can use a (Pt) microheater that deposits over a SiO2 dielectric membrane to reach the required temperature. Generally, (Pt) is used for the fabrication of micro heaters due to its excellent thermal stability at high temperatures.

Among two frames to fabricate MOS- MEMS sensors such as silicon frame and suspended structure (spider-like support), silicon membrane is commonly utilized as a substrate and sensing layer deposited on the electrodes.

In the fabrication process, the silicon wafer is oxidized by the thermal oxidation method to obtain the micrometer thickness of SiO2. Generally, the front layer is operated as the film to support the active sensing layer, while the dorsal layer is applied as an inductively coupled plasma (ICP) etching hard mask. The 10/60 nm Ti/Pt microheater is then deposited on the substrate by DC magnetron sputtering with a lift-off system. A layer of SiO2 with a thickness of 300nm, an insulating layer, is deposited on the microheater by an e-beam evaporation method that prevents any electrical

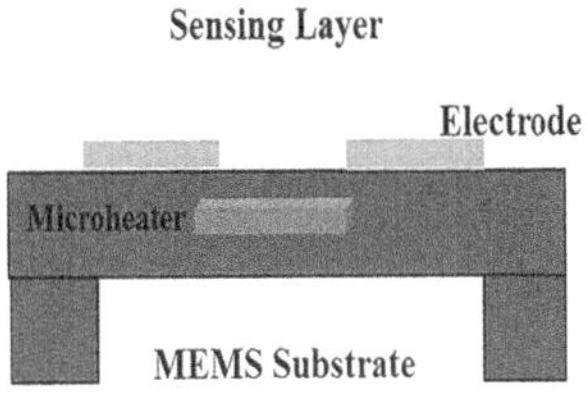

Figure 3.6 Schematic diagram of MEMS sensor.

connection between components. Generally, layer thickness 10/60 nm of Cr/Au as MOS electrode and active layer (SnO2) deposited consequently on the substrate and the fabricated electrodes with process DC magnetron sputtering, like a microheater. Finally, the back of the silicon dioxide and substrate are etched by reactive ion etching (RIE) to furnish temperature isolation and reduced energy consumption [7, 35, 36]. The Fabrication process of the gas sensor can be seen in Figure 3.6.

3.7 APPLICATION

Recently, Using MEMS technology to fabricate MOS gas sensor has been favored [37].

MOS-MEMS gas sensor technology is more practical for the future. So far, there have been many efforts to figure out new targets for gas sensing, new ways of materials processing, new types of devices, new ways of material processing, sensing materials, and emphasizing gas sensing performances. Today, the design of semiconductor gas sensors with high efficiency is desired. With increasing industrial activities, the amount of carbon dioxide as pollutant gas has been increasing. So, monitoring this gas pollution is important in environmental applications. Electrochemical potentiometric CO_2 sensors containing solid electrolytes are one of the methods which have been widely used, because of their miniaturized structure, suitable selectivity, and low cost. Choi (2013) fabricated a MEMS stacked-type potentiometric solid-state (S-PSS) sensor by using alumina as a substrate to reduce power consumption. Compared to the planar-type potentiometric solid-state (P-PSS) sensor, although the size of the sensing area was reduced, it did not affect its performance [38].

In 2017, S.E. Moon designed a micro carbon dioxide gas sensor with low energy consumption (59 mW) and high selectivity by deposition of Li_3PO_4 and Li_2CO_3 thick film as solid electrolyte and sensing material, respectively on Si substrate [39]. Yttria stabilized zirconia (YSZ) with high chemical and thermal stability as a solid electrolyte in the SnO_2 MEMS technique was deposited by RF sputtering to detect and monitor SO_2 gas molecules (Hsueh 2019). compared with using YSZ or SnO_2 separately, this fabrication (YSZ/SnO_2 MEMS SO_2 gas sensor) improves sensor response at 400 C in 250 ppb SO_2 gas concentration [40].

Also, MEMS is widely used in the field of medical and environmental using electrochemical apathosensors. MEMS-based electrochemical aptasensor is used as a recognition element for monitoring Noroviruses (NoVs) as environmental pathogen causes. In 2016, Kitajima fabricated a DNA aptamer monolayer on an (Au) working electrode by using cyclic voltammetry as well as visual observation of a fluorescent-labeled aptamer. The fabrication of a miniaturized device with MEMS technology increases detection sensitivity [41].

An incurable disease like cancer is one of the main causes of death. So, the timely diagnosis of a disease can help to cure patients. Among many methods, the electrochemical aptasensor has been widely studied to detect cancer biomarkers (platelet-derived growth factor-BB) due to its easy operation method and more sensitivity than other transduction methods. Wang (2020) fabricated electrodes with glassy-carbon microelectrochemical to prepare a sensitive, stable, and selective PDGF-BB aptasensor. They optimized the fabrication parameter by cyclic voltammetry and electrochemical impedance spectroscopy. The use of C-MEMS as a sensible method has enhanced the C-MEMS-based PDGF-BB aptasensor for the diagnosis of microdevices [42].

3.8 CONCLUSION

Generally, gas sensors provide information about the chemical composition of gases. In addition, a gas sensor can be used as a device to detect a gas leak and sound an alarm automatically. Gas sensors can detect flammable gases. These characterizations make them practical devices in various applications. Besides, gas sensors can also be arranged in different categories depending on other principles, Sensors are based on (1) fabrication technology, (2) physical properties of the gas, (3) working temperature, 4) design, etc. [4]. In this chapter, electrochemical MOS-MEMS gas sensors have been introduced. Modification and structure of working electrodes with novel nanostructures, operation method, standard techniques of fabrication, and their applications, are considered.

REFERENCES

1. Yamazoe, N. (2005). Toward innovations of gas sensor technology, *A Review, Sensors, and Actuators B*, 108, 2–14.
2. Hodgkinson, J. Saffell, J., Luff, J. Shaw, J., Ramsden, J., Huggins, C., Bogue, R., Carline, R. (2009). Gas sensor 1. The basic technologies and application, *Nanotechnology Perception*, 5(1), 71–82.

3. Matindoust, S., Baghaeinejad, M.S., Abadi, M.H., Zou, Z., Zhang, L. (2016). Food quality and safety monitoring using gas sensor array in intelligent packaging, *Sensor Review*, 36(2), 169–183.
4. Korotcenkov, G. (2013). *Handbook of Gas Sensor Materials: Properties, Advantages and Shortcomings for Applications Volume 1: Conventional Approaches*. Materials Science and Engineering Gwangju Institute of Science and Tech.Gwangju, Republic of (South Korea).
5. Fanget, S., Hentz, S., Puget, P., Arcamone, J., Matheron, M., Colinet, E., Andreucci, P., Duraffourg, L.M., Ed. Roukes, M.L. (2011). Gas sensors based on gravimetric detection—A review, *Sensors and Actuators B: Chemical*, 160(1), 804–821.
6. Comini, E., Faglia, G., Sberveglieri, G. (2009). *Solid State Gas Sensing*, Springer Science and Business Media LLC.
7. Nazemi, H., Joseph, A., Park, J., Emadi, A. (2019). Advanced micro- and nano-gas sensor technology: A review, *Sensors*, 19(6), 1285.
8. Norris, J.O.W. (2000). Optical fiber chemical sensors: Fundamentals and applications. In: Grattan, K.T.V., Meggitt, B.T. (eds) *Optical Fiber Sensor Technology*. Springer, Boston, MA.
9. Eguchi, K. (1992). Optical gas sensors, In: Sberveglieri, G. (ed.) *Gas Sensors*. Springer, Dordrecht, pp. 307–328.
10. Bársony, I., Dücső, C., Fürjes, P. (2009). Thermometric gas sensing. In: Comini, E., Faglia, G., Sberveglieri, G. (eds) *Solid State Gas Sensing*. Springer, Boston, MA.
11. Gomes, J., Rodrigues, J., Rabelo, R., Kumar, N., Kozlov, S. (2019). IoT-enabled gas sensors: Technologies, applications, and opportunities: A review, *Journal of Sensor and Actuator Networks*, 5, 57.
12. Jasinski, G., Jasinski, P. (2011). Solid electrolyte gas sensors based on cyclic voltammetry with one active electrode, *IOP Conference Series: Materials Science and Engineering*, 18, 212007.
13. Mujahid, A., Lieberzeit, P.A.L., Dickert, F. (2006). Solid-state sensors for field measurements of gases and vapors, *Encyclopedia of Analytical Chemistry: Applications, Theory and Instrumentation*, 1–30.
14. Cox, J.A., and Holmstrom, S.D. (2006). Electrochemical sensors for field measurements of gases and vapors. *Encyclopedia of Analytical Chemistry: Applications, Theory and Instrumentation.*
15. Gross, P.A., Jaramillo, T., Pruitt, B. (2018). Cyclic-voltammetry-based solid-state gas sensor for methane and other VOC detection, *Analytical Chemistry*, 90(10), 6102–6108.
16. Abdullah, M.M. Singh, P. Ikram, S. (2020). Recent developments in nanostructured metal oxide-based electrochemical sensors, *Nanofabrication for Smart Nanosensor Applications*, 123–134.
17. Wan, H., Yin, H., Mason, A.J. (2017). Rapid measurement of room temperature ionic liquid electrochemical gas sensor using transient double potential amperometry, *Sensors and Actuators B: Chemical*, 242, 658–666.
18. Tian, W. Liu, X., Yu, W. (2018). Research progress of gas sensor based on graphene and its derivatives: A review, *Applied Sciences*, 8(7), 1118.
19. Donarelli, M., Ottaviano, L (2018). 2D materials for gas sensing applications: A review on graphene oxide, MoS_2, WS_2, and phosphorene, *Sensors*, 18(11), 3638.

20. Wang, Y., Yeow, J.T.W. (2009). A review of carbon nanotubes-based gas sensors, *Journal of Sensors*, Hindawi Publishing Corporation KW, 2009.
21. Zhang, X., Cui, H., Gui, Y., Tang, J. (2017). Mechanism and application of carbon nanotube sensors in SF_6 decomposed production detection: A review, *Nanoscale Research Letters*, 12; 177.
22. Bai, H., Shi, G. (2007). Gas sensors based on conducting polymers, *Sensors*, 7, 267–307
23. Rasch, F. Postica, V. Schütt, F. Mishra, Y.K. ShayganNia, A. Lohe, M.R. Feng, X, Adelung, R., Lupan, O. (2020). Highly selective and ultra-low power consumption metal oxide-based hydrogen gas sensor employing graphene oxide as molecular sieve, *Sensors and Actuators B: Chemical*, 320, 128363.
24. Cho, In., Sim, Y. Ch., Cho, M., Cho, Y.H., Park, I. (2020). Monolithic micro light-emitting diode/metal oxide nanowire gas sensor with microwatt-level power consumption, *ACS Sensors*, 5(2), 563–570.
25. Liu, F., Lu, G. (2020). Stabilized zirconia-based solid state electrochemical gas sensor coupled with CdTiO3 for acetylene detection, *Sensors and Actuators B: Chemical*, 316, 128–199.
26. Ali, F.I., Awwad, F., Greish, Y.E., Abu-Hani, A.F., Mahmoud, S.T. (2020). Fabrication of low temperature and fast response H_2S gas sensor based on organic-metal oxide hybrid nanocomposite membrane, *Organic Electronics*, 76, 105486.
27. Wan, H., Gan, Y., Sun, J., Liang, T., Zhou, Sh., Wang, P. (2019). High sensitive reduced graphene oxide-based room temperature ionic liquid electrochemical gas sensor with carbon-gold nanocomposites amplification, *Chemical*, 299, 126952.
28. Van Tong, P., Hoa, N.D., Van Duy, N., Van Quang, V., Lam, N.T., & Van Hieu, N. (2013). In-situ decoration of Pd nanocrystals on crystalline mesoporous NiO nanosheets for effective hydrogen gas sensors, *International Journal of Hydrogen Energy*, 38(27), 12090–12100.
29. Zhou, X., Dai, Y., Muna Karanja, J., Liu, F., Yang, M. (2017). Microstructured FBG hydrogen sensor based on Pt-loaded WO3, *Optic Express*, 25(8), 8777–8786.
30. Solis, J.L. Hoel, A. Kish, L.B. Granqvist, C.G. Saukko, S., Lantto, V. (2001). Gas-sensing properties of nanocrystalline WO3 films made by advanced reactive gas deposition, *Journal of the American Ceramic Society*, 84(7), 1504–1508.
31. Aparicio-Martínez, E. Osuna, V.B. Dominguez, R. Márquez-Lucero, A. Zaragoza-Contreras, E.A. Vega-Rios, A. (2018). Room temperature detection of acetone by a PANI/Cellulose/WO3 electrochemical sensor, *Journal of Nanomaterials*, 2018, 6519694.
32. Liu, H., Zhang, L., Li, K.H.H., Tan, O.K. (2018). Microhotplates for metal oxide semiconductor gas sensor applications—Towards the CMOS-MEMS monolithic approach, *Micromachines*, 9(11), 557.
33. Wang, C., Yin, L., Zhang, L., Xiang, D., Gao, R. (2010). Metal oxide gas sensors: Sensitivity and influencing factors:A review, *Sensors*, 10, 2088–2106.
34. Barsan, N., Weimar, U. (2001). Conduction model of metal oxide gas sensors, *Journal of Electroceramics*, 7, 143–167.
35. Hsueh, T.J., Lu, C. L. (2019). A hybrid YSZ/SnO2/MEMS SO2 gas sensor, *RSC Advances*, 9, 27800–27806.

36. Kakoty, P., Bhuyan, M. (2016). Fabrication of micromachined SnO2 based MOS gas sensor with inbuilt microheater for detection of methanol, *Sensors & Transducers*, 204(9), 58–67.
37. Deng, Y. (2019). Applications of Semiconducting Metal Oxides Gas Sensors. Semiconducting Metal Oxides for Gas Sensing. Springer, Singapore, 195–241.
38. Choi, N.J., Lee, H.K., Moon, S., SeokYang, W., Kim, J. (2013). Stacked-type potentiometric solid-state CO_2 gas sensor, *Sensors and Actuators B: Chemical*, 187, 340–346.
39. Lee, J., Choi, N.J., Lee, H.K., Kim, J., Lim, S.Y., Kwon, J.Y., Lee, S.M., Moon, S.E., Jong, J.J., Yoo, D.J. (2017). Low power consumption solid electrochemical-type micro CO_2 gas sensor, *Sensors and Actuators B: Chemical*, 248, 957–960.
40. Hsueh, T-J, Lu, C-L. (2019). A hybrid YSZ/SnO2/MEMS SO_2 gas sensor, *RSC Advances*, 9, 27800–27806.
41. Kitajima, M. Wang, N.Q.X. Tay, M. Miao, J.J. Whittle, A. (2016). Development of a MEMS-based electrochemical aptasensor for norovirus detection, *Micro & Nano Letters*, 11(10), 582–585.
42. Forouzanfar, Sh., Alam, F., Pala, N., Wang, C. (2020). Highly sensitive label-free electrochemical aptasensors based on photoresist derived carbon for cancer biomarker detection, *Biosensors and Bioelectronics*, 170, 112598.

Chapter 4

Electrochemical biosensors

Mahshid Padash, Mohammadreza Faridafshin and Alireza Nouroozi
Shahid Bahonar University of Kerman, Kerman, Iran

CONTENTS

DOI: 10.1201/9781003288633-4

4.1 INTRODUCTION

The biosensor is defined based on IUPAC [1]:

> A biosensor is a self-contained integrated device which is capable of providing specific quantitative or semi-quantitative analytical information using a biological recognition element (biochemical receptor) which is in direct spatial contact with a transducer element. A biosensor should be clearly distinguished from a bioanalytical system that requires additional processing steps, such as reagent addition. Furthermore, it should be distinguished from a bio probe, which is either disposable after one measurement, i.e., single-use, or unable to continuously monitor the analyte concentration.

In 1962, Clark and Lyons first elucidated the basic concept of the biosensor. They designed the glucose enzyme electrode that used an oxygen electrode for the detection of an enzymatic reaction.

Since then, scientists in various fields such as physics, chemistry, and material science have consulted with each other to develop biosensing devices. Nowadays, these devices, due to the exquisite sensitivity and simple format for complex bioanalytical measurements, have wide applications in different fields, including monitoring of therapy and progression of diseases, drug progress, food safety, environmental monitoring, and process control. In addition to biological analytes, ions, dissolved gases, drugs, and toxins can also be detected with biosensors [2].

As the name biosensor implies, the device consists of two main parts: a bio element and a sensor element. The bio element or bioreceptor recognizes the target analyte selectively and the sensor element, as a physicochemical transducer, translates the interaction between biological molecules into a quantifiable electrical signal. The bio element that is responsible for selective recognition of the analyte can be enzymes, proteins, receptor molecules, antibodies, nucleic acids, microorganisms, cells, tissues, or aptamers. When the bioreceptor element interacts with the analyte, the generated physicochemical signal can be electrochemical, optical, thermal, piezoelectric (mass-sensitive), calorimetric, magnetic or acoustic signals [3]. So, the sensor elements such as electrode, thermistor, photon detector, and vibrational resonance can be used. Figure 4.1 shows a schematic of a biosensor.

There are several phenomena for effective adhesion of the recognition element into the sensor. The four major coupling mechanisms are; a) physical adsorption and encapsulation, b) membrane immobilization, c) covalent amalgamation, and d) matrix entrapment. When the physical intermolecular forces such as hydrophilic/hydrophobic forces, ionic forces, and van der Waals forces, lead to link the bio element to the sensor, this kind of coupling is named the physical adsorption [4]. In membrane immobilization, a semi-permeable membrane surrounds the organic element that is directly located

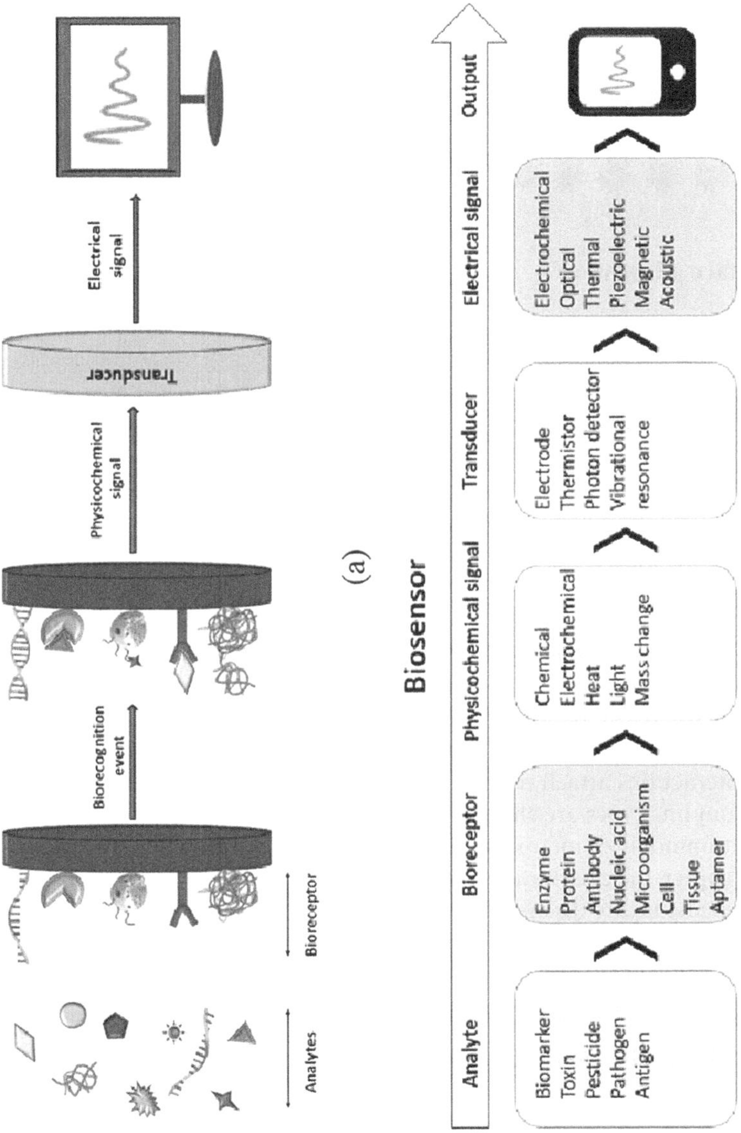

Figure 4.1 Schematic of a biosensor.

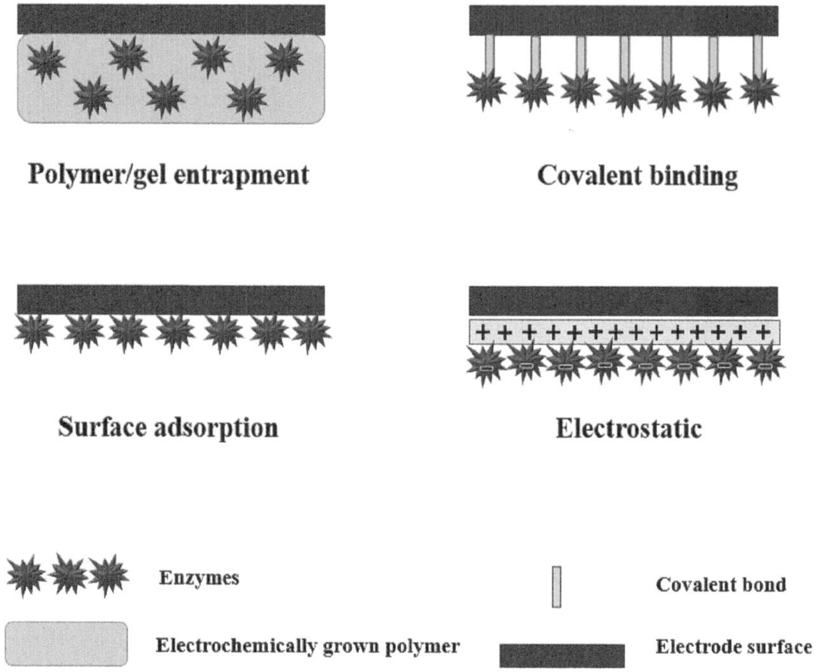

Figure 4.2 Common methods of immobilizing enzymes onto an electrode surface.

on the sensor element. The membrane plays the role of separating phase between the organic element and the target analyte [5]. For matrix entrapment coupling, porose materials like sol or gel matrixes are used as the restrictor medium for biological elements, and this matrix encapsulation forms a direct attachment with the sensor element [6]. In covalent coupling, covalent interactions attach the bio element directly to the sensor [7]. These different coupling types are shown in Figure 4.2. Table 4.1 is a Comparison of different immobilization methods.

In this chapter, the classification of biosensors is reviewed and their applications in medicine are discussed. Further, their future applications in wearable and Lab-on-a-chip (LOC) devices fields, which are widely used especially in medicine, are overviewed.

4.2 CLASSIFYING OF BIOSENSORS

Biosensors can be categorized based on the signal transduction mechanism (transducer), and biorecognition element (bioreceptor) [9].

a) Depending on the transduction mechanism, biosensors could be divided into groups of electrochemical, electrical, optical, piezoelectric (mass detection methods), calorimetric (thermometric), and thermal [3].

Table 4.1 Comparison of different standard immobilization methods

Immobilization method	*Biorecognition element*	*Principle*	*Examples*	*Advantages*	*Disadvantages*
adsorption	small molecules	van der Waals force, hydrophobic force, hydrogen bonds, ionic interaction	proteins, antibodies/ antigens	simple, quick	weak forces, leakage, low reproducibility
covalent binding	proteins	electron sharing (functional groups)	-NH2, -COOH, -SH, -OH, imidazole, phenol, phosphate	irreversible (strong), mild	conformation change of BRE, molecular architecture difficult
cross-linking	proteins, enzymes	two functional groups of cross-linker connect BRE to a transducer	carbonyl diimidazole, SAM glutaraldehyde, EDC, NHS	many cross-linker options, simple, quick, mild conditions	inter/intramolecular linking, sizing restrictions
entrapment	cells, proteins	restrain due to steric hindrance in matrix or membranes	polyacrylamide, gelatin, polyvinyl alcohol, alginate	simple, quick	leakage, permeability of interferences, diffusion barrier

Source: reprinted with permission from [8].

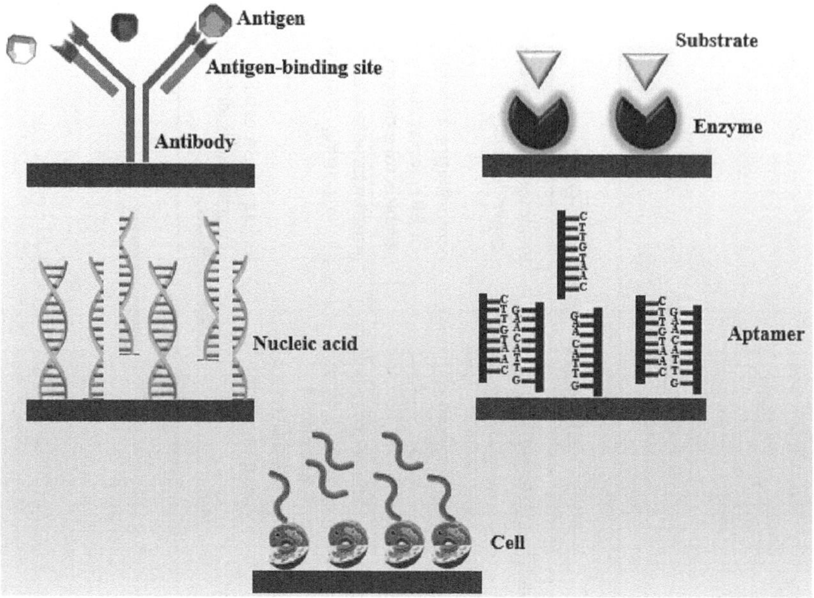

Figure 4.3 Various biological signal mechanisms of biosensing.

b) Based on bio elements, enzymes, nucleic acids, proteins, saccharides, oligonucleotides, cells, ligands, etc., there are different sets of biosensors that could be gained [10].

Each bio element in different sets of biosensors has a special mechanism to make signals. There are five major mechanisms for biological signals in biosensors that are shown in Figure 4.3. In short, among the various methods of signal transduction, the electrochemical method is of higher interest and used widely due to its benefits, such as low cost, facility of use, portability, and simplicity of manufacturing [11]. In the next section, electrochemical biosensors and their applications are discussed.

4.3 PRINCIPLES OF ELECTROCHEMICAL BIOSENSORS

IUPAC gives the definition of an electrochemical biosensor as: "a self-contained integrated device, which is capable of providing specific quantitative or semi-quantitative analytical information" [12].

The transduction element of electrochemical biosensors is a electrochemical cell that includes working, auxiliary, and reference electrodes (Figure 4.4). The working electrode is the main component that is modified for detection of analytes. The electrode as an electronic conductor transportes charge via movement of electrons and holes [13].

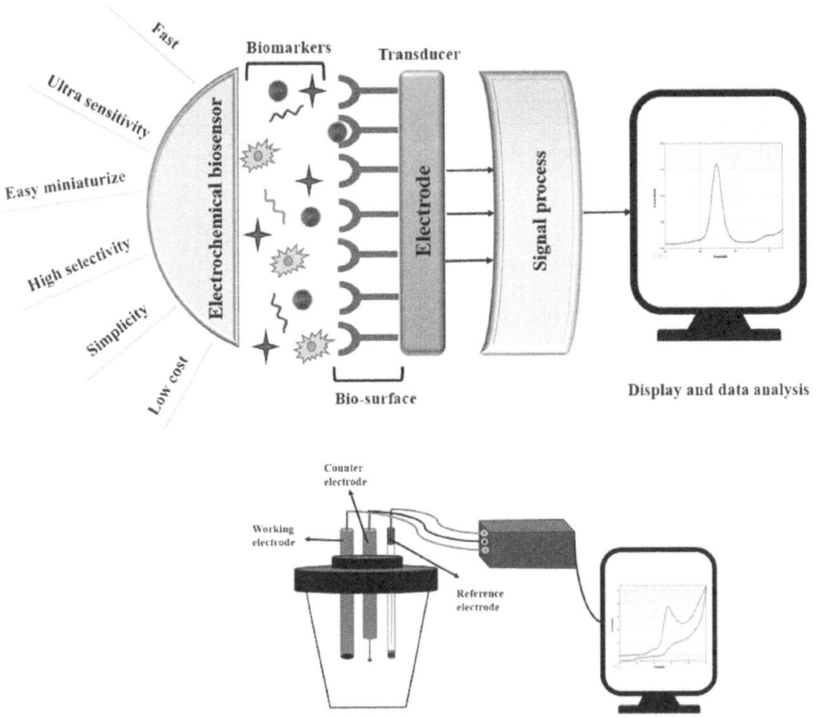

Figure 4.4 Electrochemical sensing and electrochemical working cell.

Nanomaterials and nanostructures, with their special physical and chemical features, are very good candidates for designing high-performance electrochemical biosensors. The high surface-to-volume ratio of nanostructures and nanomaterials has an important role in enhancing of analytical sensitivity. Consequently, nanomaterials and nanostructures are used widely in electrochemical biosensors. Table 4.2 shows functional nanomaterials and their effect on improving electrochemical performance [8].

The electrochemical reactions typically produce measurable current, potential, and conductive properties. So, the three main electrochemical biosensor detection strategies are amperometry, potentiometry, and conductometry [14]. The use of electrochemical impedance spectroscopy and voltammetry method is also common. In potentiometry, the difference in potential is measured while no current is flowing. This difference is produced through an ion-selective membrane that separates two solutions [8]. Amperometry measures the current related to the reduction or oxidation at the working electrode in a constant potential difference versus the reference electrode. Conductometry determines the electrical conductance as the electrochemical signal in solutions that the cations and anions are charge carriers [15]. In the following, we will briefly discuss impedance spectroscopy, potentiometry, voltammetry, and amperometry methods.

Table 4.2 Electrochemical performance enhanced through functional nanomaterials

Electrochemical performance	*Nanomaterial*	*Function*	*Improved property*
Sensitivity	Nanosized materials	Voltammetric transducer	Highly efficient diffusion profile low background current
	Highly dispersed nanomaterials, nanoporous materials, nanopillar	Tracer and carrier transducer	Enhanced surface area resulted inhigher collision rate
	Nanotubes, nanowire, nanopores	Label-free transducers	Environmental confinement can detect small changes
	Metallic nanoparticle, e.g., gold nanoparticles	Catalyst and tracer	Catalytic effect enhanced signal amplification
	Short nanotubes, nanorods, nanovesicles, nanoporous microsphere	Carrier	High loading capability
	Nanotubes, metallic nanoparticles	A molecular bridge between enzyme active site and electrode	Promote electrical communications
Selectivity (specificity)	Nanoporous structure	Transducer	Good peak separations for analytes based on diffusion and non-diffusion profiles
	Noble and reactive metal nanoparticles, g-C3N4, BN	Catalyst	Inherent catalytic property and affinity with the analyte of interest
	Metallic nanoparticles and semiconducting nanoparticles	Tracer	Distinctive signals for simultaneous detection
Low electrode fouling	Nanoporous electrodes, nanopore arrays	Filter	Large molecules repelling
	Nanomaterials contained oxygen groups	Molecular repellent	Prevent fouling from adsorption of byproducts
stability	Nanoporous electrode, nanocontainer, e.g., liposome and silica mesoporous microsphere	Entrapment function	Protect electroactive species against degradation

(Continued)

Table 4.2 (Continued)

Electrochemical performance	*Nanomaterial*	*Function*	*Improved property*
	Nanoparticle mimicking enzymes	Catalyst	Higher stability than protein
Short analysis time	Highly dispersed nanomaterials, nanoporous materials, magnetic hybrid nanomaterials	Transducer	Enhanced collision rates due to the high surface-area-to-volume ratio overcome diffusion limitation of a binding assay based solid phase immobilization
Low interference from sample matrixes	Magnetic nanoparticles, nanoporous structure, highly dispersed nanomaterials	Separator	Enrichment of target analytes
Wide dynamic range	Nanoporous network structures	Immobilized surface for biorecognition elements	Prevents saturation at high sample concentration

Source: Reprinted with permission from [8].

4.3.1 Potentiometric method

The basis of the work of potentiometric sensors is that the potential difference between reference and working electrodes changes directly with the concentration of the analyte while current is zero [16]. In potentiometric sensors, different transducers such as gas-sensitive electrodes, field effect transistors (FETs), and ion-selective, are employed based on the nature of the studied species. Some of the key properties of the potentiometric technique are selectivity, sensitivity, cost-effectiveness, and non-invasion. These feautures played an important role in the development of biosensors in biomedical applications [17–19]. Among different transducers, FET-based potentiometric sensors have achieved high consideration, especially in point-of-care devices, which might be because of FET's built-in potential for miniaturization [20]. Also, some benefits such as rapid response, ease of fabrication, and highly sensitive label-free detection by the FET, have an important role in the enthusiasm of researchers for this transducer [21]. FET is a semiconductor (transducer)- based potentiometric device that includes three following terminals: source, drain, and gate. The target-specific bioreceptors are placed on the surface of a gate terminal (dielectric material). Biorecognition changes the surface-charge on the gate terminal. The surface-charge changes can lead to changes in the gate voltage and thereby, the charge transport properties of the FET channel [22]. A typical schematic representation of FET-based biosensors is shown in Figure 4.5.

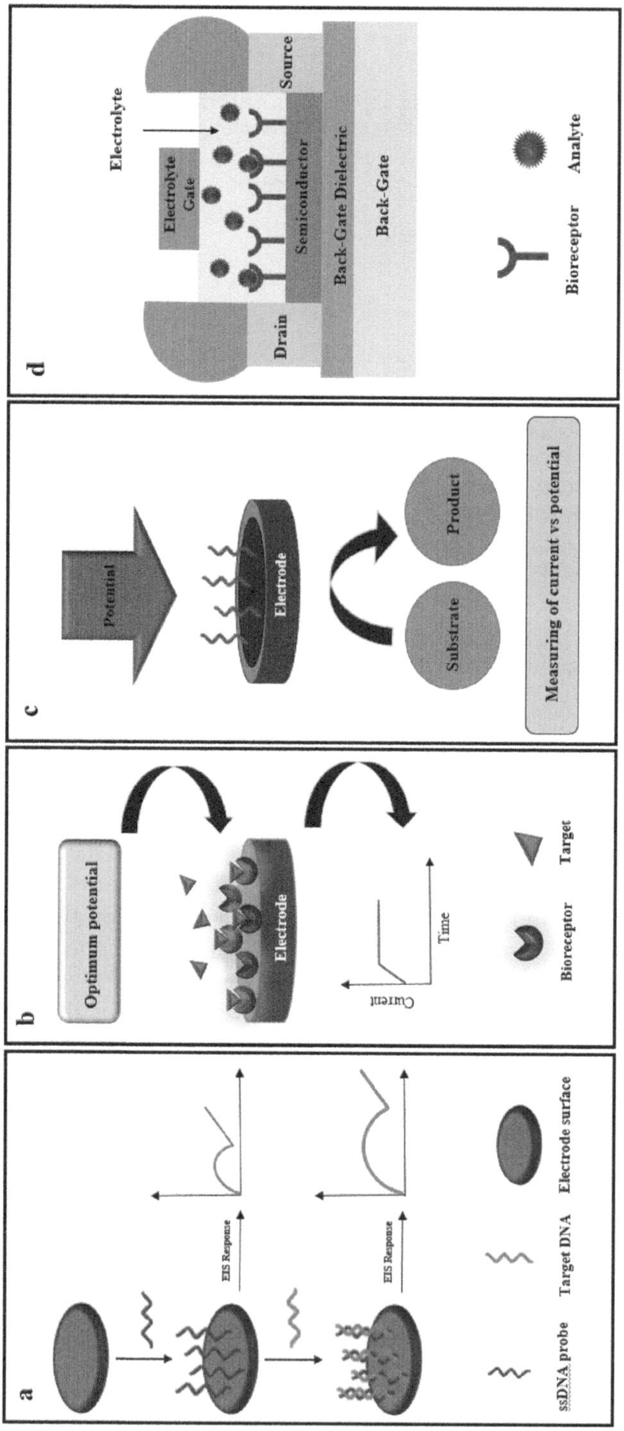

Figure 4.5 Electrochemical biosensor detection strategies: a) impedance, b) amperometry, c) voltammetry, d) potentiometric (Field Effect Transistors)

4.3.2 Voltammetric method

Voltammetric biosensors have a key role in the scientific community due to their ability to provide information about biological systems. Based on the voltammetric method, the generated current is monitored between the working electrode and a counter electrode upon sweeping potential between a reference electrode and a working electrode [11]. In voltammetry, electroactive species reach to the electrode-electrolyte interface by mass transport from the bulk of the solution, then the heterogeneous electron transfer happens at the electrode-electrolyte interface. The current is created as a result of the oxidation/reduction process of the electroactive species, which takes place at the surface of the working electrode. Depending on the applied potential waveform, voltammetric techniques are classified as square wave voltammetry (SWV), differential pulse voltammetry (DPV), cyclic voltammetry (CV), and linear sweep voltammetry (LSV) [23]. In voltammetric biosensors, the current response is the result of the process of bio-recognition between the recognition layer and the analyte that is produced by redox processes of the analyte or through labeling. The current responses are usually displayed as a peak, which correlates with the concentration of the electroactive species [24].

4.3.3 Impedance method

The electrochemical impedance spectroscopic (EIS) technique was familiar to electrochemists for over a century [8]. The EIS method is a strong tool in biosensing applications due to its effective label-free detection [25]. The biorecognition process leads some changes at the electrode-electrolyte interface in different chemical and physical properties. EIS technique uses the changes in charge transfer resistance (R_{ct}) or interfacial capacitance to indicate the biochemical changes happening at the sensor surface [8, 25]. Unlike other electrochemical methods like cyclic voltammetry, which has large amplitude perturbations, the EIS method imports small amplitude perturbations. This property makes EIS a non-destructive technique [26].

4.3.4 Amperometric method

In amperometry, the current-time response of electro-oxidation/reduction of an electroactive species at the optimum potential is checked [27]. The current response is commensurate with the concentration of the electroactive species. The excellent selectivity in detection caused by this potentiostatic technique makes the amperometry method the widely used one in the development of chemical sensors [28]. The advantages of amperometry that makes it a preferred one for sensor developers include low-cost instrumentation, wide concentration range of detection, minimization of charging current (current needed to apply potential which affects the detection limit), selectivity, and specificity [27, 29, 30]. In amperometric biosensors, specific bioreceptor-target binding creates amperometric signals either directly or indirectly. For example, a direct signal can be produced via charge transfer from an electron-rich label attached to the target molecule. On the other hand, redox processes catalyzed by enzyme labels on the target molecule can generate an indirect signal.

4.3.5 Electrochemical biosensor assay strategy: labeled vs. label-free

The electrochemical transduction generates a signal either with redox activity of a solution-based reporter or a biocatalytic reaction of an electroactive label coupled to a probe or target [31].

Label-free biosensors have attractive advantages compared to similar label-based analytical techniques (Table 4.3). Label-free electrochemical biosensors can monitor the biological reaction of specific bio-recognition elements and analytes directly. Among electrochemical techniques, the methods based on potentiometry and impedance are widely used for label-free techniques [8]. Label-free biosensors can also employ voltammetry-based techniques where oxidation or reduction of intrinsic electroactive components of (bio) analytes (e.g., nitrogenous bases in nucleic acids [33], and amino acids in proteins [34]) generate the electrochemical signal. Further, redox probes can be used. In this case, the binding of the analyte to the specific biorecognition element (BRE), changes the conformation of (BRE) that influences the accessibility of redox probes and subsequently affects the electrochemical signal. Label-free methods are typically used for biological

Table 4.3 Advantages of modern label-free biosensors over similar label-based analytical techniques

Advantages
A simplified pattern of analysis.
Reduced analysis time (rapid response time).
Lower cost of analysis.
Reduced consumption of organic solvents.
Portability and small dimensions.
No need for qualified medical personnel.
Opportunity to quantify biomolecules in real-time mode.
Target analytes are detected in natural forms, without. modifications and labels.
High sensitivity.
Direct measurement of analytes.
Opportunity to detect small molecules.
The opportunity of multiplexing.
Access to kinetic and thermodynamic parameters.

Source: [32].

reactions such as DNA hybridization [35, 36] aptamers hybridization [37], immunoassays [38], and also molecular imprinted [39] and enzymatic approaches [40].

Sometimes the interaction between the analyte and its bioreceptor may not produce an electrochemical signal, or not be even sensitive enough. In such cases, different strategies are employed in which a label is used to generate an electrochemical signal. These strategies can be typical sandwich assays [41], electrochemically active DNA intercalators [42], and sensors utilizing molecular beacons [43]. In Figure 4.6, labeled and label-free strategies are shown briefly.

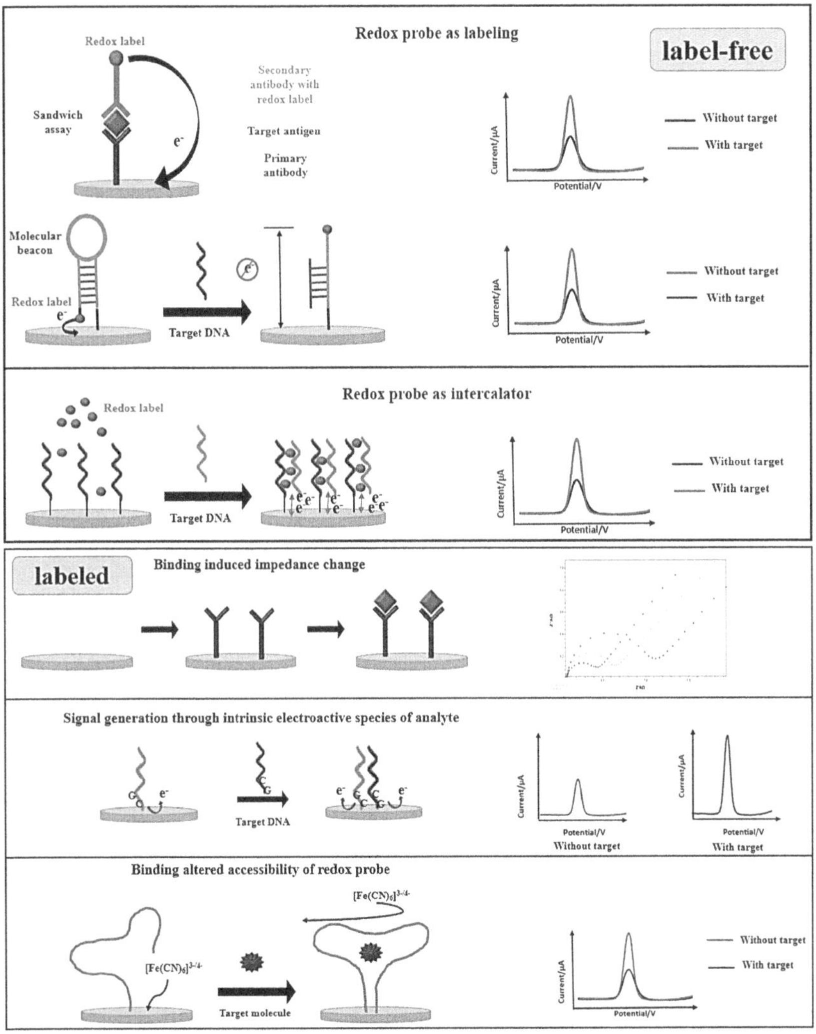

Figure 4.6 Examples of label-free and labeled assays.

4.3.6 Biocatalytic/affinity electrochemical biosensors

Electrochemical biosensors based on the nature of the biological recognition process can be divided into two main categories; a) biocatalytic and b) affinity biosensors (Figure 4.7).

4.3.6.1 *Biocatalytic biosensors*

Biocatalytic sensors use biological components like whole cells, enzymes, or tissue slices for recognizing the target analyte and making electroactive species or another detectable product [23]. Enzymatic biocatalytic sensors often have relatively simple designs, without needing expensive instrumentation. These sensors can be used with different detection configurations like stationary sample solution vs. flow conditions, or bulk sample solution vs. a micro drop detected by a microelectrode. Biocatalytic sensors can be simply adapted with industrial analysis or automatic clinical lab. For example, personal blood glucose monitoring devices are the most prosperous commercial usage of biocatalytic sensors [44].

Enzymes have high biocatalytic activity and specificity. So they are used mostly by electrochemical biosensors [45]. Enzymes are globular proteins that are made up mainly of 20 naturally occurring amino acids which catalyze biochemical reactions [23].

Enzymes as a catalyst can significantly increase the speed of an uncatalyzed reaction. The active site of the enzyme is often at the centroid of the protein. The arrangement of amino acids at the active site bind with the specific substrate. So, this strategy for bioreaction makes the enzyme selective for one type of substrate molecule [46]. Furthermore, many enzymes combine small nonprotein chemical groups, like cofactors or prosthetic groups, into the structures of their active site that assist determine substrate specificity [46]. The inherent selectivity of enzymes leads to detection of individual substances in a complex mixture, like blood or urine selectively, and also circumvents the signals produced by interfering species in complex samples. So, with this, there isn't the need for labor-intensive, time-consuming, and interference-prone sample separation and pretreatment steps used in composite methods.

Cellular materials such as plant tissues are used by some biocatalytic biosensors as the recognition component [11, 23]. These biocatalytic electrodes, work similarly to conventional enzyme electrodes (i.e., enzymes existing in the tissue or cell consume or produce electrochemically detectable species).

Live microorganisms have also been used by electrochemical biosensors to monitor biotechnological processes like food manufacturing, brewing, energy production, wastewater treatment, and pharmaceutical synthesis [8, 11–24, 28, 31, 33–177]. They can be coupled with electrochemical transducers like electrodes. Immobilization of bacteria on transducers is often via

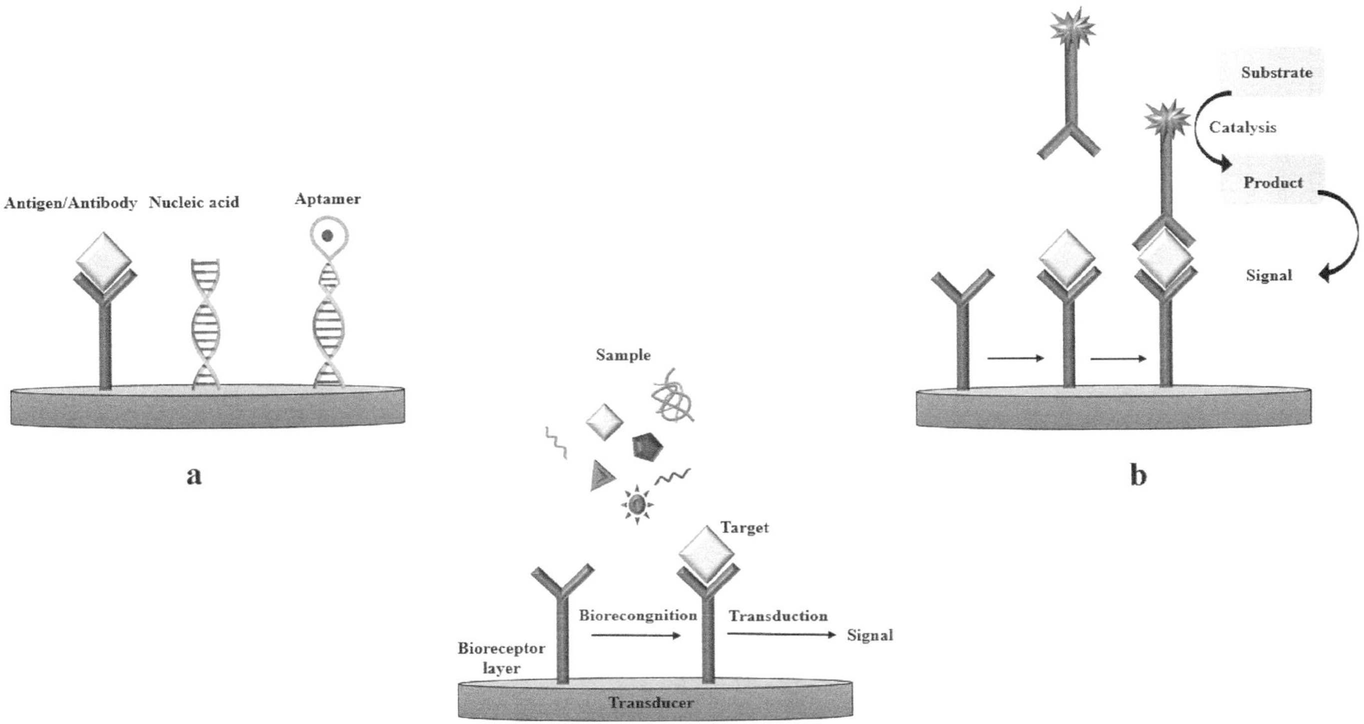

Figure 4.7 a) Affinity based biosensors, b) biocatalytic based biosensors.

microencapsulation where a inert membrane traps the microbe on the electrode surface [11]. After coupling the microorganism with the target analyte, the respiration activity of the microorganism is changed and causes a lower surface concentration of electroactive metabolites (e.g., oxygen), which can be detected by the electrochemical transducer [23, 24].

Some microorganisms, through the production of electroactive metabolites that can be directly monitored, are useful for electrochemical biosensors [11]. Using microbes in biosensors is cheaper than isolated enzymes, and also less sensitive to inhibition by other sample components. Microbes are more tolerant to slight variations in pH and temperature than enzymes [11, 25]. Some of their disadvantages that limit their application contain longer response times, hysteresis effects, longer recovery times after exposure to the analyte of interest, and possible loss in selectivity because of including many types of enzymes [11, 24, 25].

There are some limitations of biocatalytic biosensors. For example, there is no specific enzyme for many biological analytes. Therefore, affinity biosensors can act as an alternative to biocatalytic biosensors.

4.3.6.2 *Affinity biosensors*

In affinity biosensors, biomolecules such as membrane receptors, antibodies (Ab), or oligonucleotides bind with a target analyte selectively and produce a strong measurable signal [23]. The complementary size and shape of the binding site to the target analyte is critical for molecular recognition [23]. The degree of affinity and specificity of the biomolecule for its ligand determines the sensitivity and selectivity of these sensors [11]. The binding process, such as DNA hybridization or antibody-antigen (Ab-Ag) complexation is manipulated by thermodynamic considerations [23].

Ab-based affinity biosensors are called immunosensors. Immunosensors detect an analyte, an antigen or a hapten via their binding to a region of an (Ab) [47]. The binding is converted by the electrochemical transducer to the electrical response towards an output, so that it can be amplified, stored, and displayed. Complementary regions of the (Ab) bound to an (Ag) are used to furnish the antibodies in a host organism like a mouse or a rabbit with high affinity and specificity [46]. Such polyclonal (Abs) are heterogeneous regarding their binding domain, and may be clarified by a selection process to obtain a monoclonal (Abs-MAbs). All members of that particular (MAb) clone are identical. (Abs) and (MAbs) can be employed for a wide variety of substances. Theoretically, if an (Ab) can be raised against a particular analyte, an immunosensor could be utilized to detect that substance. Immunosensors are very desirable among analytical methods for their very low detection limit [47]. Immunoassays and immunosensors have been expanded for both quantitative and qualitative utilizations [11, 23]. Immunosensors with the ability to detect trace levels (ppb, ppt) of various analytes, including viruses, bacteria, hormones, drugs, pesticides, and numerous other chemicals, have lots of different applications [11, 23]. For example, detecting environmental pollutants like herbicides and pesticides in water and soil, detecting biomedical substances such as warfarin, monitoring food safety relevant to severe allergies (such as peanuts), and monitoring for biowarfare agents such as toxins, bacteria, viruses, and spores [11, 23]. Relatively inexpensive detection kits can be designed using immunosensors such as home pregnancy and fertility test kits.

Compared to antibodies, nucleic acids have been less used as the biorecognition element in affinity sensors. They use DNA or RNA nucleic acid fragments and also aptamers that are generating nucleic acid structures.

The bioreaction in affinity sensors that use DNA or RNA nucleic acid fragments is based on complementary base-pairing between the target analyte and the sensors nucleic acid sequence. Aptamers as biorecognition, recognize and bind to three-dimensional surfaces, such as those of proteins [44]. The knowledge of the nucleic acid structure and how it is manipulated for various applications is rapidly expanding and advancing, so nucleic acids are now becoming more important. Today, DNA biosensors are very operational in medicine like cancer detection, viral infections, and genetic disease [11].

4.4 ELECTROCHEMICAL LAB-ON-A-CHIP SYSTEMS

Given the evolving applications of biosensors, the creation of a portable, fast, and user-friendly biosensor is of particular importance. Therefore, miniaturizing the device through lab on a chip (LOC) technologies is essential [48–50]. LOC is a device that can scale down the laboratory functions to a chip format up to a range of only a few square centimeters [51].

Different laboratory functions can be integrated and automated by the LOC platform. With the miniaturization of laboratory functions on a LOC device, the amount of reagents for the analysis is reduced, which can lead to lower costs [52, 53]. Furthermore, the usage of LOC technologies provides the possibility of simultaneous measurements of several biomarkers. Advanced microfabrication techniques authorize the integration of microfluidics with biosensing functionalities on the same sensor chip that empowers system automation [49].

Most commonly used biosensors use electrochemical detection techniques because of their high sensitivity, selectivity, ability to operate in turbid solutions, rapid analysis, and the possibility of miniaturization [54]. Electrochemistry is a hopeful detection strategy combined with LOC systems because of its advantages such as high sensitivity, low cost, low power requirement, independence of optical pathways, and distinctive adaptability with microfabrication and micromachining technologies [55].

4.4.1 Design of LOC systems

The components of each LOC system are selected according to the analysis. Typically, LOC systems use a microfluidic manner to transport the sample. According to the analytical problems, several functionalities are combined on an LOC system. The crucial features assimilated into analytical LOC systems are sample preparation, separation, and detection systems. In LOC, biosensor systems can be integrated into microfluidics that can improve the response of sensing systems. For example, these presenting procedures can remove interferences that increase the accuracy of the measurements, or can increase the analyte concentration in detection volume which increases the measurement sensitivity.

Furthermore, LOCs due to diminished sizes and reduced volumes, only need a little amount of sample, like a drop of blood, to achieve the analysis. Besides, in this system, the distance between analyte molecule and biorecognition element is reduced via increasing diffusion-controlled electron-transfer yield, which can lead to improvement in sensor performance [56]. Microfluidics with controlling small fluid volumes have an important role in the advancement of LOC technology. It has the potential to run sample preparation, separation, reagent mixing, filtration, reaction time, and multiple analyte detection on a single chip [57].

4.4.2 Materials and fabrication

There are various materials for the fabrication of the LOC systems. Rigid inorganic substrates such as glass or silicon were used for the construction of the first microfluidic devices. Gradually, with the advancement of research in the field of microfluidics, polymers became the most usual substrate for LOC systems fabrication, and recently, paper-based LOC systems have been described. When we are designing a microfluidic system by choosing materials and fabrication methods, these three main factors have to be considered: degree of integration, required function, and application [150]. Table 4.4 briefly shows the LOC materials with their characteristics and fabrication techniques.

4.4.3 Microfluidics

The history of microfluidics is commenced from the microelectronic industries that researchers tried to use etching, photolithography, and bonding techniques in order to improve silicon-based micromachining processes. Terry et al. introduced the first silicon-based analysis system in 1979. Later, in the 1990s the application of microfluidics is advanced by Manz et al. [151].

Unlike laboratory-scale samples that are measured in milliliter-scale volumes [152], microfluidics does the measurement of nanoliter [151] and microliter-scale volumes [153]. Therefore the term microfluidics refers to each technology that moves fluid in microscopic and nanoscale volumes

Table 4.4 Materials used for the fabrication of LOC systems

Material	*Characteristics*	*Fabrication techniques*
Silicon	Easiness of surface modifications, rigid, biocompatible, high temperature resistance, not gas-permeable, low unspecific binding	Photolithography, etching, vapor deposition
Ceramics	Good electrical, mechanical, and thermal characteristics, rigid	Photolithography, etching, vapor deposition
Glass	Transparent, rigid, easiness of surface modification	photolithography, etching, vapor deposition
Poly (methyl methacrylate) (PMMA)	Rigid, transparent, low water absorption	Injection molding, hot embossing, stereolithography, laser photoablation
Cyclic olefin copolymer (COC)	Low water absorption, rigid, transparent	Injection molding, hot embossing, stereolithography
Polystyrene (PS)	Rigid, transparent	Injection molding, hot embossing, stereolithography
Polydimethylsiloxane (PDMS)	Flexible, transparent, biocompatible, extremely gas permeable, chemically inert,	Soft lithography, injection molding
Polycarbonate (PC)	Rigid, transparent, highly heat resistant	Injection molding, hot embossing, stereolithography

Source: [8, 178, 179].

through micro-sized channels on a microelectromechanical system (MEMS). Microfluidics is a combination of different disciplines, including the fluid mechanic, chemistry, surface science, biology, and in many cases, microscopy, optics, control systems, electronics, and microfabrication [154]. Microfluidic chips, (i.e., LOC systems) integrate a diverse set of biological sensors and manipulate fluids at the nanoliter and picolitre scales [155] which can automatize biological computations or experiments [156, 157].

Health-care systems need to use more accurate, faster, and more highly precise diagnostic devices. Microfluidics-based LOCs are suitable for these applications since they are capable of reducing health care costs and also provide better epidemiological data that are useful for infectious-diseases modeling. Figure 4.8 shows some applications of microfluidics.

4.4.3.1 The physics of microfluidics

For designing and optimizing microfluidics-based devices for biological applications, it is necessary to understand the underlying flow of physics and interfacial phenomena at small scales. DNA detection in microfluidics-based

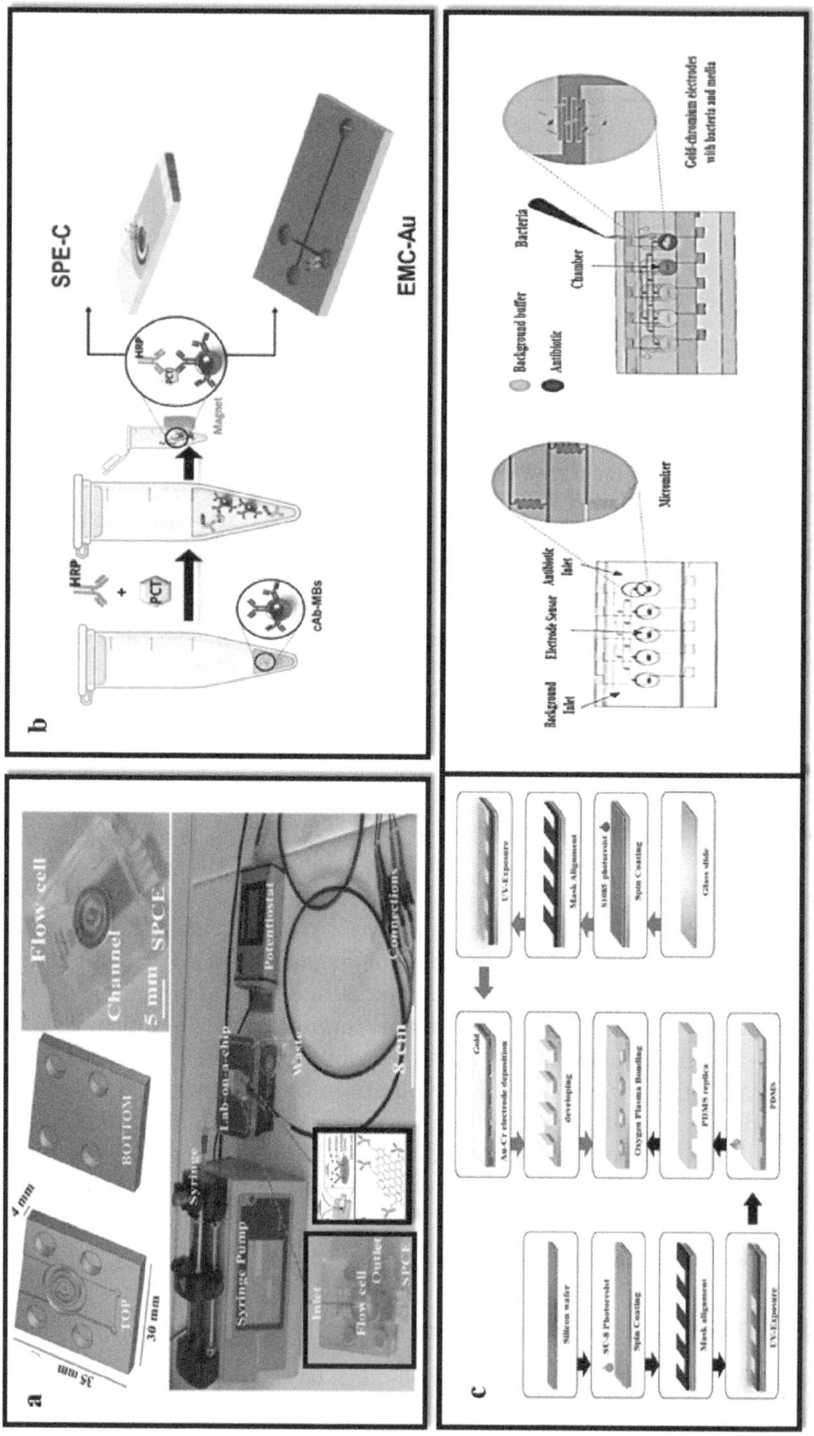

Figure 4.8 Microfluidics applications: a) microfluidic-based electrochemical immunosensor for detection of ferritin as a clinically important biomarker, b) electrochemical immunosensor on-drop and on-chip for determination of procalcitonin, c) label-free electrochemical microfluidic chip for the antimicrobial testing [180–182].

systems needs precise fluid control and flow stability. Because of the sensitive detection systems of microfluidic devices, it is critical to prevent bubble formation within the channels or chambers when the infusion of any fluids occurs. Undesired bubbles can affect the sample flow and lead to detection failures, specially in highly sensitive optical detection [159].

4.4.3.2 *Classification of microfluidic platforms*

Broadly, there are three main categories for microfluidics: digital, channel, and paper-based microfluidics. These microfluidic systems have different working principles, the device format, and driving forces to manipulate liquids, which will be discussed in their related sections [160].

4.4.3.2.1 Channel microfluidics

Channel microfluidics came out as a gadget for chemical separation (e.g., capillary electrophoresis). Later, it appeared with chip technology, commonly known as microelectromechanical systems (MEMS), to manufacture miniaturized devices. Channel microfluidics includes enclosed channels where pneumatic pressure forces liquid. The device format includes mixers, valves, channels, and pumps (manual or digital). Channel microfluidics can be useful. Continuous flow microfluidic chips have a diversity of applications in chemical and biochemical separation due to precisely handling volumes down to the nanoliter (nL) involving different molecules like proteins, DNA, cells, etc. [161].

4.4.3.2.2 Digital microfluidics (DMF)

In digital microfluidics, the liquids are manipulated in separate droplets on an array of dielectric coated electrodes [162–169]. An electrode array involving consecutive actuation of electrode receives a series of potential to mix, merge, dispense, and split droplets separately or simultaneously through digital programming. In two device formats, DMF devices can work: one-plate and two-plate. The ideal features of DMF like an ability to govern individual droplets in the absence of networks of channels, pumps, valves, or mechanical mixers, makes it desirable from channel microfluidics. DMF has a great usage in a broad range of applications such as immunoassays [170], DNA biosensing [172], cell culture [171], forensics [173], separation science [174], etc.

4.4.3.2.3 Paper-based microfluidics

Paper-based microfluidics contains open channels, while channel microfluidics involves enclosed channels. So, it seems easy to recognize them from each other. Paper-based microfluidics is a concept of liquid handling and analysis on paper by building channels on paper via hydrophobic/hydrophilic patterns [175]. Paper microfluidics has interesting advantages, such as facility of fabrication, availability, flexibility, significant low cost, and ease of liquid movement via capillary forces without need for any external driving force. These benefits encouraged researchers to develop a point-of-care tool to screen diseases in resource-deprived regions in the world [176]. Therefore, there are various formats of paper-based diagnostics technology [177].

4.5 WEARABLE ELECTROCHEMICAL BIOSENSOR

Nowadays, the prevention of disease by monitoring the early stages is considered a very cost-effective approach concerning treatment costs, once the diseases are fully manifested. This new approach also leads to better health outcomes [100, 101]. In this endeavor, wearable biosensors have gained considerable attention. The high specificity, portability, fast detection, low-cost, and low-power features of biosensors have made them very suitable as wearable devices. Wearable devices have a considerable role in accomplishing these goals since the collection of crucial information in a continuous and non-invasive manner is easily obtained [27, 93–99, 101]. The USA announced 2015 as “the year of health care for wearables” [102], while The Huffington Post stated that wearable technology is “the coming revolution in healthcare” [103]. Wearable biosensors are advancing toward non-invasive monitoring of biochemical markers and drugs and other target analytes in biofluids. In this regard, microfluidic systems are very effective and helpful. Due to the important role of microfluidics, suitable manufacturing methods are required for these goals. Therefore, an efficient, flexible, fast, and affordable manufacturing method plays a huge role in the development of wearable biosensors and human health monitoring in the future. Wearable technology is often referring to a category of gadgets that can be worn directly by a consumer for fun or just to track their physical activity and fitness. A different category of wearable technology is medical wearable devices that can be worn by patients on their skin or different parts of their body, and often includes the tracking of the body’s physiological information related to health, in some cases, at molecular levels [104–116]. Wearable devices can collect data on a 24-hour, seven-day basis, in several environmental settings, as people go through their daily routines at home or work [117]. Wearable devices can relay physiological information as the body evolves over healthy and sick states. They can help people to monitor themselves without expensive equipment, and neither educated professionals nor teams of expensive medical staff are required [107, 118]. Moreover, the characterization of non-invasive and wearable technologies for diagnosis is extremely beneficial for both continuous health monitoring and diagnostics in early and pre-disease states. They also allow quick access to clinical information by the patients, which encourages people to take more concern of their health more comfortably and cheaply, which also improves compliance [119, 120]. Some of wearable biosensors are shown in Figure 4.10.

4.5.1 Target biofluids for the wearable electrochemical biosensors

Wearable biosensors measure biochemical markers in biofluids, such as saliva, sweat, tears, and interstitial fluid for human health monitoring. In this section, we have a brief look at health monitoring in various biofluids.

4.5.1.1 Saliva analysis

Human saliva is a watery substance that includes 99.5% water with electrolytes, white blood cells, mucus, glycoproteins, enzymes, epithelial cells, among others [58]. While the monitoring of saliva is very interesting for the healthcare fields, some challenges need to be evaluated to increase the efficiency of electrochemical biosensors. For example, analysis of saliva requires a highly selective sensor because of its complicated mixture. This is also crucial that the performance of the sensor remains stable in such a high moisture environment. In addition, the devices should be quite biocompatible, due to their use in the mouth [60]. A label-free and non-invasive biosensing platform is developed for the detection of an oral cancer biomarker. For efficient detection of microRNA as a biomarker, non-toxic serine amino acid is used as linker molecules for the functionalization of nZrO2. The limit of detection of this sensor is 0.01 ng/mL, which is low enough for the smaller secretion level of targets in human saliva [59].

4.5.1.2 Tear analysis

Tears are secreted from lacrimal glands, ocular surface epithelial cells, and goblet cells. Tears include lipids, metabolites, electrolytes, and proteins/peptides [61]. For health monitoring purposes, these complex extracellular fluids can be determined by electrochemical biosensors [61–64].

4.5.1.3 Sweat analysis

Sweat is more accessible than saliva and tears, and also has abundant biochemical compounds. It is possible to measure metabolites (e.g., lactate) and electrolytes (e.g., pH, Na+) selectively by analyzing sweat with a constant flow and a temperature sensor for internal calibration [65]. The detection of trace metal in sweat can be done by using a wearable amperometric biosensor. Because mobility affects sweating, sweat sensors must be able to separate multiple components in a complex sweat mixture at different rates. Genomics and proteomics have an important role in searching for new biomarkers in sweat. For example, dermcidin (DCD) and prolactin inducible

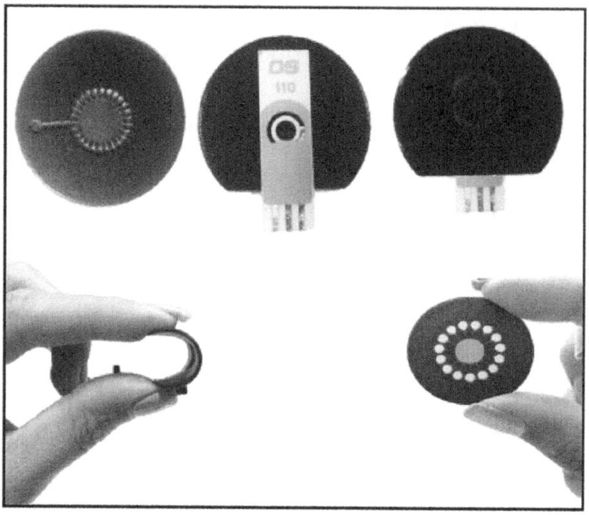

Figure 4.9 A 3D printed wearable device for electrochemical sensors for sweat analysis. (Reprinted with permission from [27].)

protein (PIP) have been found in sweat in addition to blood [121]. DCD and its receptors exist on the cell surfaces of invasive breast carcinomas, and their lymph node metastases, as well as in brain neurons and are overexpressed. PIP is present and overexpressed in prostate cancer and metastatic breast cancer instead [121]. Figure 4.9 shows a 3D printed wearable device for sweat analysis that is designed for electrochemical sensors [27].

4.5.1.4 Interstitial Fluid (ISF) analysis

Interstitial fluid (ISF) has the same composition to that of blood. It includes essential small molecules (e.g., proteins, salts, ethanol, and glucose). In addition, by using ISF, we can provide minimally invasive monitoring without the requirement for blood sampling [122]. For extraction of ions such as Na^+ in the ISF to the skin surface, a potential difference between two electrodes on the skin surface is applied, and reverse iontophoresis extracts the ions [123, 124], which has also been used by the GlucoWatch [125]. A flexible tattoo-based epidermal diagnostic device has been made by combining reverse iontophoresis with an enzyme-based amperometric biosensor [126]. Some applications of electrochemical wearable biosensors are shown in Figure 4.10.

4.5.2 Template and non-template fabrication methods

Flexible substrates are used to process the flexible sensors as a test strip [66], skin patch/tattoo [67], textile [68], etc. There are various substrates for

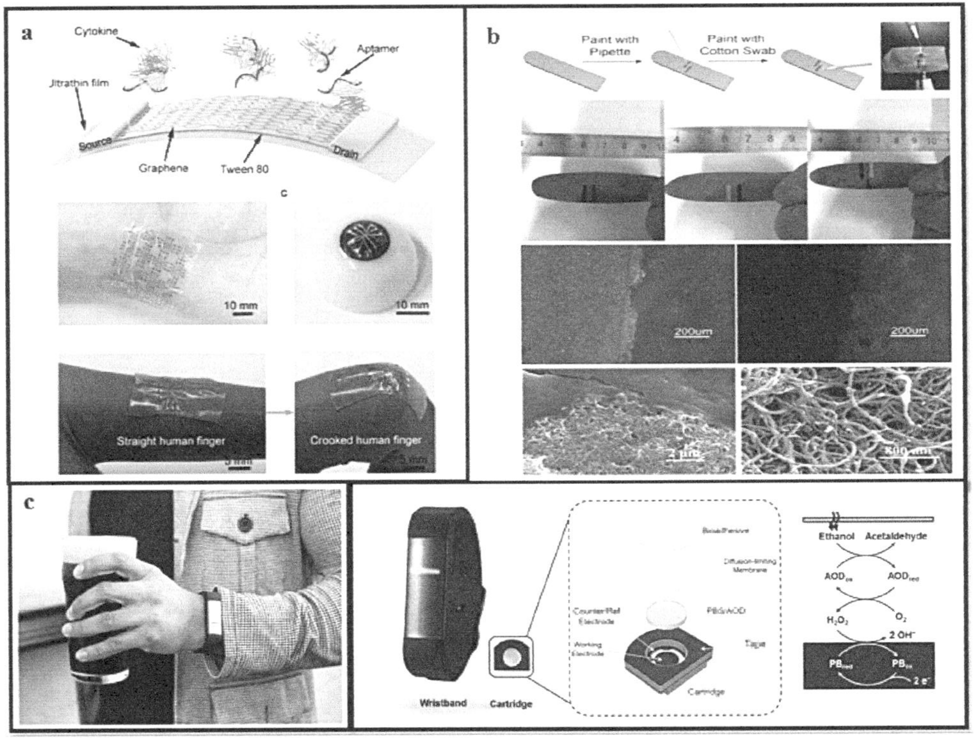

Figure 4.10 Application of electrochemical wearable biosensor: a) wearable affinity nanosensor for monitoring of cytokines in biofluids, b) wearable carbon nanotube-based amperometric biosensors on gloves for lactate determination, c) wearable transdermal enzymatic alcohol biosensor [183–185].

flexible biosensors, such as polyimide (PI), polydimethylsiloxane (PDMS), polyethylene terephthalate (PET), parylene, nylon, textile, and papers [69].

Multiple novel fabrication methods have been developed that use flexible substrates for stretchable electrochemical biosensors [127–129]. In this section, we discuss briefly these fabrication methods.

Lithographic methods (e.g., ion-beam lithography, thin-film deposition and etching, photolithography) can be applied to reproducibly fabricate high-performance devices (e.g., H_2O_2 sensors [129] and RNA sensors [130]). In these methods, despite their attractive attributes because of multiple equipment acquisitions, complex processes, the cleanroom setup, and the unique materials demand, the cost of employing them is high (115–117).

Screen printing is one of the fabrication methods that is inexpensive and simple to scale up for mass production of electrochemical biosensors with convenient electro-analytical execution [132, 133], which easily leads to the production of cheap fabric/textile-based electrochemical biosensors [134]. In this method, first, the conductive ink as a reference electrode is used as an underlayer on the textile. Next, the ink based on carbon or metal containing the detection element is coated on the substrate to act as working electrodes [135]. The ink formulation affects the spatial resolution and electrical performance of printed electrodes.

Flexography and gravure printing are the other template-based printing methods. In these methods, the ink is transferred to the substrate from a engraved (gravure) or raised (flexography) pattern on a roll. An anilox roll is used in flexography printing (Figure 4.11a) [136]. There are millions of tiny divots on the anilox roll which can bring the roll in contact with the printing cylinder. On the other hand, the anilox roll via these tiny divots can catch the ink. In flexography printing, the ink is first transferred from a bath to an anilox roll, then the ink is put on the surface of the target substrates. Gravure printing works via impressing the film into the cavities of the roll where the ink resides (Figure 4.11b) [139]. While in flexography, the ink is on the ridges of the pattern on the printing cylinder [58, 137, 138]. Gravure

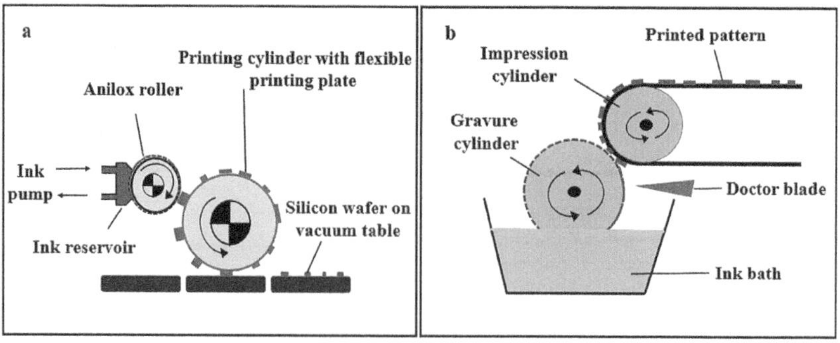

Figure 4.11 Schematic of a) flexography printing and b) gravure printing.

and flexography printing methods are inherently robust and can provide large-area manufacturing [139–145].

Some benefits of non-template-based printing technologies, like higher customization design and lower price for small-scale manufacturing, caused the development of these technologies. Non-template-based printing methods rely on dispensing the given technology. These technologies contain the use of piezoelectric material in the setup (piezoelectric), gas or pressurized air (pneumatic), driving of ink by an electric field (electrohydrodynamic), aerodynamic focusing (aerosol jet), and the heating of the material (thermal) [146, 147]. The extrusion-based 3D printing method as a representative non-template-based printing method, for manufacturing a fully 3D-printed electrode, uses a computer-controlled motion stage for applying the ink filament through a heated nozzle onto the substrate [148]. Inkjet 3D printing method unlike extrusion-based 3D printing that often needs high viscosity, uses the low viscosity inks to help ink transfer [146, 149].

4.6 HEALTHCARE APPLICATIONS

Electrochemistry technique, because of certain advantages for detection, provides a wider application for biosensors. Electrochemical measurement can be done with very small sample volumes [70]. Using this technique, low detection limits in immunoassays with simple or no sample preparation can be achieved, and atto-and zeptomole detecting electrochemical immunoassays have been made [71, 72].

Some detection methods like spectrophotometry are affected by sample components, unlike electrochemical detection and some other methods. Homogeneous immunoassays have no separation steps; hence they need a method that is not so affected by components of the sample. Therefore, electrochemical determinations can be performed on colored or turbid samples like whole blood, without interference from red blood cells, hemoglobin, bilirubin, and fat globules [73, 74]. So today, electrochemical biosensors have found a large range of medical applications, including the detection of cancer, infectious diseases, cardiac disease, and covid. We will discuss some of them below.

4.6.1 Cancer detection

Cancer has been the subject of wide scientific research. It is one of the most deadly diseases in the world. The detection of cancer in the early stages of its evolution is very important because it greatly increases the chances of treatment success [28]. We can be informed about cell differentiation, the tumor on the histogenesis, and cell functionalization by quantitative and qualitative analyzes. This information can provide the basis for cancer classification, the early detection or diagnosis of primary cancer, therapeutic guidelines, and prognosis [75].

There are many different cancer biomarkers in different classifications, like carbohydrate antigen, protein cancer markers, oncogene-related cancer biomarkers, hormone-related cancer marker, enzymatic tumor markers, isoenzymes tumor markers, and embryonic antigen markers [76]. Table 4.5 and Table 4.6 review some of the biomarkers of cancer and some the electrochemical biosensors respectively. Some of electrochemical biosensors for cancer detection are represented in Figure 4.12.

4.6.2 Infectious detection

Infectious diseases are caused by pathogenic organisms like viruses (~50–200 nm), bacteria (~0.5–10 μm), parasites, or fungi (~5–100 μm). The direct or indirect transmission of these pathogenic organisms can occur from person to person through contaminated air, contact, food or water [77].

Table 4.5 Biomarkers of cancer

Cancer	*Biomarkers*	*Reference*
Breast	BRCA1, BRCA2, MUC1, CEA, CA 15-3, CA 27, CA29, EGFR, EpCAM, HER2; miRNA-21, miRNA-373, miRNA-182, miRNA-1246, miRNA-155 and miRNA-105	[28, 29, 186, 187]
Prostate	PSA, Sarcosine; TEMPRSS2; miRNA-21, miRNA-141, miRNA-375	[30, 188, 189]
Brain	MDM2	[190]
Pancreas	CA 19-9, PAM4-protein; miRNA-21, miRNA-155, miRNA-196	[191, 192]
Gastric/Stomach	CA72-4, CA19-9, CEA, IL-6; PVT1 (DNA); miRNA-21, miRNA-331, miRNA-421	[193, 194]
Liver	AFP, DCP, GP73; miRNA-21, miRNA-122, miRNA-16	[195]
Ovarian	CA 125 (MUC-16), CEA, Claudin-4;	[196, 197]
Lung	ANXA2, CEA, Chromogranin A, CA 19-9, CYFRA 21-1 (CA-19 fragment), NSE, SCC, SAA1, HER1; P53, P16, Ras genes, Telomere length and telomere-related genes, EGFR gene (c-ErbB-1 and c-ErbB-2); miRNA-21 (in sputum)	[198–202]
Neck	MGMT gene	[203]

Abbreviations: BRCA1: breast cancer 1 gene; BRCA2: breast cancer 2 gene; MUC1: mucin1; CEA: carcinoembryonic antigen; CA: cancer antigen; EGFR: epidermal growth factor receptor; EpCAM: epithelial cell adhesion molecule; HER: human epidermal growth factor receptor; miR: micro-RNA; PSA: prostate-specific antigen; PAP: prostatic acidic phosphatase; MDM2: murine double minute 2; IL: Interleukin, AFP: α-1-fetoprotein; DCP: des-γ-carboxyprothrombin; GP73: Golgi protein 73; ANXA2: annexin A2; NSE: neuron-specific enolase; SCC: squamous cell carcinoma antigen; SAA1: serum amyloid A1; P53: protein suppressor gene; MGMT: o6-methylguanine DNA methyltransferase.

Table 4.6 Samples of electrochemical biosensors for cancer biomarkers

Cancer type	*Receptor/ Biomarker*	*Sample*	*Device/indicator*	*Label/ Label free*	*Detection methods*	*LOD*	*LR*	*Ref*
Lung	Antibody/CEA	Serum	AgNPs/THI/ICP fibers GEC/ thionine	Label free	DPV	0.5 fg	mL-1 50 fg mL-1 – 100 ng mL-1	[204]
Breast	Anti-EpCAMMNs/ MCF-7	Peripheral blood	MGE/molybdate	Label to form sandwich	SWV	1 cell mL-1	5–3 × 104 cells mL-1	[205]
Colorectal stomach pancreas	Antibody/CA19-9	Serum	Au-PGO-GCE/ HRP-H2O2-Thi	Label to form sandwich	DPV	0.006 U mL–1	0.015 – 150 U mL–1	[206]
Liver	Antibody/AFP	Serum	Cu2O@ GO-GCE/TB	Label free	SWV	0.1 fg mL-1	0.001 pg mL-1 – 100 ng mL-1	[207]
Prostate	Antibody/PSA	Serum	COOH-AgPtPd-NH2- rGO/ [Fe (CN)6]3-/4-	Label free	DPV	4 fg mL-1	4 fg mL-1 – 300 ng mL-1	[208]
Brain	Antibody/MDM2	mouse brain tissue homogenate	cysteamine – polycrystalline-GE/ [Fe (CN)6]3-/4-	Label free	electrochemical impedance spectroscopy (EIS)	1.3 pg/ ml	1 pg/ml-1 mg/ ml	[209]
Stomach	Antibody/CA72-4	Serum	rGO-TEPA/GCE/ PtPd-Fe3O4-H2O2	Label to form sandwich	Amperometry	0.3 mU mL–1	0.001–10 U mL–1	[210]

Abbreviations : CEA: carcinoembryonic antigen, Anti-EpCAM: antiepithelial cell adhesion molecule, CA19-9: Carbohydrate antigen19-9, AFP: alpha fetoprotein, PSA: prostate-specific antigen, MDM2: murine double minute 2, Au-PGO: Au nanoparticles functionalized porous, GCE: glassy carbon electrode, GE: gold electrode, rGO: reduced graphene oxide, TB: toluidine blue, MDM2: Murine double minute 2.

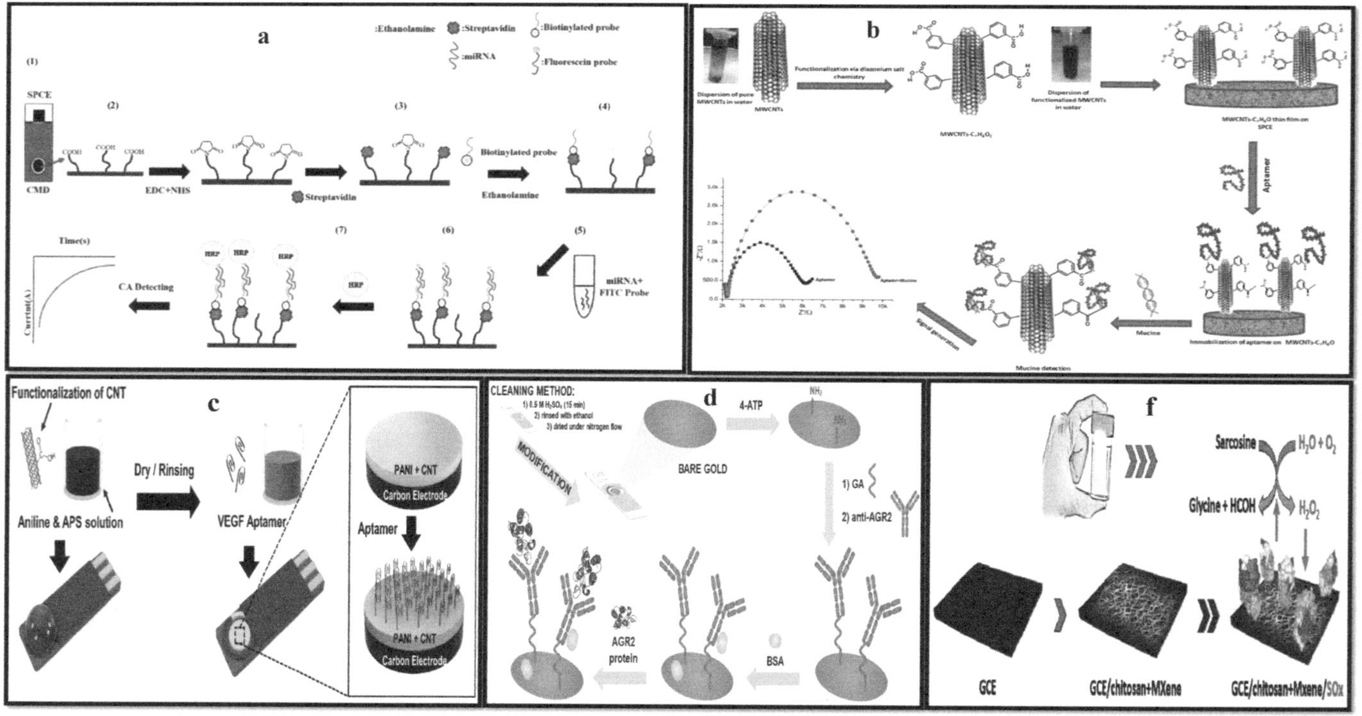

Figure 4.12 Electrochemical biosensors for cancer detection: a) high-sensitive electrochemical biosensor for dual-probe detection of miR-141 tumor marker. b) impedimetric carbon nanotubes-based biosensor for breast cancer detection. c) highly sensitive label-free electrochemical aptasensor for VEGF165 tumor marker. d) ultrasensitive biosensor for detection of the cancer antigen AGR2 using monoclonal antibody. f) ultrasensitive amperometric biosensor for detection of prostate cancer marker in urine [211–215].

Infections are the main reason for premature mortality worldwide and make a serious global public health problem. Accurate and timely laboratory diagnostic of infectious diseases is an important aim of modern medicine. Timely initiation of treatment provides a decline in the incidence of disease and also prevents the outbreak of dangerous epidemics. So, it is critical to detect pathogenic viruses accurately and rapidly for clinical point-of-care (POC) purposes. Electrochemical biosensors, because of their benefits such as rapid response, sensitivity, low cost, and selectivity, have been proposed for infectious disease detection applications.

During the last few decades, viral infectious diseases have become the main concern of the world. These viruses include Zika virus (ZIKV), Ebola virus (EBV), human immune deficiency virus (HIV), hepatitis virus, influenza virus, dengue virus (DENV) and most recently COVID [78–80]. These viruses make continuous threats to public health because of their genetic stability and rapid-spreading capacity. Since late 2019, the pandemic dimensional spread of COVID-19 has created a serious threat to the lives and health of people worldwide. Currently, diagnosis of COVID-19 infected patients is being done by polymerase chain reaction (PCR)-based tests [81]. However, the PCR method, due to the slow diagnostic time (3 to 4 hours), gives rise to false-positive and false-negative results, lack of sensitivity, and requirement of extra kits, has limited its application [80]. Therefore, fast detection of COVID-19 to prevent subsequent secondary spread requires rapid, selective, and sensitive diagnostic devices. So electrochemical biosensors, due to their benefits, have been considered for the detection of COVID-19 [83, 84]. Suryasnata Tripathy and Shiv Govind Singh developed a label-free electrochemical transduction for COVID-19-specific viral RNA/c-DNA detection, intending to provide a comfortable point-of-care testing. Gold nanoparticles were used as the transducing elements. Smart healthcare can be provided via developing a readout that can be interfaced with smartphones and operated via software applications [84]. As can be seen in Figure 4.13, a schematic is presented that shows the envisaged interfacing of the miniaturized device with a smartphone. Figure 4.14 shows some samples of electrochemical biosensors for infectious disease diagnosis. Table 4.7 shows some biosensors for infectious diseases and Table 4.8 enlists the advantages and disadvantages of electrochemical techniques over widely used conventional methods.

4.6.3 Cardiac detection

Cardiovascular disease (CVD) as a main reason of human death, has attracted much attention in the medical community. For patient survival and also for cost and time saving in the successful prognosis of cardiovascular disease, the early and quick diagnosis of the disease is crucial. For diagnosis of CVD according to World Health Organization (WHO) criteria, patients must face at least two of three csonditions: increase of the biochemical

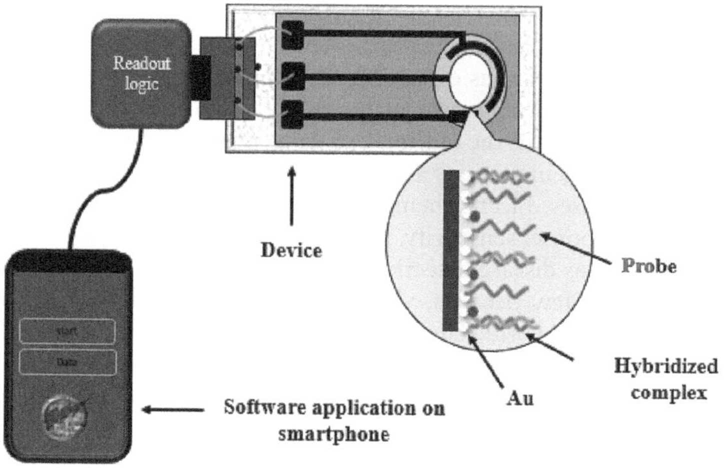

Figure 4.13 Label-Free electrochemical detection of DNA hybridization: a method for COVID-19 diagnosis [84].

markers in blood sample, characteristic pain in the chest, and changes in diagnostic electrocardiogram (ECG). [93]. Sometimes an ECG makes primary diagnosis of CVD more difficult because the electrocardiograms are normal or not diagnostic in about half of the CVD patients referred to the emergency department. [94–97]. Therefore, although ECG is an important management tool for guiding therapy [94, 95], it works poorly in early diagnostic of CVD. So, measurement of cardiac markers has an important role in helping the diagnosis of CVD. A more rapid and sensitive method is required to make faster diagnosis of CVD.

The development of biosensors makes the conditions for overcoming some of the diagnosis problems smoother by providing cost-effective, rapid, and sensitive measurements [98].

Biosensors can make a fast diagnosis by reducing the waiting time for results dissemination, and providing better health care. Lately, lab-on-chip and microfluidics-based biosensor technology is reviewed for cardiac markers detection [99]. Here, a summary of different markers and different sensor platforms that are available for the detection of CVD is presented and summarized in Table 4.9 and Figure 4.15.

Electrochemical sensors were the first scientifically suggested biosensors for multiple analytes. In the last few years, electrochemical biosensors for the detection of cardiac markers containing cardiac troponin I or T (cTnI/T), C-reactive protein (CRP), myoglobin, interlukin-6 (IL-6), LDL, heart fatty acid binding protein (H-FABP), B-type natriuretic peptide (BNP), lipoprotein-associated phospholipase A, creatine kinase MB subform (CK-MB), and myeloperoxidase (MPO) have been provided [85–92, 238].

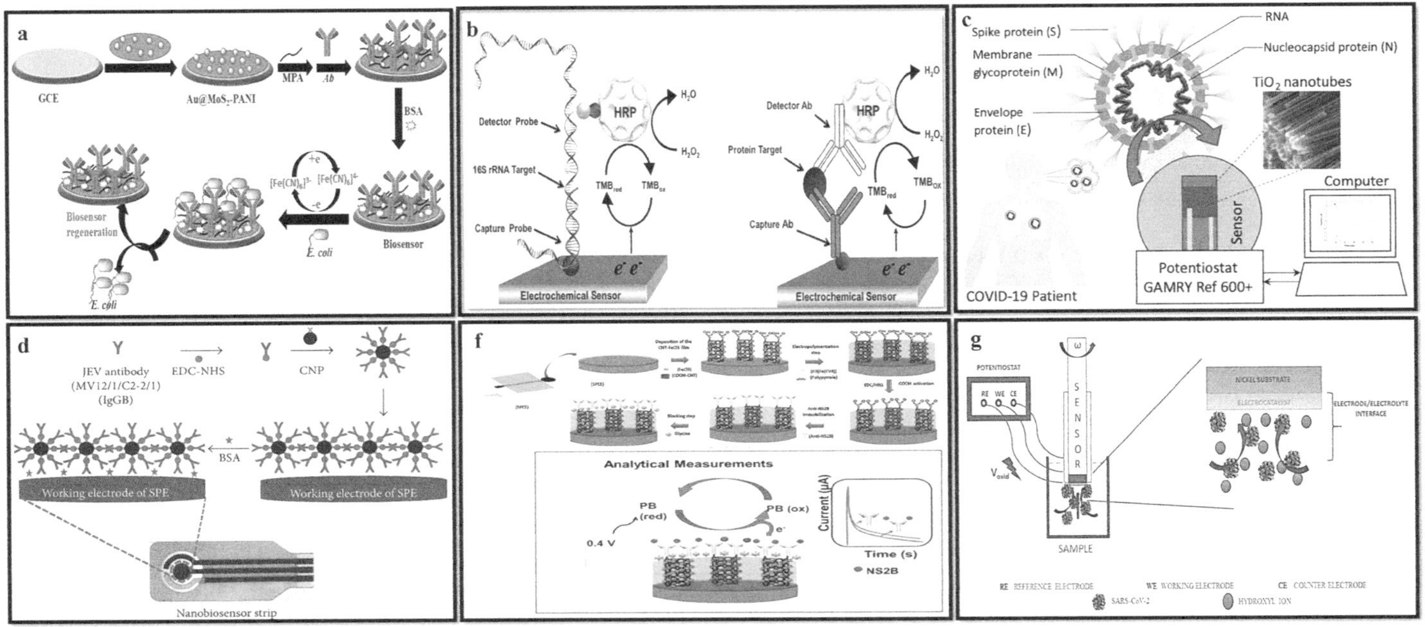

Figure 4.14 Some electrochemical biosensors for infectious diseases:a) label-free electrochemical biosensor for escherichia coli detection, b) electrochemical biosensor array for urinary tract infection diagnosis, c) electrochemical biosensing of SARS-CoV-2, d) electrochemical biosensor strip for Japanese encephalitis virus detection, f) label-free Zika virus immunosensor for amperometric detection of the NS2B protein, g) electrochemical sensor for point-of-care covid-19 diagnosis using saliva [216–221].

Table 4.7 Some biosensors for infectious diseases

Target	*Biosensor format*	*Transducer*	*Linear range*	*Detection Limit*	*Reference*
Zika virus	DNA biosensor	Impedance	25–340 nM	25 nM	[222]
B/C genotyping of hepatitis B virus	DNA based sensor	Amperometry	100–800 f. (B type) 100–800 f. (C type)	1.12 f. (B type)	[223]
Ebola Virus	DNA biosensor	Differential Pulse Voltammetry	10–75 nM	4.7 nM	[224]
Influenza virus	Glycan based FET biosensor	Potentiometry	$10^{0.5}$–$10^{8.5}$ TCID $_{50}$/mL	$10^{0.5}$ TCID $_{50}$/mL	[225]
Dengue Virus Serotype 3	DNA sensor	Differential Pulse Voltammetry	-	3.09 nM	[226]
Legionella pneumophilia	DNA sensor	Square Wave Voltammetry	1 ZM–1 μM	1 ZM	[227]
Human immune deficiency virus -1	DNA biosensor	Impedance	1.0×10^{-13} M–1.0×10^{-10} M	2.3 × 10-14 M	[228]
Avian Influenza	DNA sensor	Square Wave Voltammetry	1–10 pM	1.39 pM	[229]
Mycobacterium Tuberculosis	DNA aptasensor	Differential Pulse Voltammetry	1.00–1 × 105 fg/mL	0.9 fg/mL	[230]

Table 4.8 Advantages and disadvantages of electrochemical techniques over widely used conventional methods

Method	*Advantages*	*Disadvantages*	*References*
Cell culture systems	separation of different viruses in mixed cultures or from unexpected agents, more sensitive than antigen tests, antiviral susceptibility	long commune period for almost all viruses, need technical expertise for cytopathic effect check	[231, 232]
Molecular approach (Nucleic acid detection)	great sensitivity and specificity, useful for viruses that are not able to be cultivated by conventional cell cultures	expensive instrumentation, only detection of the desired virus and probably lack of identification of unexpected agents and mixed infections, availability of most method only at research laboratories	[233, 234]
Immunofluorescence (IF) method	good sensitivity and specificity	not applicable for all viruses, need training and experience for reading result, less sensitive than cell culture method	[235]
Electrochemical biosensors	saving time, fast response, cost-effective, excellent detection limits, easy miniaturization, need the sample volume, simple usage without requirement of professional training	lower shelf life, Lower durability, affected by sample matrix	[236, 237]

Table 4.9 Different caradic biomarkers

Cardiac biomarker	*Relevant type of cardiovascular disease detected*	*Cut-off levels*
Troponin I (cTnI)	Diagnosis of acute myocardial infarction (AMI)	0.01–0.1 ng mL^{-1}
Troponin T (cTnT)	Detection of AMI	0.05–0.1 ng mL^{-1}
Myoglobin	Early Diagnosis of AMI	70–200 ng mL^{-1}
C-reactive protein (CRP)	Early detection of inflammation/cardiac risk factor	$< 10^3$ ng mL^{-1}: low risk $1–3 \times 10^3$ ng mL^{-1}: mid $> 3–15 \times 10^3$ ng mL^{-1}: high risk
Lipoprotein-associated phospholipase A	Detection of coronary heart disease	235 ng/mL
Interlukin-6 (IL-6)	Detection of Inflammation/cardiac risk factor	< 0.0013 ng mL^{-1}: low risk 0.00138–0.002 ng mL^{-1}: mid > 0.002 ng mL^{-1}: high risk
Low-density lipoprotein (LDL)	Diagnosis of atherosclerotic cardiovascular disease (ASCVD)	< 70 mg/dL for heart or blood vessel disease < 100 mg/dL for high risk patients < 130 mg/dL otherwise
B-type natriuretic peptide (BNP)	Detection of acute coronary syndromes/diagnosis of heart failure/ventricular overload	
Creatine kinase MB subform (CK-MB)	Early diagnosis of AMI	10 ng mL^{-1}
Heart fatty acid binding protein (H-FABP)	Diagnosis of myocardial necrosis	≥ 6 ng mL^{-1}
Myeloperoxidase (MPO)	Detection of inflammation	> 350 ng mL^{-1}

Source: [238–242]

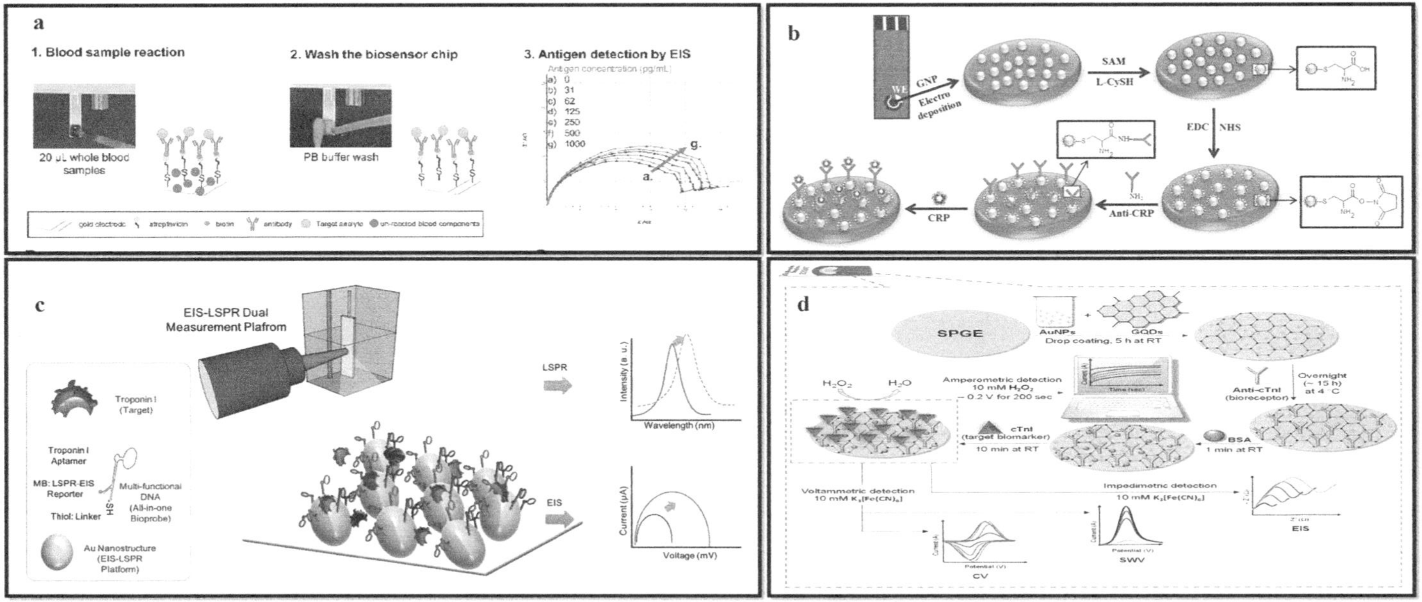

Figure 4.15 Some of electrochemical biosensors for the detection of cardiac biomarkers: a) electrochemical immunosensor for detection of cardiac troponin I in blood samples, b) label-free electrochemical immunosensor for C-reactive protein, c) a dual-mode biosensor for troponin I biomarker, d) enzyme-free electrochemical nano-immunosensor for cardiac troponin I [243–246].

4.7 CONCLUSION

Biosensor technology has achieved high interest in the last decade because of its advantages, such as rapid analysis time, low detection limit, and low cost. Great developments in the field of biosensors have been made. By using microfluidics and wearable devices, biosensors have greater application in healthcare. This chapter reviews the basis of biosensors and electrochemical transducers and their application in healthcare. By integrating with a microfluidic channel and using wearable devices, the biosensors can be promising candidates for real time monitoring, reagentless, label free, and miniaturized applications. In coming years, biosensors in the shape of biochips can be used for in-home medical diagnostics of diseases without the necessity of sending samples to a laboratory that save time.

REFERENCES

1. Thevenot, D. R.; Toth, K.; Durst, R. A., Wilson, G. S. Electrochemical biosensors: recommended definitions and classification. *Pure and Applied Chemistry.* 1999 Jan 1;71(12):2333–2348.
2. Kirsch, J.; Siltanen, C.; Zhou, Q.; Revzin, A.; Simonian, A. Biosensor technology: recent advances in threat agent detection and medicine. *Chemical Society Reviews.* 2013;42(22):8733–8768.
3. Monosik, R.; Stredanský, M.; Sturdik, E. Biosensors-classification, characterization and new trends. *Acta chimica slovaca.* 2012 Apr 1;5(1):109.
4. Wang, J.; Musameh, M. Carbon-nanotubes doped polypyrrole glucose biosensor. *Analytica Chimica Acta.* 2005 May 10;539(1–2):209–213.
5. Yabuki, S.; Mizutani, F.; Hirata, Y. Fabrication of an acetylcholine sensor using an enzyme-immobilized polyion composite membrane. *Analytical Chemistry.* 2001 Dec 5; 50(12): 899–901.
6. Gupta, R.; Chaudhury, N. K. Entrapment of biomolecules in sol–gel matrix for applications in biosensors: Problems and future prospects. *Biosensors and Bioelectronics.* 2007 May 15;22(11):2387–2399.
7. Rajesh, B. V.; Takashima, W.; Kaneto, K. 2005. An amperometric urea biosensor based on covalent immobilization of urease onto an electrochemically prepared copolymer poly (N-3-aminopropyl pyrrole-co-pyrrole) film. *Biomaterials* 26(17):3683–3690.
8. Wongkaew, N.; Simsek, M.; Griesche, C.; Baeumner, A. J. Functional nanomaterials and nanostructures enhancing electrochemical biosensors and lab-on-a-chip performances: recent progress, applications, and future perspective. *Chemical Reviews.* 2018 Sep 24;119(1):120–194.
9. Perumal, V.; Hashim, U. Advances in biosensors: Principle, architecture and applications. *Journal of Applied Biomedicine.* 2013;12:1–5.
10. Mohanty, S. P.; Kougianos, E. Biosensors: A tutorial review. *IEEE Potentials.* 2006 Jul 5;25(2):35–40.
11. Eggins, B. R. *Chemical Sensors and Biosensors*, West Sussex: John Wiley & Sons, 2002.

12. Thevenot, D. R.; Toth, K.; Durst, R. A.; Wilson, G. S. Electrochemical biosensors: Recommended definitions and classification. *Biosensors and Bioelectronics*. 2001;16:121–131.
13. Bard, A. J.; Faulkner, L. R. 2000. *Electrochemical Methods: Fundamentals and Applications*, 2nd ed. New York: Wiley.
14. Cesewski, E.; Johnson, B. N. Electrochemical biosensors for pathogen detection. *Biosensors and Bioelectronics*. 2020 Jul 1;159:112.
15. Noh, J.; Kim, H. C.; Chung, T. D. Biosensors in microfluidic chips. *Microfluidics*. 2011:117–152.
16. Bi H, Han X. Editors; Kohji Mitsubayashi, Osamu Niwa and Yuko Ueno. Chemical sensors for environmental pollutant determination. In Chemical, Gas, and Biosensors for Internet of Things and Related Applications (pp. 147–160). Elsevier, Amsterdam, Netherlands, 2019.
17. Liang, P.; Wang, H.; Xia, X.; Huang, X.; Mo, Y.; Cao, X.; Fan, M. Carbon nanotube powders as electrode modifier to enhance the activity of anodic biofilm in microbial fuel cells. *Biosensors and Bioelectronics*. 2011 Feb 15;26(6): 3000–3004.
18. Koncki, R. Recent developments in potentiometric biosensors for biomedical analysis. *Analytica Chimica Acta*. 2007 Sep 5;599(1):7–15.
19. Wang, Y., Chen, Q., Zeng, X. Potentiometric biosensor for studying hydroquinone cytotoxicity in vitro. *Biosensors and Bioelectronics*. 2010 Feb 15;25(6): 1356–1362.
20. Kaisti, M. Detection principles of biological and chemical FET sensors. *Biosensors and Bioelectronics*. 2017 Dec 15;98:437–448.
21. Singh, N.K.; Thungon, P.D.; Estrela, P.; Goswami, P. Development of an aptamer-based field effect transistor biosensor for quantitative detection of Plasmodium falciparum glutamate dehydrogenase in serum samples. *Biosensors and Bioelectronics*. 2019;123:30–35.
22. Pachauri, V.; Ingebrandt, S. Biologically sensitive field-effect transistors: From ISFETs to NanoFETs. *Essays in Biochemistry*. 2016 Jun 30;60(1):81–90.
23. Wang, J. *Analytical Electrochemistry*. Hoboken, NJ: John Wiley & Sons VCH, 2006.
24. Karube, I.; Suzuki, M. Microbial biosensors In *Biosensors: A Practical Approach*, ed. E. G. Cass, The Practical Approach Series 7. Oxford: Oxford University Press, 1990.
25. Bartlett, P.N. *Bioelectrochemistry: Fundamentals, Experimental Techniques and Applications*. John Wiley & Sons, Hoboken, NJ, 2008.
26. Sang, S.; Wang, Y.; Feng, Q.; Wei, Y.; Ji, J.; Zhang, W. Progress of new label-free techniques for biosensors: A review. *Critical Reviews in Biotechnology*. 2016 May 3;36(3):465–481.
27. Padash, M.; Carrara, S. A 3D printed wearable device for sweat analysis. In *2020 IEEE International Symposium on Medical Measurements and Applications (MeMeA)* 2020 Jun 1 (pp. 1–5). IEEE.
28. El Aamri, M.; Yammouri, G.; Mohammadi, H.; Amine, A.; Korri-Youssoufi, H. Electrochemical biosensors for detection of MicroRNA as a cancer biomarker: Pros and Cons. *Biosensors*. 2020 Nov;10(11):186.
29. Bohunicky, B.; Mousa, S.A. Biosensors: The new wave in cancer diagnosis. *Nanotechnology, Science and Applications*. 2010;4:1–10.

30. Yang, H.; Wang, J.; Yang, C.; Zhao, X.; Xie, S.; Ge, Z. Nano Pt@ ZIF8 modified electrode and its application to detect sarcosine. *Journal of the Electrochemical Society*. 2018 Apr 11;165(5):H247.
31. Grieshaber, D.; MacKenzie, R.; Vörös, J.; Reimhult, E. Electrochemical biosensors—Sensor principles and architectures. *Sensors*. 2008;8:1400–1458.
32. Andryukov, B.G.; Besednova, N.N.; Romashko, R.V.; Zaporozhets, T.S.; Efimov, T.A. Label-free biosensors for laboratory-based diagnostics of infections: Current achievements and new trends. *Biosensors*. 2020;10(2):11.
33. Palecek, E.; Bartosik, M. Electrochemistry of nucleic acids. *Chemical Reviews*. 2012 Jun 13;112(6):3427–3481.
34. Palecek, E.; Tkac, J.; Bartosik, M.; Bertók, T.; Ostatna, V.; Paleček, J. Electrochemistry of nonconjugated proteins and glycoproteins. Toward sensors for biomedicine and glycomics. *Chemical Reviews*. 2015 Mar 11;115(5):2045–2108.
35. Wang, J.; Kawde, A. N.; Erdem, A.; Salazar, M. Magnetic bead-based label-free electrochemical detection of DNA hybridization. *Analyst*. 2001;126(11):2020–2024.
36. Xiao, Y.; Qu, X.; Plaxco, K. W.; Heeger, A. J. Label-free electrochemical detection of DNA in blood serum via target-induced resolution of an electrode-bound DNA pseudoknot. *Journal of the American Chemical Society*. 2007 Oct 3;129(39):11896–11897.
37. Farjami, E.; Campos, R.; Nielsen, J. S.; Gothelf, K. V.; Kjems, J.; Ferapontova, E. E. RNA aptamer-based electrochemical biosensor for selective and label-free analysis of dopamine. *Analytical Chemistry*. 2013 Jan 2;85(1):121–128.
38. He, P.; Wang, Z.; Zhang, L.; Yang, W. Development of a label-free electrochemical immunosensor based on carbon nanotube for rapid determination of clenbuterol. *Food Chemistry*. 2009 Feb 1;112(3):707–714.
39. Zhao, W.; Li, B.; Xu, S.; Huang, X.; Luo, J.; Zhu, Y.; Liu, X. Electrochemical protein recognition based on macromolecular self-assembly of molecularly imprinted polymer: a new strategy to mimic antibody for label-free biosensing. *Journal of Materials Chemistry B*. 2019;7(14):2311–2319.
40. Wang, H. F.; Ma, R. N.; Sun, F.; Jia, L. P.; Zhang, W.; Shang, L.; Xue, Q. W.; Jia, W. L.; Wang, H. S. A versatile label-free electrochemical biosensor for circulating tumor DNA based on dual enzyme assisted multiple amplification strategy. *Biosensors and Bioelectronics*. 2018 Dec 30;122:224–230.
41. Duan, C.; Meyerhoff, M. E. Immobilization of proteins on gold coated porous membranes via an activated self-assembled monolayer of thioctic acid. *Microchimica Acta*. 1995 Sep 1;117(3–4):195–206.
42. Sun, X.; He, P.; Liu, S.; Ye, J.; Fang, Y. Immobilization of single-stranded deoxyribonucleic acid on gold electrode with self-assembled aminoethanethiol monolayer for DNA electrochemical sensor applications. *Talanta*. 1998 Oct 1;47(2):487–495.
43. Radi, A. E.; Acero Sánchez, J. L.; Baldrich, E.; O'Sullivan, C. K. Reagentless, reusable, ultrasensitive electrochemical molecular beacon aptasensor. *Journal of the American Chemical Society*. 2006 Jan 11;128(1):117–124.
44. Ricci, F.; Adornetto, G.; Palleschi, G. A review of experimental aspects of electrochemical immunosensors. *Electrochimica Acta*. 2012 Dec 1;84:74–83.
45. Grieshaber, D.; MacKenzie, R.; Vörös, J.; Reimhult, E. Electrochemical biosensors-sensor principles and architectures. *Sensors*. 2008 Mar;8(3):1400–1458.
46. Copeland, R. A. *Enzymes*. New York: John Wiley & Sons VCH, 2000.

47. Ronkainen-Matsuno, N. J.; Thomas, J. H.; Halsall, H. B.; Heineman, W. R. Electrochemical immunoassay moving into the fast lane. *TrAC Trends in Analytical Chemistry*. 2002 Apr 1;21(4):213–225.
48. Eicher, D.; Merten, C. A. Microfluidic devices for diagnostic applications. *Expert Review of Molecular Diagnostics*. 2011;11(5):505–519.
49. Gervais, L.; de Rooij, N.; Delamarche, E. Microfluidic chips for point-of-care immunodiagnostics. *Advnced Materials*. 2011;23(24): H151–H176.
50. Trietsch, S. J.; Hankemeier, T.; van der Linden, H. J. Lab-on-a-chip technologies for massive parallel data generation in the life sciences: A review. *Chemometrics and Intelligent Laboratory Systems*. 2011;108(1):64–75.
51. Gupta, S.; Ramesh, K.; Ahmed, S.; Kakkar, V. Lab-on-chip technology: A review on design trends and future scope in biomedical applications. *International Journal of Bio Technology and Research*. 2016;8:311–322.
52. Temiz, Y.; Lovchik, R. D.; Kaigala, G. V.; Delamarche, E. Lab-on-a-chip devices: How to close and plug the lab? *Microelectronic Engineering*. 2015;132:156–175.
53. Dulay, S. B.; Gransee, R.; Julich, S.; et al. Automated microfluidically controlled electrochemical biosensor for the rapid and highly sensitive detection of Francisella tularensis. *Biosensors and Bioelectronics*. 2014;59:342–349.
54. Shah, J.; Wilkins, E. Electrochemical biosensors for detection of biological warfare agents. *Electroanalysis*. 2003;15(3):157–167.
55. Wang, J. Portable electrochemical systems. *TRAC: Trends in Analytical Chemistry*. 2002;21:226–232.
56. Lynn, N.S., Homola, J. Biosensor enhancement using grooved micromixers: part I, numerical studies. *Analytical Chemistry*. 2015;87, 5516–5523.
57. Choi, S., Goryll, M., Sin, L. Y. M., Wong, P. K., Chae, J. Microfluidic-based biosensors toward point-of-care detection of nucleic acids and proteins. *Microfluidics and Nanofluidics*. 2011;10:231–247.
58. Wang, J.; Timchalk, C.; Lin, Y. Carbon nanotube-based electrochemical sensor for assay of salivary cholinesterase enzyme activity: An exposure biomarker of organophosphate pesticides and nerve agents. *Environmental Science & Technology*. 2008;42:2688–2693.
59. Kumar, S.; Sharma, J. G.; Maji, S.; Malhotra, B. D. A biocompatible serine functionalized nanostructured zirconia based biosensing platform for non-invasive oral cancer detection. *RSC Advances*. 2016;6(80):77037–77046.
60. Yang, X, Cheng, H. Recent developments of flexible and stretchable electrochemical biosensors. *Micromachines*. 2020 Mar;11(3):243.
61. Jalbert, I. Diet, nutraceuticals and the tear film. *Experimental Eye Research*. 2013 Dec 1;117:138–146.
62. Thomas, N.; Lähdesmäki, I.; Parviz, B. A. A contact lens with an integrated lactate sensor. *Sensors and Actuators B: Chemical*. 2012 Feb 20;162(1):128–134.
63. Yao, H., Shum, A. J., Cowan, M., Lähdesmäki, I., Parviz, B. A. A contact lens with embedded sensor for monitoring tear glucose level. *Biosensors and Bioelectronics*. 2011 Mar 15;26(7):3290–3296.
64. Yao, H.; Liao, Y.; Lingley, A. R.; Afanasiev, A.; Lähdesmäki, I.; Otis, B. P.; Parviz, B. A. A contact lens with integrated telecommunication circuit and sensors for wireless and continuous tear glucose monitoring. *Journal of Micromechanics and Microengineering*. 2012 Jun 7;22(7):075007.
65. Bariya, M.; Nyein, H. Y. Y.; Javey, A. Wearable sweat sensors. *Nature Electronics*. 2018;1:160–171.

66. Gao, W.; Emaminejad, S.; Nyein, H. Y.; Challa, S.; Chen, K.; Peck, A.; Fahad, H. M.; Ota, H.; Shiraki, H.; Kiriya, D.; Lien, D. H. Fully integrated wearable sensor arrays for multiplexed in situ perspiration analysis. *Nature*. 2016 Jan;529(7587):509–514.
67. Bandodkar, A. J., Jia, W., Wang, J. Tattoo-based wearable electrochemical devices: a review. *Electroanalysis*. 2015 Mar;27(3):562–572.
68. Papaiordanidou, M.; Takamatsu, S.; Rezaei-Mazinani, S.; Lonjaret, T.; Martin, A.; Ismailova, E. Cutaneous recording and stimulation of muscles using organic electronic textiles. *Advanced Healthcare Materials*. 2016 Aug;5(16):2001–2006.
69. Wang, S.; Chinnasamy, T.; Lifson, M. A.; Inci, F.; Demirci, U. Flexible substrate-based devices for point-of-care diagnostics. *Trends in Biotechnology*. 2016 Nov 1;34(11):909–921.
70. Ronkainen-Matsuno, N. J.; Thomas, J. H.; Halsall, H. B.; Heineman, W. R. *TrAC Trends in Analytical Chemistry*. 2002;21(4):213.
71. Bauer, C. G., Eremenko, A. V., Ehrentreich-Forster, E., Bier, F. F., Makower, A., Halsall, H. B., Heineman, W. R.; Scheller, F. W. Zeptomole-detecting biosensor for alkaline phosphatase in an electrochemical immunoassay for 2, 4-dichlorophenoxyacetic acid. *Analytical Chemistry*. 1996;68(15):2453.
72. Jenkins, S. H.; Heineman, W. R.; Halsall, H. B. Extending the detection limit of solid-phase electrochemical enzyme immunoassay to the attomole level. *Analytical Biochemistry*. 1988;168(2):292.
73. Wijayawardhana, C. A.; Halsall, H. B.; Heineman, W. R. *Electroanalytical Methods of Biological Materials*, ed. A. BrajterToth, J. Q. Chambers. New York: Marcel Dekker, 2002, pp. 329–365.
74. Yao, H., Jenkins, S. H., Pesce, A. J., Halsall, H.B., Heineman, W. R. Electrochemical homogeneous enzyme immunoassay of theophylline in hemolyzed, icteric, and lipemic samples. *Clinical Chemistry (Washington, D. C.)*. 1993;39:1432.
75. Finn, O. J. Immune response as a biomarker for cancer detection and a lot more. *The New England Journal of Medicine*. 2005;353:1288.
76. Diamandis, E. P. Cancer biomarkers: can we turn recent failures into success? *The Journal of the National Cancer Institute*. 2010;102:1462.
77. Qureshi, A.; Niazi, J. H. Biosensors for detecting viral and bacterial infections using host biomarkers: A review. *Analyst*. 2020;145(24):7825–7848.
78. Khater, M.; De La Escosura-Muñiz, A.; Merkoçi, A. Biosensors for plant pathogen detection. *Biosensors and Bioelectronics*. 2017 Jul 15;93:72–86.
79. Kwon, P. S.; Ren, S.; Kwon, S. J.; Kizer, M. E.; Kuo, L.; Xie, M.; Zhu, D.; Zhou, F.; Zhang, F.; Kim, D.; Fraser, K. Designer DNA architecture offers precise and multivalent spatial pattern-recognition for viral sensing and inhibition. *Nature Chemistry*. 2020 Jan;12(1):26–35.
80. Lu, R. et al. Caracterización genómica y epidemiología del nuevo coronavirus 2019: implicaciones para los orígenes del virus y la unión al receptor. *The Lancet*. 2020;395(10224):566–568.
81. Chu, D. K.; Pan, Y.; Cheng, S. M.; Hui, K. P.; Krishnan, P.; Liu, Y.; Ng, D. Y.; Wan, C. K.; Yang, P.; Wang, Q.; Peiris, M. Molecular diagnosis of a novel coronavirus (2019-nCoV) causing an outbreak of pneumonia. *Clinical Chemistry*. 2020 Apr 1;66(4):549–555.
82. An, J.; Liao, X.; Xiao, T.; Qian, S.; Yuan, J.; Ye, H.; Qi F.; Shen C.; Wang L.; Liu Y.; Cheng, X. Clinical characteristics of recovered COVID-19 patients

with re-detectable positive RNA test. *Annals of Translational Medicine*. 2020 Sep;8(17):1084.

83. Kim, H. E., Schuck, A., Lee, S. H., Lee, Y., Kang, M., Kim, Y. S. Sensitive electrochemical biosensor combined with isothermal amplification for point-of-care COVID-19 tests. *Biosensors and Bioelectronics*. 2021 Jun 15;182:113168.
84. Tripathy, S., Singh, S. G. Label-free electrochemical detection of DNA hybridization: a method for COVID-19 diagnosis. *Transactions of the Indian National Academy of Engineering*. 2020 Jun;5:205–209.
85. Tweedie, M.; Subramanian, R.; Lemoine, P.; Craig, I.; McAdams, E.T.; McLaughlin, J.A.; Maccraith, B.; Kent, N. Fabrication of impedimetric sensors for label-free point-of-care immunoassay cardiac marker systems, with passive microfluidic delivery. *Conference Proceedings – IEEE Engineering in Medicine and Biology Society*, 2006;1:4610–4614.
86. Buch, M.; Rishpon, J. An electrochemical immunosensor for C-reactive protein based on multi-walled carbon nanotube-modified electrodes. *Electroanalysis: An International Journal Devoted to Fundamental and Practical Aspects of Electroanalysis*. 2008 Dec;20(23):2592–2594.
87. Chen, X.; Wang, Y.; Zhou, J.; Yan, W.; Li, X.; Zhu, J. J. Electrochemical impedance immunosensor based on three-dimensionally ordered macroporous gold film. *Analytical Chemistry*. 2008 Mar 15;80(6):2133–2140.
88. Centi, S.; Bonel Sanmartin, L.; Tombelli, S.; Palchetti, I.; Mascini, M. Detection of C reactive protein (CRP) in serum by an electrochemical aptamer-based sandwich assay. *Electroanalysis: An International Journal Devoted to Fundamental and Practical Aspects of Electroanalysis*. 2009 Jun;21(11):1309–1315.
89. Hennessey, H., Afara, N., Omanovic, S., Padjen, A. L. Electrochemical investigations of the interaction of C-reactive protein (CRP) with a CRP antibody chemically immobilized on a gold surface. *Analytica Chimica Acta*. 2009 Jun 8;643(1–2):45–53.
90. Lee, J. K.; Noh, G. H.; Pyun, J. C. Capacitive immunoaffinity biosensor by using diamond-like carbon (DLC) electrode. *BioChip Journal*. 2009 Jan 1;3(4): 287–292.
91. Quershi, A.; Gurbuz, Y.; Kang, W. P.; Davidson, J. L. A novel interdigitated capacitor based biosensor for detection of cardiovascular risk marker. *Biosensors and Bioelectronics*. 2009 Dec 15;25(4):877–882.
92. Venkatraman, V. L.; Reddy, R. K.; Zhang, F.; Evans, D.; Ulrich, B.; Prasad, S. Iridium oxide nanomonitors: Clinical diagnostic devices for health monitoring systems. *Biosensors and Bioelectronics*. 2009 Jun 15;24(10):3078–3083.
93. Yang, Z.; Zhou, D. M. Cardiac markers and their point-of-care testing for diagnosis of acute myocardial infarction. *Clinical Biochemistry*. 2006 Aug 1;39(8):771–780.
94. Yusuf, S.; Pearson, M.; Sterry, H.; Parish, S.; Ramsdale, D.; Rossi, P.; Sleight, P. The entry ECG in the early diagnosis and prognostic stratification of patients with suspected acute myocardial infarction. *European Heart Journal*. 1984 Sep 1;5(9):690–696.
95. Foy, S. G., Kennedy, I. C., Ikram, H., Low, C. J., Shirlaw, T. M., Crozier, I. G. The early diagnosis of acute myocardial infarction. *Australian and New Zealand Journal of Medicine*. 1991 Jun;21(3):335–337.
96. Stubbs, P., Collinson, P. O. Point-of-care testing: A cardiologist's view. *Clinica chimica acta*. 2001 Sep 15;311(1):57–61.

97. Kost, G. J., Tran, N. K. Point-of-care testing and cardiac biomarkers: the standard of care and vision for chest pain centers. *Cardiology Clinics*. 2005 Nov 1;23(4):467–790.
98. Mascini, M., Tombelli, S. Biosensors for biomarkers in medical diagnostics. *Biomarkers*. 2008 Jan 1;13(7–8):637–657.
99. Mohammed, M. I., Desmulliez, M. P. Lab-on-a-chip based immunosensor principles and technologies for the detection of cardiac biomarkers: A review. *Lab on a Chip*. 2011;11(4):569–595.
100. World Health Organization. Cardiovascular diseases World Health Organization. Cardiovascular diseases. Fact Sheet. 2007;317.
101. Allender, S.; Scarborough, P.; Peto, V.; Rayner, M.; Leal, J.; Luengo-Fernandez, R.; Gray, A. European cardiovascular disease statistics. *European Heart Network*. 2008 Feb;3:11–35.
102. Martín-Ventura, J. L., Blanco-Colio, L. M., Tuñón, J., Muñoz-García, B., Madrigal-Matute, J., Moreno, J. A., de Céniga, M. V., Egido, J. Biomarkers in cardiovascular medicine. *Revista Española de Cardiología (English Edition)*. 2009 Jun 1;62(6):677–688.
103. McDonnell, B.; Hearty, S.; Leonard, P.; O'Kennedy, R. Cardiac biomarkers and the case for point-of-care testing. *Clinical Biochemistry*. 2009 May 1;42(7–8): 549–561.
104. Vasan, R. S. Biomarkers of cardiovascular disease: Molecular basis and practical considerations. *Circulation*. 2006 May 16;113(19):2335–2362.
105. Anderson, L. Candidate-based proteomics in the search for biomarkers of cardiovascular disease. *The Journal of Physiology*. 2005 Feb;563(1):23–60.
106. Yang, Y. N., Lin, H. I., Wang, J. H., Shiesh, S. C., Lee, G. B. An integrated microfluidic system for C-reactive protein measurement. *Biosensors and Bioelectronics*. 2009 Jun 15;24(10):3091–3096.
107. Casas, J. P., Shah, T., Hingorani, A. D., Danesh, J., Pepys, M. B. C-reactive protein and coronary heart disease: A critical review. *Journal of Internal Medicine*. 2008 Oct;264(4):295–314.
108. Tang, L.; Kang, K. A. Editors: Giuseppe Cicco, Duane F. Bruley, Marco Ferrari, David K. Harrison. Preliminary study of fiber optic multi-cardiac-marker biosensing system for rapid coronary heart disease diagnosis and prognosis. In *Oxygen Transport to Tissue XXVII* (pp. 101–106). Boston, MA: Springer, 2006.
109. Daniels, J. S.; Pourmand, N. Label-free impedance biosensors: Opportunities and challenges. *Electroanalysis: An International Journal Devoted to Fundamental and Practical Aspects of Electroanalysis*. 2007 Jun;19(12):1239–1257.
110. Kwon, Y. C.; Kim, M. G.; Kim, E. M.; Shin, Y. B.; Lee, S. K.; Lee, S. D.; Cho, M. J.; Ro, H. S. Development of a surface plasmon resonance-based immunosensor for the rapid detection of cardiac troponin I. *Biotechnology Letters*. 2011 May 1;33(5):921–927.
111. Krishnamoorthy, S., Iliadis, A. A., Bei, T., Chrousos, G. P. An interleukin-6 ZnO/SiO2/Si surface acoustic wave biosensor. *Biosensors and Bioelectronics*. 2008 Oct 15;24(2):313–318.
112. Monson, C. F.; Driscoll, L. N.; Bennion, E.; Miller, C. J.; Majda, M. Antibody-Antigen exchange equilibria in a field of an external force: design of reagentless biosensors. *Analytical Chemistry*. 2009 Sep 1;81(17):7510–7514.
113. Fan, X.; White, I. M.; Shopova, S. I.; Zhu, H.; Suter, J. D.; Sun, Y. Sensitive optical biosensors for unlabeled targets: A review. *Analytica Chimica Acta*. 2008 Jul 14;620(1–2):8–26.

114. Wolf, M.; Juncker, D.; Michel, B.; Hunziker, P.; Delamarche, E. Simultaneous detection of C-reactive protein and other cardiac markers in human plasma using micromosaic immunoassays and self-regulating microfluidic networks. *Biosensors and Bioelectronics*. 2004 May 15;19(10):1193–3202.
115. Hill, H. R., Martins, T. B. The flow cytometric analysis of cytokines using multi-analyte fluorescence microarray technology. *Methods*. 2006 Apr 1;38(4): 312–316.
116. Hun, X., Zhang, Z. Functionalized fluorescent core-shell nanoparticles used as a fluorescent labels in fluoroimmunoassay for IL-6. *Biosensors and Bioelectronics*. 2007 May 15;22(11):2743–2748.
117. Ganesh, N., Block, I. D., Mathias, P. C., Zhang, W., Chow, E., Malyarchuk, V., Cunningham, B. T. Leaky-mode assisted fluorescence extraction: application to fluorescence enhancement biosensors. *Optics Express*. 2008 Dec 22;16(26): 21626–21640.
118. Jung, J. W., Jung, S. H., Yoo, J. O., Suh, I. B., Kim, Y. M., Ha, K. S. Label-free and quantitative analysis of C-reactive protein in human sera by tagged-internal standard assay on antibody arrays. *Biosensors and Bioelectronics*. 2009 Jan 1; 24(5):1469–1473.
119. Pultar, J.; Sauer, U.; Domnanich, P.; Preininger, C. Aptamer–antibody on-chip sandwich immunoassay for detection of CRP in spiked serum. *Biosensors and Bioelectronics*. 2009 Jan 1;24(5):1456–1461.
120. Raj, V.; Hari, P. R.; Antony, M.; Sreenivasan, K. Selective estimation of C-reactive protein in serum using polymeric formulations without antibody. *Sensors and Actuators B: Chemical*. 2010 Apr 8;146(1):23–27.
121. Albrecht, C.; Kaeppel, N.; Gauglitz, G. Two immunoassay formats for fully automated CRP detection in human serum. *Analytical and Bioanalytical Chemistry*. 2008 Jul;391(5):1845–1852.
122. Fogh-Andersen, N.; Altura, B.M.; Altura, B.T.; Siggaard-Andersen, O. Composition of interstitial fluid. *Clinical Chemistry*. 1995;41:1522–1525.
123. Rao, G.; Guy, R.H.; Glikfeld, P.; LaCourse, W.R.; Leung, L.; Tamada, J.; Potts, R.O.; Azimi, N. Reverse iontophoresis: Noninvasive glucose monitoring in vivo in humans. *Pharmaceutical Research: An Official journal of The American Association of Pharmaceutical Scientists*. 1995;12:1869–1873.
124. Leboulanger, B.; Guy, R.H.; Delgado-Charro, M.B. Reverse iontophoresis for non-invasive transdermal monitoring. *Physiological Measurement*. 2004;25: R35–R50.
125. Edelman, S.V. Watching your glucose with the glucowatch. *Diabetes Technology & Therapeutics*. 2001;3:283–284.
126. Kim, J.; Sempionatto, J.R.; Imani, S.; Hartel, M.C.; Barfidokht, A.; Tang, G.; Campbell, A.S.; Mercier, P.P.; Wang, J. Simultaneous monitoring of sweat and interstitial fluid using a single wearable biosensor platform. *Nature Materials*. 2018;5:1800880.
127. Kim, D.H.; Lu, N.; Ma, R.; Kim, Y.S.; Kim, R.H.; Wang, S.; Wu, J.; Won, S.M.; Tao, H.; Islam, A.; et al. Epidermal electronics. *Science*. 2011;333:838–843.
128. Wang, C.; Hwang, D.; Yu, Z.; Takei, K.; Park, J.; Chen, T.; Ma, B.; Javey, A. User-interactive electronic skin for instantaneous pressure visualization. *Nature Materials*. 2013;12:899–904.
129. Hondred, J.A.; Stromberg, L.R.; Mosher, C.L.; Claussen, J.C. High-resolution graphene films for electrochemical sensing via inkjet maskless lithography. *ACS Nano*. 2017;11:9836–9845.

130. Purwidyantri, A.; Chen, C.H.; Hwang, B.J.; Luo, J.D.; Chiou, C.C.; Tian, Y.C.; Lin, C.Y.; Cheng, C.H.; Lai, C.S. Spin-coated Au-nanohole arrays engineered by nanosphere lithography for a Staphylococcus aureus 16S rRNA electrochemical sensor. *Biosensors and Bioelectronics*. 2016;77:1086–1094.
131. Park, J.; Kim, J.; Kim, S.Y.; Cheong, W.H.; Jang, J.; Park, Y.G.; Na, K.; Kim, Y.T.; Heo, J.H.; Lee, C.Y.; et al. Soft, smart contact lenses with integrations of wireless circuits, glucose sensors, and displays. *Science Advances*. 2018;4: eaap9841.
132. Honeychurch, K.C.; Hart, J.P. Screen-printed electrochemical sensors for monitoring metal pollutants. *TrAC-Trends in Analytical Chemistry*. 2003;22: 456–469.
133. Alonso-Lomillo, M.A.; Domínguez-Renedo, O.; Arcos-Martínez, M.J. Screen-printed biosensors in microbiology: A review. *Talanta*. 2010;82:1629–1636.
134. Chuang, M.C.; Windmiller, J.R.; Santhosh, P.; Ramírez, G.V.; Galik, M.; Chou, T.Y.; Wang, J. Textile-based electrochemical sensing: Effect of fabric substrate and detection of nitroaromatic explosives. *Electroanalysis*. 2010;22:2511–2518.
135. Windmiller, J.R.; Wang, J. Wearable electrocemical sensors and biosensors: A review. *Electroanalysis*. 2013;25:29–46.
136. Fung, C.M.; Lloyd, J.S.; Samavat, S.; Deganello, D.; Teng, K.S. Facile fabrication of electrochemical ZnO nanowire glucose biosensor using roll to roll printing technique. *Sensors & Actuators, B: Chemical*. 2017;247:807–813.
137. Parrilla, M.; Ferré, J.; Guinovart, T.; Andrade, F.J. Wearable potentiometric sensors based on commercial carbon fibres for monitoring sodium in sweat. *Electroanalysis*. 2016;28:1267–1275.
138. Jalbert, I. Diet, nutraceuticals and the tear film. *Experimental Eye Research*. 2013 Dec 1;117:138–146.
139. Bariya, M.; Shahpar, Z.; Park, H.; Sun, J.; Jung, Y.; Gao, W.; Nyein, H.Y.Y.; Liaw, T.S.; Tai, L.C.; Ngo, Q.P.; et al. Roll-to-roll gravure printed electrochemical sensors for wearable and medical devices. *ACS Nano*. 2018;12:6978–6987.
140. Overgaard, M.H.; Sahlgren, N.M.; Hvidsten, R.; Kühnel, M.; Dalby, K.N.; Vosch, T.; Laursen, B.W.; Nørgaard, K. Facile synthesis of mildly oxidized graphite inks for screen-printing of highly conductive electrodes. *Advanced Engineering Materials*. 2019;21:1801304.
141. Kim, S.; Sojoudi, H.; Zhao, H.; Mariappan, D.; McKinley, G.H.; Gleason, K.K.; Hart, A.J. Ultrathin high-resolution flexographic printing using nanoporous stamps. *Science Advances*. 2016;2:e1601660.
142. Jung, Y.; Park, H.; Park, J.A.; Noh, J.; Choi, Y.; Jung, M.; Jung, K.; Pyo, M.; Chen, K.; Javey, A.; et al. Fully printed flexible and disposable wireless cyclic voltammetry tag. *Scientific Reports*. 2015;5:1–6.
143. Lau, P.H.; Takei, K.; Wang, C.; Ju, Y.; Kim, J.; Yu, Z.; Takahashi, T.; Cho, G.; Javey, A. Fully printed, high performance carbon nanotube thin-film transistors on flexible substrates. *Nano Letters*. 2013;13:3864–3869.
144. Lee, W.; Koo, H.; Sun, J.; Noh, J.; Kwon, K.S.; Yeom, C.; Choi, Y.; Chen, K.; Javey, A.; Cho, G. A fully roll-to-roll gravure-printed carbon nanotube-based active matrix for multi-touch sensors. *Scientific Reports*. 2015;5:17707.
145. Yeom, C.; Chen, K.; Kiriya, D.; Yu, Z.; Cho, G.; Javey, A. Large-area compliant tactile sensors using printed carbon nanotube active-matrix backplanes. *Advanced Materials*. 2015;27:1561–1566.

146. Kim, J.; Kumar, R.; Bandodkar, A.J.; Wang, J. Advanced materials for printed wearable electrochemical devices: A review. *Advanced Electronic Materials*. 2017;3:1600260.
147. Suganuma, K. *Introduction to Printed Electronics*, Vol. 74. New York: Springer, 2014.
148. He, Y.; Wu, Y.; Fu, J.Z.; Gao, Q.; Qiu, J.J. Developments of 3D printing microfluidics and applications in chemistry and biology: A review. *Electroanalysis*. 2016;28:1658–1678.
149. Moya, A.; Gabriel, G.; Villa, R.; Javier del Campo, F. Inkjet-printed electrochemical sensors. *Current Opinion in Electrochemistry*. 2017;3:29–39.
150. Nge, P.N.; Rogers, C.I.; Woolley, A.T. Advances in microfluidic materials, functions, integration, and applications. *Chemical Reviews*. 2013;113:2550–2583.
151. Wei-Cheng, T.; Erin, P. *Microfluidics for Biological Applications* (pp. 323–384). New York: Springer-Verlag, 2008.
152. Ringeisen, B.R.; Henderson, E.; Wu, P.K.; Pietron, J.; Ray, R.; Little, B.; Biffinger, J.C.; Jones-Meehan, J.M. High power density from a miniature microbial fuel cell using Shewanella oneidensis DSP10. *Environmental Science & Technology*. 2006;40:2629–2634.
153. Qian, F.; Baum, M.; Gu, Q.; Morse, D.E. A 1.5 mL microbial fuel cell for on-chip bioelectricity generation. *Lab on a Chip*. 2009;9:3076–3081.
154. Gomez, F.A. *Biological Applications of Microfluidics*. Hoboken, NJ: John Wiley & Sons, Inc., 2008.
155. Chakrabarty, K. Design automation and test solutions for digital microfluidic biochips. *IEEE Transactions on Circuits and Systems*. 2010;57:4–17.
156. David, N.B.; Philip, J.L.; Luke, P.L. Microfluidics-based systems biology. *Molecular BioSystems*. 2006;2:97112.
157. Michael, S.L.; Ron, W.; Laura, F.L. Automated design and programming of a microfluidic DNA computer. *Natural Computing*. 2006;5:1–13.
158. Yager, P.; Edwards, T.; Fu, E.; Helton, K.; Nelson, K.; Tam, M.R.; Weigl, B.H. Microfluidic diagnostic technologies for global public health. *Nature Biotechnology*. 2006;442:412–418.
159. Clarissa, L.; Nathaniel, C.C.; Carl, A.B. Nucleic acid-based detection of bacterial pathogens using integrated microfluidic platform systems. *Sensors*. 2009;9:3713–3744.
160. Chen, S.; Shamsi, M.H. Biosensors-on-chip: A topical review. *Journal of Micromechanics and Microengineering*. 2017 Jul 20;27(8):083001.
161. Tamal, D.; Suman, C. Biomicrofluidics recent trends and future challenges. *Sādhanā*. 2009;34:573–590.
162. Mandell, G.L.; Bennett, J.E.; Dolin, R. *Principles and Practice of Infectious Diseases*, 6th ed. Maryland Heights, MO: Elsevier, 2005.
163. Liu, R.H.; Lodes, M.J.; Nyuyen, T.; Siuda, T.; Slota, M.; Fuji, H.S.; McShea, A. Validation of a fully integrated microfluidic array device for influenza a subtype identification and sequencing. *Analytical Chemistry*. 2006;78:5184–5193.
164. Myers, K.M.; Gaba, J.; Al-Khaidi, S.F. Molecular identification of Yersinia enterocolitica from pasteurized whole milk using DNA microarry chip hybridization. *Molecular and Cellular Probes*. 2006;20:71–80.
165. Wang, L.; Li, P.C.H.; Yu, H.-Z.; Parameswaran, A.M. Fungal pathogenic nucleic acid detection achieved with a microfluidic microarray device. *Analytica Chimica Acta*. 2008;610:97–104.

166. Wang, Y.; Stanzel, M.; Gumbrecht, W.; Humenik, M.; Spinzl, M. Estrase 2 oligodeoxynucleotide conjugate as sensitive reporter for electrochemical detection of nucleic acid hybridization. *Biosensors and Bioelectronic*. 2007;22: 1798–1806.
167. Nebling, E.; Grunwald, T.; Albers, J.; Schafer, P.; Hintsche, R. Electrical detection of viral DNA using ultramicroelectrode arrays. *Analytical Chemistry*. 2004;76:689–696.
168. Baeumner, A.J.; Cohen, R.N.; Miksic, V.; Min, J. RNA biosensor for the rapid detection of viable Escherichia coli in drinking water. *Biosensors and Bioelectronic*. 2003;8:405–413.
169. Baeumner, A.J.; Leonard, B.; McElwee, J.; Montagna, R.A. A rapid biosensor for viable B. anthracis spores. *Analytical and Bioanalytical Chemistry*. 2004;380:15–23.
170. Baeumner, A.J.; Humiston, M.C.; Montagna, R.A.; Durst, R. Detection of viable oocysts of Cryptosporidium parvum following nucleic acid sequence based amplification. *Analytical Chemistry*. 2001;73:1176–1180.
171. Zaytseva, N.V.; Montagna, R.A.; Lee, E.M.; Baeumer, A.J. Multianalyte single-membrane biosensor for the serotype-specific detection of dengue virus. *Analytical and Bioanalytical Chemistry*. 2004;380:46–53.
172. Piliarik, M.; Paravoa, L.; Homola, J. High-throughput SPR sensor for food safety. *Biosensors and Bioelectronics*. 2009;24:1399–1404.
173. Zaytseva, N.V.; Goral, V.N.; Montagna, R.A.; Baeumner, A.J. Development of a microfluidic biosensor module for pathogen detection. *Lab on a Chip*. 2005;5:801–811.
174. Cheong, K.H.; Yi, D.K.; Lee, J.-G.; Park, J.-M.; Kim, M.J.; Edel, J.B.; Ko, C. Gold nanoparticles for one step DNA extraction and real-time PCR of pathogens in a single chamber. *Lab on a Chip*. 2008;8:810–813.
175. Tamal, D.; Suman, C. Biomicrofluidics: Recent trends and future challenges. *Sãdhanã*. 2009;34:573–590.
176. Teon, E.C.; de Michele, K; Marelize, B.; Manuel, J.L. Editors: T. Eugene Cloete, Michele de Kwaadsteniet, Marelize Botes and J. Manuel López-Romero. Nano technology in water treatment application. In *Molecular Tools for Microbial Detection*. Norwich: Caister Academic Press, 2010.
177. Dmitri, I.; Daniel, J.O.; Anthony, G.; Roger, S.; Michael, C.; Rodney, F. Nucleic acid approaches for detection and identification of biological warfare and infectious diseases agents. *Biotechniques*. 2003;35:862–869.
178. Zhang, C.; Xu, J.; Ma, W.; Zheng, W. PCR microfluidic devices for DNA amplification. *Biotechnology Advances*. 2006;24(3):243–284.
179. Nguyen, N.T.; Shaegh, S.A.M.; Kashaninejad, N.; Phan, D.T. Design, fabrication and characterization of drug delivery systems based on lab-on-a-chip technology. *Advanced Drug Delivery Reviews*. 2013;65(11–12):1403–1419.
180. Garg, M.; Christensen, M.G.; Iles, A.; Sharma, A.L.; Singh, S.; Pamme, N. Microfluidic-based electrochemical immunosensing of ferritin. *Biosensors*. 2020; 10(8):91.
181. Molinero-Fernández, Á.; Moreno-Guzmán, M.; López, M.Á.; Escarpa, A. Magnetic bead-based electrochemical immunoassays on-drop and on-chip for procalcitonin determination: disposable tools for clinical sepsis diagnosis. *Biosensors*. 2020;10(6):66.

182. Jeon, H.; Khan, Z.A.; Barakat, E.; Park, S. Label-free electrochemical microfluidic chip for the antimicrobial susceptibility testing. *Antibiotics*. 2020;9(6):348.
183. Wang, Z.; Hao, Z.; Yu, S.; Huang, C.; Pan, Y.; Zhao, X. A wearable and deformable graphene-based affinity nanosensor for monitoring of cytokines in biofluids. *Nanomaterials*. 2020;10(8):1503.
184. Luo, X.; Shi, W.; Yu, H.; Xie, Z.; Li, K.; Cui, Y. Wearable carbon nanotube-based biosensors on gloves for lactate. *Sensors*. 2018;18(10):3398.
185. Lansdorp, B.; Ramsay, W.; Hamid, R.; Strenk, E. Wearable enzymatic alcohol biosensor. *Sensors*. 2019;19(10):2380.
186. Hakimian, F.; Ghourchian, H. Ultrasensitive electrochemical biosensor for detection of microRNA-155 as a breast cancer risk factor. *Analytica Chimica Acta*. 2020 Nov 1(1136):1–8.
187. Chen, X.; Zhang, Q.; Qian, C.; Hao, N.; Xu, L.; Yao, C. Electrochemical aptasensor for mucin 1 based on dual signal amplification of poly (o-phenylenediamine) carrier and functionalized carbon nanotubes tracing tag. *Biosensors and Bioelectronics*. 2015 Feb 15;64:485–492.
188. Filella, X.; Fernández-Galán, E.; Bonifacio, R.F.; Foj, L. Emerging biomarkers in the diagnosis of prostate cancer. *Pharmacogenetics and Precision Medicine*. 2018;11:83–94.
189. Vaarala, M.H.; Porvari, K.; Lukkarinen, O.; Vihko, P. TheTMPRSS2 gene encoding transmembrane serine protease is overexpressed in a majority of prostate cancer patients: Detection of mutatedTMPRSS2 form in a case of aggressive disease. *International Journal of Cancer*. 2001;94:705–710.
190. Elshafey, R.; Tlili, C.; Abulrob, A.; Tavares, A.C.; Zourob, M. Label-free impedimetric immunosensor for ultrasensitive detection of cancer marker Murine double minute 2 in brain tissue. *Biosensors and Bioelectronics*. 2013;39:220–225.
191. Laocharoensuk, R. Development of electrochemical immunosensors towards point-of-care cancer diagnostics: Clinically relevant studies. *Electroanalysis*. 2016;28:1716–1729.
192. Hasan, S.; Jacob, R.; Manne, U.; Paluri, R. Advances in pancreatic cancer biomarkers. Oncology *Reviews*. 2019;13:410.
193. Matsuoka, T.; Yashiro, M. Biomarkers of gastric cancer: Current topics and future perspective. *World Journal of Gastroenterology*. 2018;24:2818–2832.
194. Ashizawa, T.; Okada, R.; Suzuki, Y.; Takagi, M.; Yamazaki, T.; Sumi, T.; Aoki, T.; Ohnuma, S.; Aoki, T. Clinical significance of interleukin-6 (IL-6) in the spread of gastric cancer: Role of IL-6 as a prognostic factor. *Gastric Cancer*. 2005;8:124–131.
195. Lou, J.; Zhang, L.; Lv, S.; Zhang, C.; Jiang, S. Biomarkers for hepatocellular carcinoma. biomark. *Cancer*. 2017;9:1–9.
196. Farzin, L.; Shamsipur, M. Recent advances in design of electrochemical affinity biosensors for low level detection of cancer protein biomarkers using nanomaterial-assisted signal enhancement strategies. *Journal of Pharmaceutical and Biomedical Analysis*. 2018;147:185–210.
197. Li, J.; Sherman-Baust, A.C.; Tsai-Turton, M.; Bristow, R.E.; Roden, R.B.S.; Morin, P.J. Claudin-containing exosomes in the peripheral circulation of women with ovarian cancer. *BMC Cancer*. 2009;9:244.

198. Altintas, Z.; Tothill, I. Biomarkers and biosensors for the early diagnosis of lung cancer. *Sensors & Actuators, B: Chemical*. 2013;188:988–998.
199. Arya, S.K.; Bhansali, S. Lung Cancer and Its Early Detection Using Biomarker-Based Biosensors. *Chemical Reviews*. 2011;111:6783–6809.
200. Xie, Y.; Todd, N.W.; Liu, Z.; Zhan, M.; Fang, H.; Peng, H.; Alattar, M.; Deepak, J.; Stass, S.A.; Jiang, F. Altered miRNA expression in sputum for diagnosis of non-small cell lung cancer. *Lung Cancer*. 2010;67:170–176.
201. Lim, E.H.; Zhang, S.-L.; Li, J.-L.; Yap, W.-S.; Howe, T.-C.; Tan, B.-P.; Lee, Y.-S.; Wong, D.; Khoo, K.-L.; Seto, K.-Y.; et al. Using whole genome amplification (WGA) of low-volume biopsies to assess the prognostic role of EGFR, KRAS, p53, and CMET mutations in advanced-stage non-small cell lung cancer (NSCLC). *Journal of Thoracic Oncology*. 2009;4:12–21.
202. Staden, R.-I.S.-V.; Comnea-Stancu, I.R.; Surdu-Bob, C.C. Molecular screening of blood samples for the simultaneous detection of CEA, HER-1, NSE, CYFRA 21-1 using stochastic sensors. *Journal of the Electrochemical Society*. 2017;164:B267–B273.
203. Carr, O.; Raymundo-Pereira, P.A.; Shimizu, F.M.; Sorroche, B.P.; Melendez, M.E.; Pedro, R.D.O.; Miranda, P.B.; Carvalho, A.L.; Reis, R.M.; Arantes, L.M.; et al. Genosensor made with a self-assembled monolayer matrix to detect MGMT gene methylation in head and neck cancer cell lines. *Talanta*. 2020; 210:120609.
204. Lu, W.; Cao, X.; Tao, L.; Ge, J.; Dong, J.; Qian, W. A novel label-free amperometric immunosensor for carcinoembryonic antigen based on Ag nanoparticle decorated infinite coordination polymer fibres. *Biosensors and Bioelectronics*. 2014:57:219.
205. Shen, C.; Liu, S.; Li, X.; Yang, M. Electrochemical detection of circulating tumor cells based on dna generated electrochemical current and rolling circle Amplification. *Analytical Chemistry*. 2019;91(18):11614.
206. Yang, F.; Yang, Z.; Zhuo, Y.; Chai, Y.; Yuan, R. Ultrasensitive electrochemical immunosensor for carbohydrate antigen 19-9 using Au/porous graphene nanocomposites as platform and Au@Pd core/shell bimetallic functionalized graphene nanocomposites as signal enhancers. *Biosensors and Bioelectronics*. 2015;66:356.
207. Wang, H.; Zhang, Y.; Wang, Y.; Ma, H.; Du, B.; Wei, Q. Facile synthesis of cuprous oxide nanowires decorated graphene oxide nanosheets nanocomposites and its application in label-free electrochemical immunosensor. *Biosensors and Bioelectronics*. 2017;87:745.
208. Sharifuzzaman, M.; Barman, S.C.; Rahman, M.T.; Zahed, M.A.; Xuan, X.; Park, J.Y. Green synthesis and layer-by-layer assembly of amino-functionalized graphene oxide/carboxylic surface modified trimetallic nanoparticles nanocomposite for label-free electrochemical biosensing. *Journal of the Electrochemical Society*, 2019;166:B983.
209. Elshafey, R.; Tlili, C.; Abulrob, A.; Tavares, A.C.; Zourob, M. Label-free impedimetric immunosensor for ultrasensitive detection of cancer marker Murine double minute 2 in brain tissue. *Biosensors and Bioelectronics*. 2013 Jan 15;39(1):220–225.
210. Wu, D.; Guo, Z.; Liu, Y.; Guo, A.; Lou, W.; Fan, D.; Wei, Q. Sandwich-type electrochemical immunosensor using dumbbell-like nanoparticles for the determination of gastric cancer biomarker CA72-4. *Talanta*. 2015 Mar 1;134:305–309.

211. Leung, W.H.; Pang, C.C.; Pang, S.N.; Weng, S.X.; Lin, Y.L.; Chiou, Y.E.; Pang, S.T.; Weng, W.H. High-sensitivity dual-probe detection of urinary miR-141 in cancer patients via a modified screen-printed carbon electrode-based electrochemical biosensor. *Sensors*. 2021;21(9):3183.
212. Nawaz, M.A.H.; Rauf, S.; Catanante, G.; Nawaz, M.H.; Nunes, G.; Louis Marty, J.; Hayat, A. One step assembly of thin films of carbon nanotubes on screen printed interface for electrochemical aptasensing of breast cancer biomarker. *Sensors*. 2016;16(10):1651.
213. Park, Y.; Hong, M.S.; Lee, W.H.; Kim, J.G.; Kim, K. Highly sensitive electrochemical aptasensor for detecting the VEGF165 tumor marker with PANI/CNT nanocomposites. *Biosensors*. 2021;11(4):114.
214. Białobrzeska, W.; Dziąbowska, K.; Lisowska, M.; Mohtar, M.A.; Muller, P.; Vojtesek, B.; Krejcir, R.; O'Neill, R.; Hupp, T.R.; Malinowska, N.; Bięga, E. An ultrasensitive biosensor for detection of femtogram levels of the cancer antigen AGR2 using monoclonal antibody modified screen-printed gold electrodes. *Biosensors*. 2021;11(6):184.
215. Hroncekova, S.; Bertok, T.; Hires, M.; Jane, E.; Lorencova, L.; Vikartovska, A.; Tanvir, A.; Kasak, P.; Tkac, J. Ultrasensitive Ti3C2TX MXene/chitosan nanocomposite-based amperometric biosensor for detection of potential prostate cancer marker in urine samples. *Processes*. 2020;8(5):580.
216. Raj, P.; Oh, M.H.; Han, K.; Lee, T.Y. Label-free electrochemical biosensor based on Au@ MoS_2–PANI for Escherichia coli detection. *Chemosensors*. 2021;9(3):49.
217. Mohan, R.; Mach, K.E.; Bercovici, M.; Pan, Y.; Dhulipala, L.; Wong, P.K.; Liao, J.C. Clinical validation of integrated nucleic acid and protein detection on an electrochemical biosensor array for urinary tract infection diagnosis. *PLoS One*. 2011;6(10):e26846.
218. Vadlamani, B.S.; Uppal, T.; Verma, S.C.; Misra, M. Functionalized TiO2 nanotube-based electrochemical biosensor for rapid detection of SARS-CoV-2. *Sensors*. 2020;20(20):5871.
219. Lai, H.C.; Chin, S.F.; Pang, S.C.; Henry Sum, M.S.; Perera, D. Carbon nanoparticles based electrochemical biosensor strip for detection of Japanese encephalitis virus. *Journal of Nanomaterials*. 2017;2017, 7 pages.
220. Silva, B.V.; Cordeiro, M.T.; Rodrigues, M.A.; Marques, E.T.; Dutra, R.F. A label and probe-free Zika virus immunosensor prussian blue@ carbon nanotube-based for amperometric detection of the NS2B protein. *Biosensors*. 2021;11(5):157.
221. Ramanujam, A.; Almodovar, S.; Botte, G.G. Ultra-fast electrochemical sensor for point-of-care covid-19 diagnosis using non-invasive saliva sampling. *Processes*. 2021;9(7):1236.
222. Antonio, H.; Faria, M.; Zucolotto, V. Biosensors and Bioelectronics Label-free electrochemical DNA biosensor for zika virus identification. Biosensors *&* *Bioelectronics*. 2019;131:149–155.
223. Chen, J.; Liu, Z.; Zheng, Y.; Lin, Z.; Sun, Z.; Liu, A.; Chen, W.; Lin, X. B/C genotyping of hepatitis B virus based on dual-probe electrochemical biosensor. *Journal of Electroanalytical Chemistry*. 2017;785:75–79.
224. Ilkhani, H.; Farhad, S. A novel electrochemical DNA biosensor for Ebola virus detection, *Analytical Biochemistry*. 2018;557:151–155.

225. Hideshima, S.; Hayashi, H.; Hinou, H.; Nambuya, S.; Kuroiwa, S.; Nakanishi, T.; Momma, T.; Nishimura, S.I.; Sakoda, Y.; Osaka, T. Glycan-immobilized dual-channel field effect transistor biosensor for the rapid identification of pandemic influenza viral particles, *Scientific Reports*. 2019;9:1–10.
226. Oliveira, N., Souza, E., Ferreira, D., Zanforlin, D., Bezerra, W., Borba, M.A., Arruda, M., Lopes, K., Nascimento, G., Martins, D., Cordeiro, M. A sensitive and selective label-free electrochemical DNA biosensor for the detection of specific dengue virus serotype 3 sequences. *Sensors*. 2015 Jul;15(7):15562–15577.
227. Mobed, A.; Hasanzadeh, M.; Babaie, P.; Agazadeh, M.; Mokhtarzadeh, A.; Rezaee, M.A. DNA-based bioassay of legionella pneumonia pathogen using gold nanostructure: a new platform for diagnosis of legionellosis. *International Journal of Biological Macromolecules*. 2019;128:692–699.
228. Gong, Q.; Wang, Y.; Yang, H. A sensitive impedimetric DNA biosensor for the determination of the HIV gene based on graphene-Na fi on composite film, *Biosensors and Bioelectronics*. 2017;89:565–569.
229. Kurzatkowska, K.; Sirko, A.; Zagórski-Ostoja, W.; Dehaen, W.; Radecka, H.; Radecki, J., Electrochemical label-free and reagentless genosensor based on an ion barrier switch-off system for DNA sequence-specific detection of the avian influenza virus. *Analytical Chemistry*. 2015;87:9702–9709
230. Thakur, H.; Kaur, N.; Sabherwal, P.; Sareen, D.; Prabhakar, N. Aptamer based voltammetric biosensor for the detection of Mycobacterium tuberculosis antigen MPT64. *Microchimica Acta*. 2017 Jul;184(7):1915–1922.
231. Leland, D.S.;Ginocchio, C.C. Role of cell culture for virus detection in the age of technology. *Clinical Microbiology Reviews*. 2007;20(1):49–78.
232. Hudu, S.A.; Alshrari, A.S.; Syahida, A.; Sekawi, Z. Cell culture, technology: enhancing the culture of diagnosing human diseases. *Journal of Clinical and Diagnostic Research: JCDR*. 2016;10(3):DE01.
233. Menon, S.; Mathew, M.R.; Sam, S.; Keerthi, K.; Kumar, K.G. Recent advances and challenges in electrochemical biosensors for emerging and re-emerging infectious diseases. *Journal of Electroanalytical Chemistry*. 2020 Aug 25:114596.
234. Shieh, W.J. Editors: Yi-Wei Tang, Charles W. Stratton. Advanced pathology techniques for detecting emerging infectious disease pathogens. In *Advanced Techniques in Diagnostic Microbiology* (pp. 543–561). Cham: Springer, 2018.
235. Kumar, P. Methods for rapid virus identification and quantification. *Materials and Methods*. 2013;3(207):10–13070.
236. Malecka, K.; Menon, S.; Palla, G.; Kumar, K.G.; Daniels, M.; Dehaen, W.; Radecka, H.; Radecki, J. Redox-active monolayers self-assembled on gold electrodes—effect of their structures on electrochemical parameters and DNA sensing ability. *Molecules*. 2020;25(3):607.
237. Hammond, J.L.; Formisano, N.; Estrela, P.; Carrara, S.; Tkac, J. Electrochemical biosensors and nanobiosensors. *Essays in Biochemistry*. 2016;60(1):69–80.
238. Qureshi, A.; Gurbuz, Y.; Niazi, J.H. Biosensors for cardiac biomarkers detection: A review. *Sensors and Actuators B: Chemical*. 2012 Aug 1;171:62–76.
239. Sofogianni, A.; Alkagiet, S.; Tziomalos, K. 2018. Lipoprotein-associated phospholipase A2 and coronary heart disease. *Current Pharmaceutical Design*. 24(3): 291–296.
240. Ofori-Asenso, R.; Zoungas, S.; Tonkin, A.; Liew, D. LDL-cholesterol is the only clinically relevant biomarker for atherosclerotic cardiovascular disease (ASCVD) risk. *Clinical Pharmacology & Therapeutics*. 2018 Aug;104(2):235–238.

241. Lanman, R.B.; Wolfert, R.L.; Fleming, J.K.; Jaffe, A.S.; Roberts, W.L.; Warnick, G.R.; McConnell, J.P. Lipoprotein-associated phospholipase A2: review and recommendation of a clinical cut point for adults. *Preventive Cardiology*. 2006;9(3):138–143.
242. Cleveland Clinic, Last reviewed by a Cleveland Clinic medical professional on 10/27/2022, accessed 20 February 2020, ‹https://my.clevelandclinic.org/health/articles/16866-cholesterol-guidelines--heart-health›
243. Lee, T.H.; Chen, L.C.; Wang, E.; Wang, C.C.; Lin, Y.R.; Chen, W.L. Development of an electrochemical immunosensor for detection of cardiac troponin I at the point-of-care. *Biosensors*. 2021;11(7):210.
244. Thangamuthu, M.; Santschi, C.; Martin, O.J.F. Label-free electrochemical immunoassay for C-reactive protein. *Biosensors*. 2018;8(2):34.
245. Lee, T.; Kim, J.; Nam, I.; Lee, Y.; Kim, H.E.; Sohn, H.; Kim, S.E.; Yoon, J.; Seo, S.W.; Lee, M.H.; Park, C. Fabrication of troponin I biosensor composed of multi-functional DNA structure/Au nanocrystal using electrochemical and localized surface plasmon resonance dual-detection method. *Nanomaterials*. 2019;9(7):1000.
246. Mansuriya, B.D.; Altintas, Z. Enzyme-free electrochemical nano-immunosensor based on graphene quantum dots and gold nanoparticles for cardiac biomarker determination. *Nanomaterials*. 2021;11(3):578.

Chapter 5

Graphene/carbon nanotubes-based biosensors for glucose, cholesterol, and dopamine detection

S. S. Jyothirmayee Aravind
Indian Institute of Technology Madras, Chennai, India

S. Assa Aravindh
University of Oulu, Oulu, Finland

CONTENTS

DOI: 10.1201/9781003288633-5

5.1 INTRODUCTION TO CARBON NANOMATERIALS

5.1.1 Carbon nanotubes

Brought to Earth by a massive astronomical body, carbon is a unique element such that it combines itself or with other atoms without much expense of energy [1–7]. The carbon nanomaterial era has begun with the discovery of Buckminster fullerenes in 1985 [8]. In the beginning of the last decade, Japanese electron microscopist Sumio Iijima made another remarkable discovery–the carbon nanotubes [9]. The "helical microtubules of graphitic carbon," as he called multi-walled carbon nanotubes (MWNT), comprises of a number of concentrically arranged cylinders of hexagonally arranged carbon atoms. Soon, the single walled carbon nanotube (SWNT), single graphite layer rolled up into a hollow cylinder, is also invented (Figure 5.1). Double-walled carbon nanotubes (DWNT) comprising two concentric cylinders are a special case of MWNT with properties intermediate to single and multiwalled carbon nanotubes.

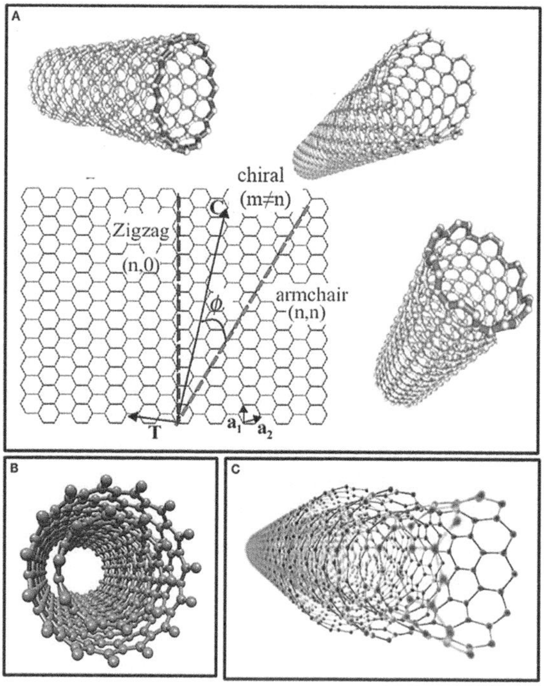

Figure 5.1 (A) SWNT structures as a function of chirality (B) Double walled nanotube and (C) Multiwalled carbon nanotubes. (Reprinted with permissions from [10] under the Creative Commons Attribution License (CC BY), Frontiers, 2015.)

5.1.2 Structure

SWNT possesses a simple morphology of a rolled-up graphene sheet consisting of hexagonal lattice of carbon atoms in the side walls and capped at both ends by one half of a fullerene-like molecule. The structure of nanotubes can be defined in terms of its chirality, that is, the direction in which the tubes get rolled up. The chiral vector with indices (n, m) are shown in Figure 5.1 [10]. Depending on the geometric arrangement of carbon atoms at the seam of the cylinders, SWNT can be categorized as chiral (m not equal to n), zigzag (m = 0) or arm chair (m = n) nanotubes. The chirality of the nanotube is important as it predicts the nanotube's mechanical, optical, and electronic properties. Nanotubes with $|n - m| = 3k$ are metallic whereas those with $|n - m| = 3k \pm 1$ are semiconducting in nature (where k is an integer). The diameter of SWNTs can be <3nm, whereas MWNTs can be of tens of nm in diameter. The lattice constant and inter-tube spacing are required to generate a SWNT, DWNT, and MWNT. These two parameters vary with tube diameter or in radial direction. Most of the experimental measurements and theoretical calculations predict the average C-C bond length as, 0.142 nm and inter-tube spacing as 0.34 nm [11].

5.1.3 Properties

The advantages of CNTs compared to other nanomaterials stems from a unique combination of electrical, thermal, optical, magnetic, mechanical, and chemical properties, which makes it attractive for a wide range of applications, including biosensing. Electronic properties of CNT have received the greatest attention in nanotube research and applications. With its nanometer dimensions and highly symmetric structure, nanotubes are endowed with remarkable quantum effects and electronic, magnetic, and thermal properties [12, 13]. Furthermore, the electronic properties have been correlated with mechanical, chemical, biological, thermal, and magnetic interactions with nanotubes. This outputs an extended electromechanical, electrochemical, thermal, electronic, and electromagnetic properties with applications of CNT in sensors, actuators, field emission, batteries, fuel cells, capacitors, and many others. Carbon atoms arranged via sp^2 bonding outturns the stiffer and stronger material known. Both experimental and theoretical measurements predict that CNT is as stiff as or stiffer than diamond, with the high Young's modulus and tensile strength. The Young's modulus of SWNT ranges from 320 to 1470 GPa and from 270 to 950 GPa for MWNT, obtained from direct tensile loading tests [14, 15]. Small radius, large specific surface area and σ-π rehybridization make CNT very attractive in chemical and biological applications because of their strong sensitivity to chemical or environmental interactions.

5.1.4 Synthesis and functionalization of CNTs

CNTs can be prepared via both physical and chemical routes. Laser ablation and arc discharge are the physical methods, whereas the most widely used chemical method is the chemical vapor deposition (CVD).

In laser ablation, intense laser pulse is employed to ablate carbon target (graphite) to a temperature of 3000–4000°C and then condensation of the evaporated carbon atoms. Large scale growth of SWNT has been achieved by Thess et al., by this method [16]. The byproducts during CNT growth are fullerenes, amorphous carbon, and graphitic polyhedrons. However, this method has a disadvantage of high temperature need, which limits the large-scale production. Also, the entangled nanotubes obtained by this method are difficult to process and hence employ in large scale applications.

The arc discharge method was investigated for the production of both SWNT and MWNT. In arc discharge, carbon atoms are evaporated by plasma of helium gas ignited by high currents passed through carbon electrodes. SWNT were first observed in the arc discharge apparatus by co-evaporating iron [17] or cobalt [18] as metal catalysts in a methane atmosphere. By tuning the pressure of inert gas in the discharge chamber and arcing current, MWNT can be grown. Scaling up of high quality MWNT by this method has been achieved for the first time in 1992 [19]. The synthesized MWNT have average length and diameter of the order of 10 microns and 5–30 nm respectively. The as-grown samples may contain impurities such as multi layered graphitic particles in polyhedron shapes, which can be overcome by heating the sample in oxygen atmosphere. Smalley and coworkers have exploited a method for purification of as grown SWNT by refluxing it in a nitric acid solution, which will oxidize away amorphous carbon species and will remove some of the metal catalyst species [20].

Chemical Vapor deposition involves heating a catalyst in a tubular furnace in the presence of a hydro-carbon gas. An employed hydrocarbon can be ethylene or acetylene along with H_2, N_2 or Ar in a CVD chamber. An employed catalyst is typically transition metal nanoparticles. The hydrocarbon dissociates at around 700°C and dissolves into the catalyst, which is in a molten state. The carbon atoms saturate in the metal nanoparticles and subsequently get precipitated, leading to the formation of tubular carbon solids in sp^2 structure. A tubular form is preferred to the open sheets of graphite because of the lower energy arising from no or fewer dangling bonds. Iron, nickel, molybdenum, or cobalt are often employed as catalysts in CVD, because of the finite solubility of carbon in these metals at high temperature. This leads to the formation of metal-carbon solution and hence CNT growth. Enhanced plasma CVD (PECVD), with the advantage of compatibility with semiconductor processing, has also attracted extensive attention in CNT growth for device fabrication. PECVD is employed for growing aligned CNT over substrates [21].

Certain advantages of the CVD technique over laser ablation or arc discharge methods are that CVD is suitable for producing good quality nanotubes on a large scale. Further, controlled CVD growth on catalytically patterned substrates is suitable for arrayed nanotube architectures useful for potential applications, which is not possible with arc discharge or laser ablation techniques. Another aspect is the lower growth temperature in CVD, which is important for industrial applications because it allows CNT to grow on systems containing materials that cannot be heated to high temperature, such as glass or transistors. Lower temperature also implies reduced fabrication costs with lower power consumption of the CVD reaction and reduced constraints on their design.

5.1.5 Purification of CNTs

Purification of as prepared nanotubes are necessary as they may contain unwanted amorphous carbon impurities, metal catalysts, other carbon derivative such as fullerenes, etc. Several post processing purification techniques, such as thermal oxidation in air [16], hydrothermal treatment [22], H_2O-plasma treatment [23], acid treatment [24], etc., are reported. Purification of MWNT can be achieved by heating the as-grown material in an oxygen environment to oxidize away the graphitic particles [19] since they exhibit a higher oxidation rate than MWNT. Typical purification procedures consist of an acid treatment to remove the metal, followed by filtration, washing in water, and then drying in a vacuum [24]. A purification process for SWNT has been developed by Tohji et al. [22]. The method involves refluxing the as-grown SWNT in a nitric acid solution for an extended period of time, oxidizing away amorphous carbon impurities and eliminating some of the metal catalyst species.

5.1.6 Graphene

Carbon, being the sixth element of the periodic table, has a ground state electronic configuration of $1s^2 2s^2 2P_x^1 2P_y^1 2P_z^0$. The four valence electrons in the outermost shells of the carbon atom can form sp, sp^2, and sp^3 hybridizations. When carbon atoms share sp^2 electrons with three other neighboring carbon atoms, a layer of honeycomb structure is formed. A single layer of carbon atoms arranged in such a honeycomb structure is commonly known as mono layer graphene. Graphene can be considered as the parent of other carbon forms, such as graphite, fullerene, and CNTs.

5.2 GRAPHENE SYNTHESIS METHODS

Commonly used synthesis methods of graphene include epitaxial growth, chemical vapor deposition, arc discharge and exfoliation techniques. A brief summary of the CVD process and exfoliation techniques are outlined here.

5.2.1 Chemical vapor deposition techniques

CVD is a widely used synthesis technique for both single and multi-layered graphene films. The growth of graphene films in CVD usually occurs at low pressure or ultra-high vacuum conditions, in the presence of hydrocarbon gas atop transition metal catalyst substrates. CVD graphene on metal substrates has shown potential for large area single layer growth compared to all other techniques [25].

The chemical reactions inside the chamber are determined by a number of factors such as gas ratios and gas partial pressure, reactor pressure, temperature, growth time, etc. Ever since the first isolation of single layer graphene in 2004, CVD technique has gained popularity due to its ease of set up in research laboratories, successful long-term use in industrial settings, and the potential to scale up fabrication. Moreover, it's a relatively inexpensive technique for large area, high quality graphene production.

One of the earliest and simplest methods for the production of graphene consisted in micromechanical exfoliation or cleavage of graphite [26]. In this method, a scotch tape was employed to peel off layers of highly oriented pyrolytic graphite (HOPG) and then transfer to a SiO_2 substrate. This simple yet efficient method could produce graphene of good electrical and structural quality. However, this method suffers from non-scalability.

5.2.2 Exfoliation

Various exfoliation techniques for production of graphene have been experimented with over time in this still an evolving area. Exfoliation is a two-step method wherein graphite is oxidized first to graphite oxide (GO), and then the GO is reduced to form few layered graphene. As graphite is an inert mineral, it can undergo oxidation only with very strong oxidation agents. The very first preparation of GO was in 1859, by the English chemist Benjamin Brodie. In this method, potassium chlorate and nitric acid were used with the mixture of graphite as the oxidant and intercalating agents, respectively. However, this technique had certain drawbacks such as long reaction time (four days), low yield, evolution of toxic acid vapors and NO_2/N_2O_4 gases, and the generation of highly explosive ClO_2 when chlorate was mixed with strong acids [27].

Later on, Staudenmaier and Hummers methods as well as variation of these methods have been reported. Staudenmaier proposed the use of H_2SO_4 with $KClO_3$ and HNO_3, but the explosive ClO_2 gas still remained, along with the prolonged reaction time. Hummers method involves treatment of graphite with potassium permanganate ($KMnO_4$), $NaNO_3$ and sulfuric acid (H_2SO_4). Here, $KMnO_4$ is a strong oxidant which guarantees the completion of the reaction within several hours and is safe in that there is no production of explosive ClO_2. Of the two oxidizing agents reported, preparations using chlorates lead to the synthesis of graphite oxide with mainly hydroxyl and epoxy functionalities, and the planar structure of the material is preserved. On the other hand, GO prepared with permanganate showed carboxyl functionalities, and the wrinkled structure is observed. Despite its high efficiency and the safety matter, Hummer's method still has some drawbacks. So several modifications were attempted. The main strategies include the removal of $NaNO_3$. In a technique called Tour method, graphite powder is oxidized in a 9:1 mixture of sulfuric and phosphoric acids, using potassium permanganate as an oxidizing agent [28]. Phosphoric acid serves as a stabilizing agent. The optimal ratio of graphite and potassium permanganate is 1:6. Variations of this method have also been reported, wherein the heating time can be considerably reduced from 12 hours to nearly one hour [29]. Figure 5.2 shows a structural model for GO [30].

Multiple research efforts have been reported for the reduction of GO to graphene. It can be mainly categorized into three groups: physical (thermal), chemical, and hybrid, wherein both physical and chemical factors come into play. In thermal reduction, GO is heated directly or irradiated with UV or microwave in a vacuum, inert, or reducing atmosphere. The mechanism of thermal exfoliation of GO can be attributed to the prompt release of CO or CO_2 gases, which come from the decomposition of functional groups of GO during rapid heating [31]. The advantage of thermal reduction is that it removes the oxygen containing functional groups and at the same time, restores the structure of GO by thermal annealing. It advances the restoration of oxidation defects of the GO basal plane afloat the sp^3-sp^2 rehybridization of carbon atoms [32].

One of the authors have reported a solar exfoliation technique of preparing graphene, wherein graphite (G) was first oxidized to graphite oxide (GO) by

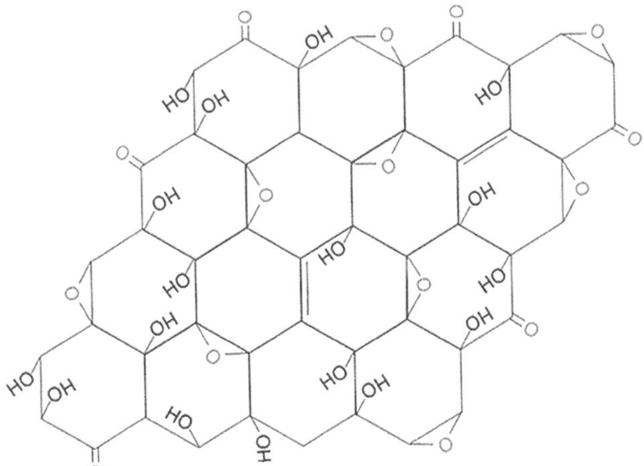

Figure 5.2 A model showing the graphite oxide structure (Reprinted (adapted) with permission from [30]. Copyright 2010 American chemical society).

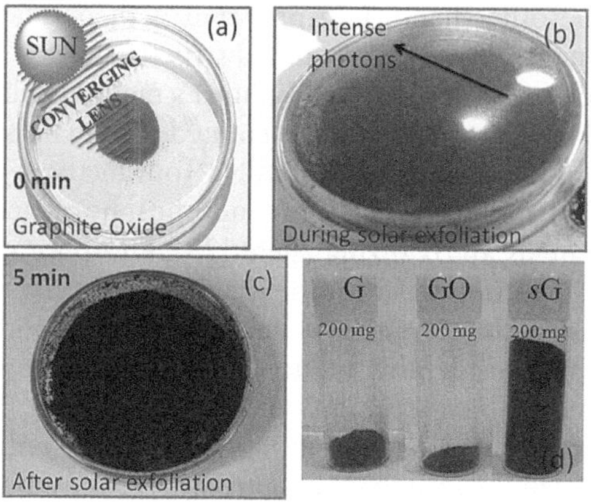

Figure 5.3 Different stages of solar exfoliation of GO to graphene (sG) and comparison of initial and final product. (Reproduced from [32] with permission from the Royal Society of Chemistry.)

modified Hummer's method. Irradiating GO with focused natural sunlight under ambient conditions resulted in a high-quality graphene (sG) with a reduced number of layers. The reaction took place at a much lower temperature (150–200°C) than thermal exfoliation, with the decomposition of functional groups and evolution of CO_2 immediately after irradiation. In other words, the pressure developed between graphitic layers during solar exfoliation could easily overcome the van der Waals force between them, thus resulting in rapid exfoliation. Figure 5.3 shows different stages of the solar exfoliation technique, as well as the comparison of initial and final products [33].

Chemical reduction involves the use of chemicals in reducing GO to graphene. Ever since the first known chemical H_2S was used in 1934 [34], several chemicals have been investigated, including hydrazine, hydroquinone, hexamethylenetetramine, hydroiodic acid, or $NaBH_4$ as reducing agents. The reduction could be carried out in liquid or in a vapor atmosphere at moderate or at room temperature [35]. Among these, hydrazine is widely researched as it can reduce GO with improved electrical and structural properties similar to pristine graphene to a large extent [36]. However, the toxic nature of hydrazine poses environmental hazards that are harmful to living organisms. As an alternative, several green routes have been emerged, including organic acids, plant extracts, microorganisms, sugars, antioxidants, amino acids, proteins, and ascorbic acid [37]. The chemical reduction method has the advantage that it is possible to simultaneously conduct chemical modifications of synthesizing material.

The hybrid production method involving the combination of chemical and thermal methods under supercritical conditions aims to improve upon the level of reduction obtained by chemical methods. Hydrothermal reduction is a simple, fast, and environmentally friendly route since it involves only water.

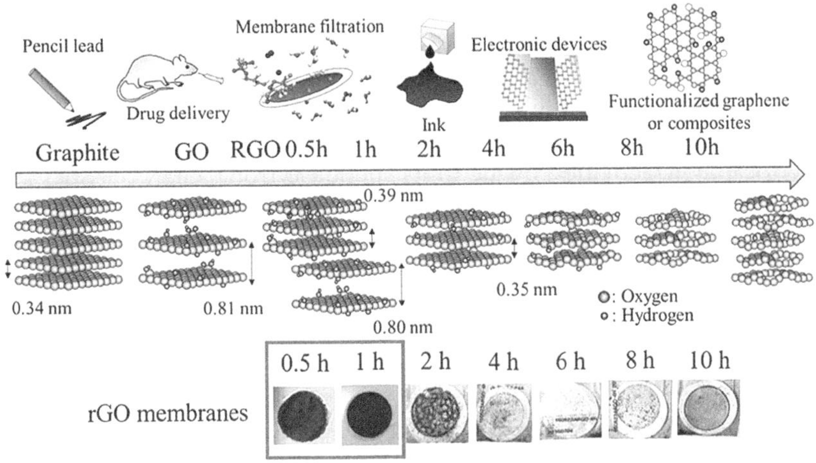

Figure 5.4 Schematic showing overall structural evolution of hydrothermally reduced GO and its potential applications. (Reprinted with permission from [38] Springer Nature, 2018 http://creativecommons.org/licenses/by/4.0/.)

Further, the experimental setup is rather cost effective and easy. Studies reported that a closed system with a certain temperature and internal pressure promotes the restoration of the aromatic structure, which is favorable for minimizing the defects. Moreover, it has a good scalability and suits industrial large-scale production. Huang et al. conducted a detailed investigation of the structural changes occurring in GO during the hydrothermal reduction process and the major outcomes of the study can be seen in Figure 5.4 [38].

5.2.3 Carbon nanomaterials for electrochemical detection

Nanomaterials can serve as excellent supporting materials for enzyme immobilization in biosensors because they offer ideal characteristics for balancing the key factors that determine the efficiency of biocatalysts, including surface area, mass transfer resistance, and effective enzyme loading. Carbon nanomaterials have been explored as a recognition element of the sensor wherein they offer binding sites for target biomarkers or molecules capturing target biomarkers. It also finds applications in a transducer component that converts the detected molecular interaction on the electrode surface into a measurable signal [39]. Carbon nanotubes combine high electrical conductivity, good chemical stability, robust mechanical strength, as well as biocompatibility. These unique properties offer great promise for CNTs in the area of electrochemical sensing. Both single walled and multiwalled nanotubes as well as their modified composites have been investigated for sensing biomolecules.

Graphene is another widely researched carbon nanomaterial with unique physical and chemical properties. The large, theoretical specific surface area of graphene (~2630 m^2/g for single layer) endows it with high electrocatalytic activities and ultra-high loading capacity for biomolecules and drugs [40]. Graphene has extraordinary electronic transport properties and high electrocatalytic activities, owing to its subtle electronic characteristics and attractive π–π interaction [41]. The high room temperature electrical conductivity and thermal conductivity of graphene is another important property making it a potential candidate for many practical applications [40, 42]. Since every atom in a graphene sheet is a surface atom, it possess high adsorptive capability. Thus, molecular interaction and electron transport through graphene can be highly sensitive to adsorbed molecules [41, 43]. The distinctive electrochemical responses of graphene to target molecules are from the planar geometric structure and special electronic character of graphene [44]. Moreover, the easy synthesis, low cost, and non-toxicity of graphene maximizes its potential applications [45].

5.2.4 Glucose biosensors

Highly sensitive and selective detection of glucose is critical in clinical diagnosis, fermentation engineering, chemical industry, as well as the food industry. Diabetes mellitus is a major health concern affecting millions of people globally, leading to several disabilities and death. It happens either due to a lack of insulin production or due to the inability of the body to utilize the available insulin. The frequent and precise monitoring of physiological glucose levels in the body is essential in order to confirm proper treatment and to ease long term consequences. The conventional finger pricking method of checking blood glucose levels to adjust the insulin dosage doesn't normally consider the night time variations, and therefore lacks an accurate blood glucose prediction. In this scenario, continuous glucose monitoring (CGM) is a promising platform for measuring the glucose level throughout the day and night. Thus, the acquisition of a low-cost, miniature glucose measuring device can help patients and clinicians manage the disease. CGM devices are based on the electrochemical measurement and quantification of glucose, which is an attractive approach for glucose determination since these devices have multiple advantages, such as specificity, speed of use, simplicity of construction, accuracy, portability, possibility of miniaturization, and easy operation. In particular, electrochemical glucose biosensors offer advantages such as the specific recognition of glucose, low cost, and rapid quantification [46]. These provide attractive means to analyze the content of biological samples from a direct conversion between a biological event and an electronic signal [47, 48]. Therefore, these electrochemical biosensors have been considered as one of the emerging technologies for point-of-care testing detection platform [49].

Electrochemical biosensing techniques can be the amperometric, potentiometric, or conductometric type. In amperometric devices, the faradic current resulting from the electron transfer between a biological system and an electrode held at an appropriate potential is measured. Potentiometric biosensors relate electric potentials to the concentration of analyte by using an ion-selective electrode or a gas-sensing electrode as the physical transducer. Conductometric biosensors record the changes in the conductance of the biological component arising between a pair of metal electrodes. Electrochemical biosensors work on a principle that two or three electrode electrochemical cells with reference, working, and counter electrodes, convert or transfer a biological event into an electrochemical signal.

5.2.5 CNT and graphene-based glucose biosensor

The first enzyme-based glucose biosensor was reported by Clark and Lyons in 1962 [50]. In 1975, Yellow spring instrument company developed a commercial product based on Clark's technology, with the successful launch of the first glucose analyzer based on the amperometric detection of H_2O_2 from the samples of the whole blood, the so-called model 23A YSI analyzer [51] as in Figure 5.5.

This first-generation biosensor was based on the H_2O_2 detection for glucose monitoring, since H_2O_2 is also produced by the enzymatic reaction at a concentration which is proportional to the glucose concentration. This biosensor was constructed by immobilizing glucose oxidase between two membrane layers. The first polycarbonate membrane layer permits only glucose molecules to move towards the enzyme layer, thus preventing interference from other larger substances such as proteins and enzymes from the whole blood. Hence, only glucose reaches the enzyme layer, where it is oxidized. H_2O_2 produced from this reaction then passes through a cellulose acetate membrane, and finally is detected at the platinum electrode surface. The two reactions are as below:

$$\text{Reaction 1:} \quad \text{Glucose} + O_2 \xrightarrow{GO_x} \text{Gluconic acid} + H_2O_2$$

$$\text{Reaction 2:} \quad H_2O_2 \xrightarrow{+0.7\text{ V vs. Ag/AgCl}} O_2 + 2H^+ + 2e^-$$

The disadvantages of first-generation biosensors were the relatively high cost arising from the platinum, as well as the interference from other species at high applied voltage, in the absence of membranes. Second generation biosensors have had a boost from the advancements in the fields of screen-printing technologies and semiconductor integration technologies. Also, they have employed synthetic electron acceptors, which helps to shuttle electrons between the redox center of the enzyme and the surface of the electrode. This system is efficient at lower applied potentials, and thereby reducing the interference effects from other biological compounds. Based on these principles, the first mediated amperometric biosensor for glucose detection was reported in 1984 [53]. In 1987, the first home-use blood glucose biosensor based on

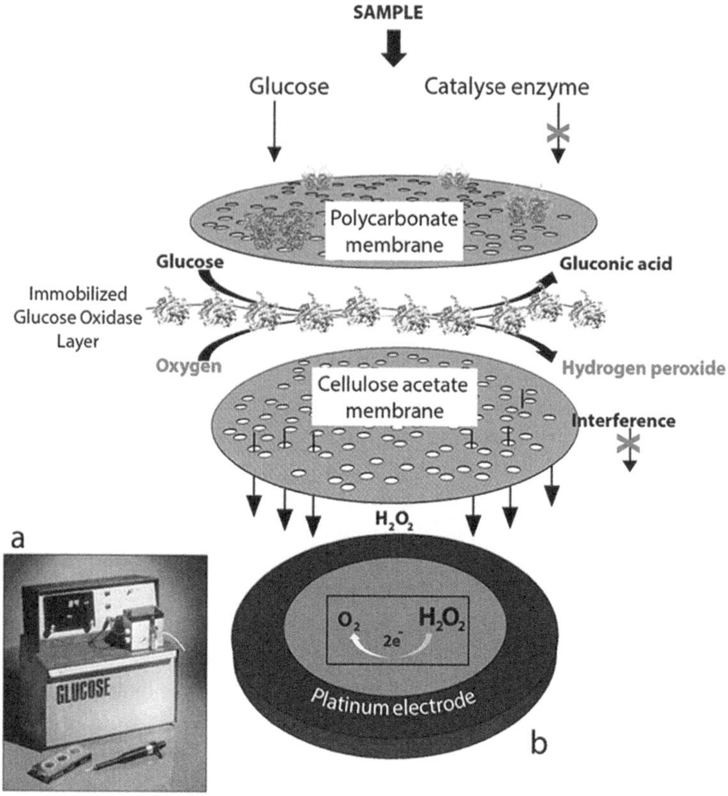

Figure 5.5 (a) YSI 23A glucose biosensor and (b) sensor probe with immobilized enzyme membrane for the Yellow Springs Instruments. (Reprinted with permission from [52] under the Creative Commons Attribution License, MDPI, 2020.)

mediators and screen-printed electrodes was launched under the brand name of ExacTech by the MediSense Company, which was a company founded jointly by the universities of Cranfield and Oxford. This is considered as a great step forward for the development of amperometric biosensors and devices for home use. Figure 5.6 shows a schematic of this biosensor and its working principle.

Though commercial mediated glucose biosensors were a success, the nature of the biochemical structure of the glucose oxidase enzyme, the relative solubility and toxicity of the mediators, and the overall poor stability of these mediators toward extended continuous operation led researchers to focus on the third-generation sensors, based on direct electron transfer between the enzyme redox center and the electrode. Minimization of electron transfer distance between the enzyme redox center and the electrode was the key. Initial efforts to achieve this included modification of the enzyme

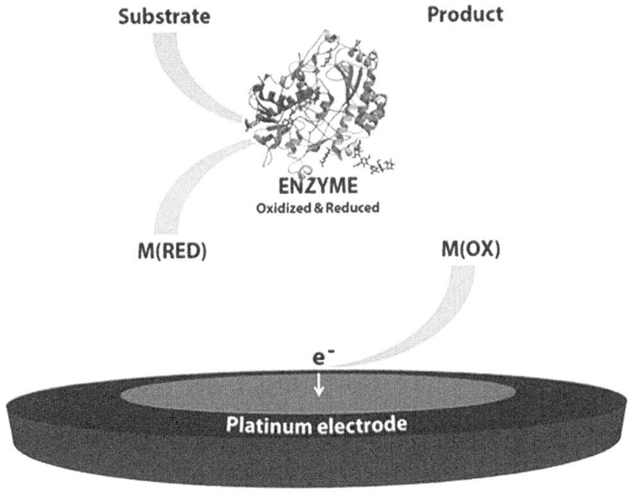

Figure 5.6 Schematic image of the second generation mediated biosensor. (Reprinted with permission [52] under the Creative Commons Attribution License, MDPI, 2020.)

with chemicals such as PVP and ferrocene carboxylic acid [54, 55]. With the advancement in nanotechnology, various nanoparticles, including carbon nanomaterials, have been employed to create a successful matrix for the transfer of electrons between the enzyme redox center and the electrode.

Nanomaterials such as gold and carbon nanotubes exhibit similar size and dimensions to redox enzymes, and hence been explored as a way to establish a bridge between the electrode and the redox center as electrical connectors [56–58]. In one such study reported in 2003, carboxylic acid functionalized Au nanoparticles of 1.4 nm diameter along with the enzyme was found to act as relay units facilitating electron transport from the redox center of the glucose oxidase – flavin adenine dinucleotide (FAD) to the electrode surface, thus activating the catalytic function of the enzymes [59]. The same group in 2004 reported the use of carbon nanotubes as molecular wires, wherein glucose oxidase on the edge of a single walled carbon nanotube was successfully linked to the electrode surface, thus immobilizing the electrode into the enzyme. Electrons could then be transferred along the length of the carbon nanotubes [60]. A similar study of an enhanced direct connection between an enzyme and an electrode via aligned carbon nanotubes arrays was also reported in 2005 [61]. These pioneering studies led researchers to explore the nanostructures as the components of biosensors, and the advancements in nanotechnology have accelerated further the developments in the field of biosensors.

Enzyme immobilization is a promising biotechnological application of CNT [62], especially for the fabrication of various biosensors and biofuel cells. The sensitivity of CNTs towards their surface makes them an ideal material for highly sensitive nanoscale biosensor devices. However, the

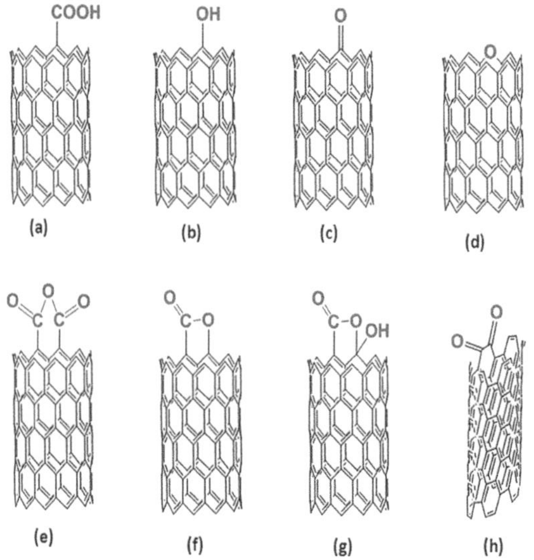

Figure 5.7 Possible structures of oxygen containing groups on CNT surface. (Reprinted with permission from [67] under the Creative Commons Attribution 3.0 Unported License (CC BY 3.0) IntechOpen, 2011.)

strong intermolecular pi-pi interactions, and hydrophobic nature of CNTs must be overcome in order to develop CNTs-based biosensors. The successful realization of CNT-based biosensors needs proper control of their chemical and physical properties, as well as their functionalization and surface immobilization. In this aspect, many functionalization strategies to enhance the solubility and stability of CNTs have been reported [63–65]. It has been shown that functionalized CNTs also offer solutions to concerns related to biocompatibility and toxicology [66]. In fact, the functionalization of CNTs has become a prerequisite for facile fabrication of nanodevices. Figure 5.7 shows the schematic of CNT with different possible oxygen containing functional groups.

Covalent and non-covalent functionalization are the two main methodologies for linking enzymes on CNTs. The noncovalent method of immobilization of enzymes is preferred to covalent methods, since it preserves the conformational structure of the immobilized enzymes [68]. The adsorption of enzymes on the surface of CNT is via π-π stacking interaction between the sidewalls of CNT and the aromatic rings of enzymes [69]. Though enzymes linked with a non-covalent approach provide high surface loading due to the weak interaction between enzyme and CNTs, further modification with polymer and nanoparticles is often needed. Nafion is explored as a binding agent to attach the CNT composite mixture onto the electrode to improve the direct transfer of glucose oxidase enzyme [70]. An electrochemical glucose biosensor fabricated using Pt nanoparticles with a diameter of

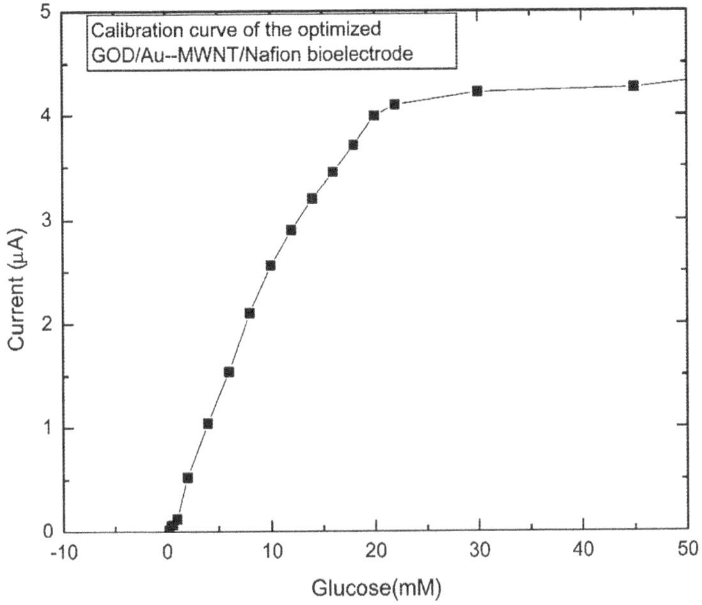

Figure 5.8 Calibration curve of GOD/Au-MWNT/Nafion electrode. (Reprinted with permission from [73] Copyright 2009 American Chemical Society.)

2-3 nm deposited onto a nafion-containing GOx/CNTs film showed a response time below 3 sec and a detection limit of up to 0.5 l M [71]. In another study, ionic liquids and glucose oxidase are made to attach onto the CNT coated poly (sodium 4-styrenesulfonate (PSS) surface, resulting in a high glucose oxidation rate [72]. Rakhi et al. reported an amperometric glucose biosensor by depositing glucose oxidase (GOD) enzyme over a Nafion-solubilized Au-multiwalled nanotube (MWNT) electrode, which offers fast and sensitive glucose quantification [73]. Figure 5.8 shows the calibration curve for glucose determination using their electrode.

An analytical model using a single walled carbon nanotube field effect transistor for glucose detection and a comparison with the experiments reported to have a good consensus between the model and the experimental data [74]. The effects of glucose adsorption on CNT electrical properties such as gate voltage versus a wide range of glucose concentration are formulated in this study. The simulated data illustrate that the analytical model can be employed with an electrochemical glucose sensor to forecast the behavior of the sensing mechanism in biosensors. Claussen et al., employed a bottom-up nanoelectrode array fabrication approach that utilizes low-density and horizontally oriented single-walled carbon nanotubes (SWCNTs) as a template for the growth and precise positioning of Pt nanospheres. They also developed a computational model to optimize the nanosphere spatial arrangement and interpret the trade-offs among kinetics, mass transport, and charge transport in an enzymatic biosensing scenario. This interesting study revealed the superior glucose sensing of tightly packed Pt nanosphere/SWCNT nanobands over low-density Pt nanosphere/SWCNT arrays, demonstrating the significance of nanoparticle placement on biosensor performance [75].

The uses of graphene-based materials for biosensing involve two aspects. One is based on charge-biomolecule interactions at π-π domains, electrostatic forces, and charge exchange leading to electrical variations in the pristine graphene. Also, the effect of defects and disorder helps in the immobilization of molecular receptors onto the graphene surface.

Pristine graphene offers an infinite surface at a molecular level. With large specific surface area, there is a high possibility of active sites for charge-biomolecular interactions, resulting in an enhanced sensing, as well as supporting the desired functionalization to target biomolecules to improve the selectivity. Figure 5.9 demonstrates the possible interactions of the graphene-based material system, wherein different areas are shown to serve different functions [76]. It is interesting to see that the functionalized graphene area is able to directly detect the biomolecules by its own oxide components, due to the synthesis in which oxygen containing functional groups are formed on the edge and surface sites of graphene sheet. Also, the functionalized graphene allows binding of enzymes, nanoparticles, etc., as shown.

Graphene based electrodes have shown superior performance, in terms of electrocatalytic activity and macroscopic scale conductivity, to carbon nanotubes based electrodes [77], indicating that the opportunities in electrochemistry confronted by carbon nanotubes are open to graphene as well. In addition, compared to carbon nanotubes, graphene exhibits potential advantages of low cost, high surface area, ease of processing and safety [78]. Another advantage of graphene is its high purity, as transition metals such as Fe, Ni, etc., are absent in graphene from GO reduction [79]. With its

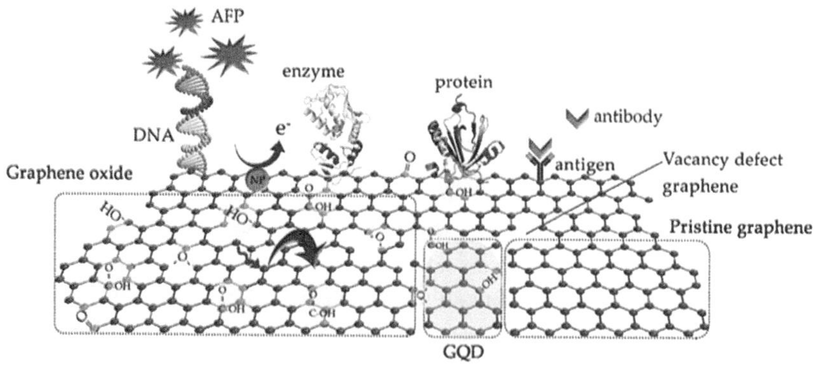

Figure 5.9 Schematic of graphene-based materials that can be immobilized with biomolecules as the receptor. (Reprinted with permission from [76] Copyright 2011, American Chemical Society.)

excellent electron transfer promoting ability for some enzymes and excellent catalytic behavior toward small biomolecules such as H_2O_2, NADH, graphene became extremely attractive for enzyme-based biosensors, e.g. glucose and cholesterol biosensors [80].

The first graphene-based glucose biosensor was reported by Shan et al. in 2009, using a graphene/polyethylenimine-functionalized ionic liquid nanocomposites modified electrode. The biosensor exhibits wide linear glucose response (2 to 14 mM, with an R value 0.994), good reproducibility and high stability [81]. A chemically reduced graphene oxide (CR-GO) based biosensor exhibited an enhanced glucose sensing with a wide linear range (0.01–10 mM), high sensitivity (20.21 μA mM cm^{-2}) and low detection limit of 2.00 μM (S/N 3) [82]. Kang et al. used biocompatible chitosan to disperse graphene and construct glucose biosensors, and the fabricated graphene-based enzyme sensor exhibited excellent sensitivity (37.93 μA mM^{-1} cm^{-2}) and long-term stability for measuring glucose [83].

The synergistic effect of graphene and metal nanoparticles such as Au and Pt have been explored for the fabrication of glucose biosensors [84]. Studies using Pd [85] and Ag [86] nanoparticles and graphene have reported enhanced electron transfer with a response time of below one second. In another study, a biosensor electrode fabricated using graphene/nano-Au composite by one-step electrochemical co-reduction method and immobilized with GOx enzyme displayed a low detection limit of 17 μM (S/N = 3), a high sensitivity of 56.93 μA $mM^{-1}cm^{-2}$, acceptable reproducibility, very good stability, selectivity, and anti-interference ability [87].

The glucose biosensors based on Pt-Au and Au nanoparticle spacers decorated graphene nanosheets are also reported. The GOx immobilized nanoparticle/graphene decorated bioelectrode retains its biocatalytic activity and offers fast and sensitive glucose quantification. A gold nanoparticles dispersed functionalized graphene-based glucose biosensor displayed a linear response up to 30 mM and a detection limit of 1 μM of glucose. The mechanism of improved sensing in the presence of the nanoparticles modified graphene was attributed to elimination of restacking of functionalized graphene in the presence of nanoparticle spacers, thereby contributing to increased surface area [88].

5.2.6 Non-enzymatic glucose biosensors

Conventional glucose sensors depend on the immobilization of glucose oxidase (GOx) as molecular recognition elements on various substrates and usually have good selectivity and high sensitivity in glucose detection. These enzyme-based sensors also display good performance in detecting glucose. However, their instability, and vulnerability to temperature and interfering chemicals, limits their further development. Non-enzymatic glucose biosensors exhibit higher stability than the enzyme-based glucose biosensor, mainly because glucose can directly generate current signals on the surface of nanomaterial-modified electrodes in non-enzymatic biosensors [89–91]. Hence, several studies have been reported on non-enzymatic glucose sensors to avoid the instability problem of enzyme-based sensors. Different materials used to fabricate non-enzymatic biosensors are shown in Figure 5.10.

Non-enzymatic glucose sensors based on metals (Pt, Cu, Ni), alloys (Pt-Pd) or metal oxides (CuOx and MnO_2) as electrocatalysts [92–97] are reported. Gold nanomaterials also attained wide attention due to their great conductivity, catalytic capability, and biocompatibility for glucose oxidation in a neutral solution [98]. Graphene can be decorated with metal nanoparticles, polymers, and such resulting nanocomposites have been extensively developed for glucose detection. For example, He et al., fabricated an Au-graphene electrode via one-step electrodeposition of gold and graphene oxide for the non-enzymatic determination of glucose in sweat. It is reported that graphene helps in better dispersion of Au nanoparticles, resulting in enhanced electrochemical response. The sensor showed a linear response in the 0.05 – 14 mM and 14 – 42 mM glucose concentration range with a high sensitivity (604 and 267 μA cm^{-2} mM^{-1}) and a low detection limit (12 μM) [99]. In another study using 3D graphene foam modified with Co_3O_4 nanowires, a remarkable detection limit of 25 nM of glucose is reported [100]. Mohammadi et al., compared the glucose sensing performance of an Au/Fe_2O_3/f-MWCNTs/GCE electrode with that of one modified with GOx enzyme as well, and the mechanism of sensing of enzymatic and non-enzymatic electrodes were elucidated. Their non-enzymatic and enzymatic electrodes show sensitivity of 512.4 and 921.4 mA/mM.cm^2 and detection limit of 1.7 and 0.9 mM, respectively [101].

5.2.7 Cholesterol biosensors

Cholesterol and its fatty acids, being the structural components of biological membranes as well as nerve and brain cells, are crucial to the human body. Cholesterol is synthesized in liver as well as furnished from dietary intake and contributes to the production of hormones, bile acids, vitamin D, and other vital molecules [102]. Cholesterol is carried in the blood by molecules called lipoproteins. A lipoprotein is a complex consisting of lipid (fat) and protein. An improper level of cholesterol is associated with severe problems. Hypercholesterolemia is a condition caused by elevated cholesterol levels (>200 mg/dL^{-1}) which can lead to damaging of the blood vessels and also an increased chance of developing a range of life-threatening cardiac and brain vascular diseases, such as hypertension, arteriosclerosis, coronary heart disease, lipid metabolism dysfunction, brain thrombosis, etc. [103, 104]. Other major conditions associated with high cholesterol include nephrosis, diabetes, jaundice, and cancer. On the other hand, low cholesterol levels may possibly cause hyperthyroidism, anemia, and malabsorption [105]. In order to diagnose and detect these health disorders, determination of cholesterol levels in blood is necessary. Evaluation of food cholesterol levels is also vital

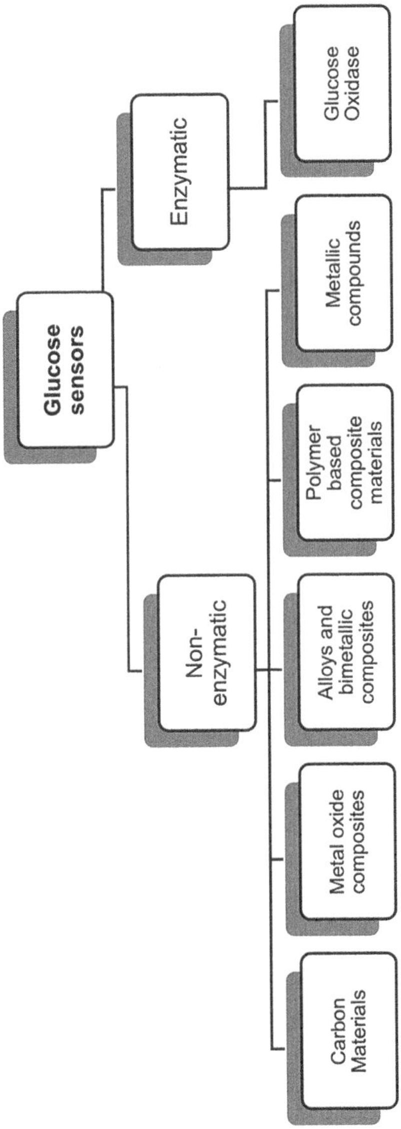

Figure 5.10 Types of enzymatic and non-enzymatic electrochemically active materials for glucose biosensors.

to select a low cholesterol diet. Various methods, such as colorimetric [106], enzymic colorimetric [107], spectrophotometric [108], and microphotometric [109], are available for serum cholesterol determination. However, these methods have certain disadvantages, such as they are time consuming, cumbersome, labor intensive, less specific, and less sensitive.

The above-mentioned drawbacks have been reduced with the emergence of biosensors. Biosensors with immobilized enzymes have attracted huge interest in analytical technology as they demonstrate the exclusive selectivity of the biological molecules and the processing power of modern microelectronics [110]. Most of the reported cholesterol biosensors are based on the amperometric technique of detection because it offers advantages such as better real-time monitoring and ease of miniaturization. Cholesterol biosensors based on photometric behaviors namely, luminescence, fluorescence, and surface plasmon resonance (SPR) are also reported [111–113]. Enzymatic amperometric biosensors with immobilized enzymes on the electrode surface constitute a redox reaction between the enzyme and an analyte. This reaction leads to a subsequent electron transfer between the enzyme and the electrode, which is then transduced to an electrical signal (current), proportional to the analyte present. The stoichiometric consumption of O_2 and production of H_2O_2 upon catalysis of cholesterol by the enzyme, as well as the transfer of electrons during the reaction, has laid the basis of amperometric cholesterol biosensors.

Amperometric determination of H_2O_2 production or O_2 consumption during enzymatic cholesterol oxidation is the frequently used strategy in cholesterol biosensors [114]. However, this reaction requires high anodic potential (0.7 V versus Ag/AgCl), at which other electrochemically active interfering species present in the physiological samples may undergo simultaneous oxidation, resulting in overestimation of the response [115]. As an alternative, the third generation amperometric biosensors proposed the direct electron transfer between the enzyme and the electrode without the involvement of O_2. Figure 5.11 shows the schematic of three generations of amperometric cholesterol biosensors.

Electrical contact between redox enzymes and electrodes is crucial in the construction of third-generation biosensors. Numerous approaches have been proposed to achieve fast electron transfer kinetics [116]. Artificial electron transfer mediators with high electrical conductivity can be exploited to shuttle electrons between the enzyme redox center and the surface of the electrode (Figure 5.2B). The poor biocompatibility of the surface on which the enzyme is immobilized may sometimes deface the enzyme activity, resulting in unstable sensing activity. In the last few years, one of the main aspects of the research on biosensors thus focused on developing a suitable support matrix employing novel materials that can result in a stable biosensor with improved electrical communication between the enzyme and the electrode.

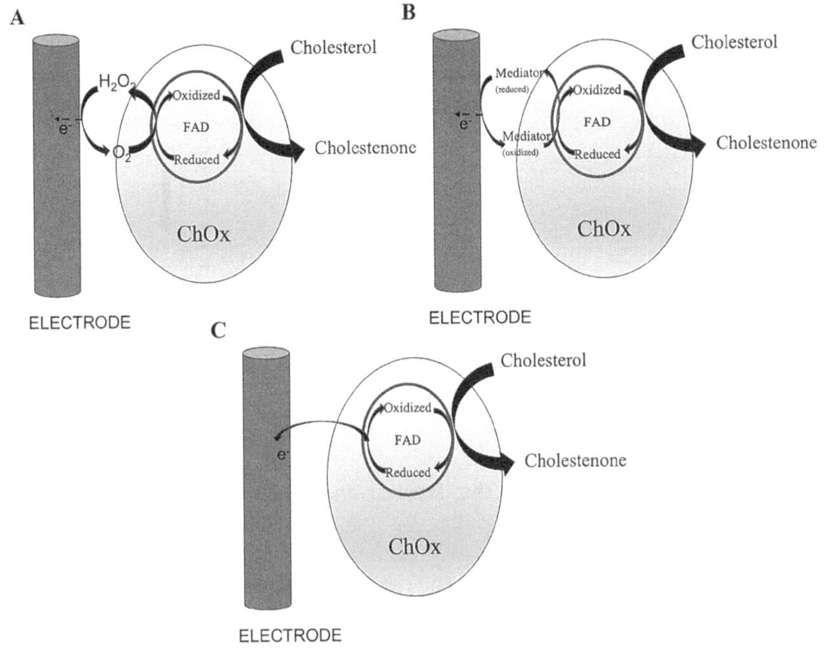

Figure 5.11 Three generations of amperometric cholesterol biosensors: (A) First generation biosensors, wherein H_2O_2 detection is on the electrode surface. (B) Second generation cholesterol biosensors use artificial electron mediators to transfer electrons between FAD and an electrode. (C) Third generation cholesterol biosensors where direct electron transfer occurs between FAD and an electrode.

As already mentioned, nanomaterials show great promise as enzyme immobilization matrices for the development of biosensors. The nanomaterials exhibit unique chemical, physical, and electronic properties, owing to their nanoscale dimension. In addition, their dimensional similarity with biomolecules makes it possible to conjugate them into novel functional hybrid systems that show promising analytical behaviors [117, 118].

Conducting polymers [119], CNT [120], nanoparticles [121], and sol-gel/hydrogels [122], modified metals, and carbon surfaces are commonly used to prepare solid electrode systems and supporting substrates. For electrochemical detection, carbon-based materials such as graphite, carbon black, and carbon fiber are preferred, due to their excellent electrical and mechanical properties. These materials have a high chemical inertness and provide a wide range of anode working potentials with low electrical resistivity. Noble metal nanoparticles are known to be excellent catalysts, due to their high ratio of surface atoms with free valences to the cluster of total atoms [123]. Deposition of the noble metal nanoparticles over carbon nanomaterials

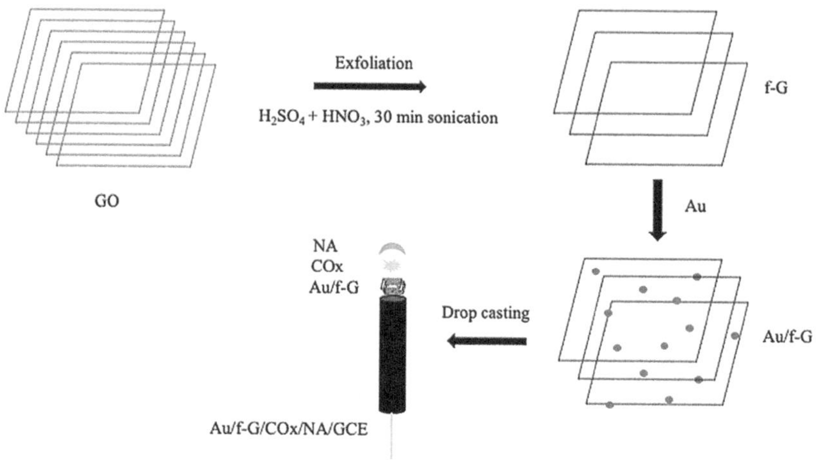

Figure 5.12 Schematic showing the fabrication of Au/f-G based Cholesterol Biosensor.

reduces their loading by increasing the catalyst utilization and improving the catalyst activity and performance. Graphene sheets prepared through the exfoliation of graphite oxide (GO) leave behind some defects and vacancies [124], and these defects can act as good anchoring sites for the deposition of metal nanoparticles that can be used for cholesterol biosensor applications.

The synergistic effect of Au nanoparticles and functionalized-graphene nanoplatelets (f-G) in the catalysis and detection of cholesterol from solution has been observed by Jyothirmayee Aravind et al. [125]. A schematic of their fabricated sensor (Figure 5.12) depicts a three-step process of exfoliation and functionalization of GO to f-G, deposition of Au nanoparticles on f-G and finally the fabrication of a biosensor electrode by drop casting of Au/f-G, the enzyme (COx) and nafion (NA) solution.

A uniform loading of nanoparticles over the substrate is crucial for a better sensing performance, and microscopic studies are often employed to analyze the particle distribution. Figure 5.13 shows one such example of transmission electron microscopic (TEM) and scanning electron microscopic (SEM) images of f-G and Au nanoparticles dispersed f-G.

Electrochemical activity of the carbon nanostructures based electrocatalysts can be studied by recording cyclic voltammograms (CV) using a three-electrode electrochemical cell wherein a constant voltage is applied between the reference and working electrodes and the current is measured between the working electrode and the counter electrode. The electrolyte is usually added to the test solution in order to ensure sufficient conductivity. The cell can be used to study the electrochemical reactions, either at the electrode/electrolyte interface or in the electrolyte. The voltage is applied to the working electrode at a fixed scan rate and current is measured and plotted versus the applied voltage to get the cyclic voltammogram.

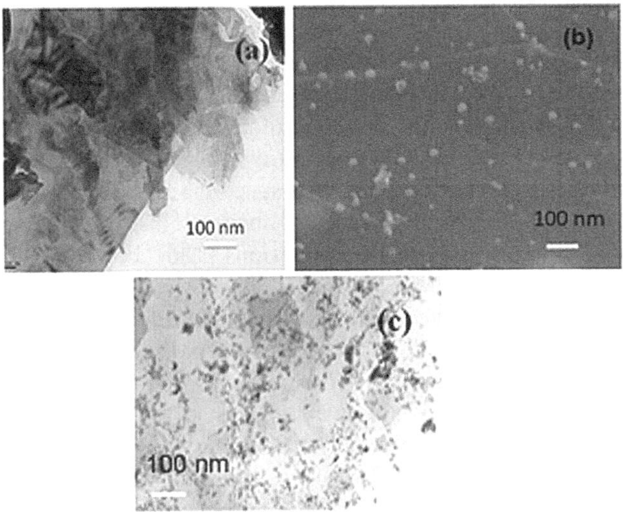

Figure 5.13 (a) TEM image of f-G (b) FESEM and (c) TEM images of Au/f-G. (Re-printed with permission from [125].)

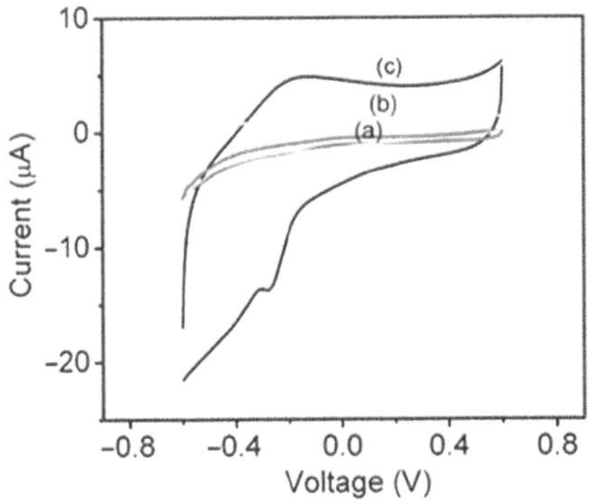

Figure 5.14 (a) Au/f-G/COx/NA/GCE in 0.1 M phosphate buffer solution (in the absence of cholesterol) (b) f-G/COx/NA/GCE and (c) Au/ f-G/ COx/ NA/GCE in 0.1 M phosphate buffer (PB) solution (pH 7) containing100 μM cholesterol at a scan rate of 50 mV/s. (Re-printed with permission from [125].)

As an example of a cholesterol detection study using nanoparticle modified biosensor, Figure 5.14 shows the cyclic voltammograms (CV) of the f-G and Au/f-G modified electrodes. In the presence of cholesterol, the Au/ f-G modified electrode displayed a couple of redox peaks, implying the potential characteristic of Au nanoparticles in facilitating direct electron transfer between the active sites of immobilized COx and the electrode surface. The fabricated Au/f-G electrode exhibited a sensitivity of 314 nA/μM.cm^2 for the detection of cholesterol with a linear response up to 135 μM.

5.2.8 Dopamine biosensors

Dopamine (DA) is a neurotransmitter which plays an important role in the central nervous, hormonal, renal, and cardiovascular systems. Parkinson's disease is a long-term degenerative, neurological disorder of the central nervous system, associated with many motor and non-motor actions, affecting body functions. Given the lack of a specific analytical test, it is usually diagnosed through reviewing patients' medical history, and conducting some physical and neurological examinations [126]. In 1960, Ehringer and Hornykiewicz, discovered that abnormal concentrations of DA in the striatum of brain could be related to neurological disorders such as Parkinson's [127, 128]. DA has also been implicated in the pathogeneses and treatment of psychiatric disorders such as schizophrenia, depression, and addiction [129–131]. In addition, DA is associated with the regulation of cognitive functions such as attention, stress, rewarding behavior, and reinforcing effects of certain stimuli [132–134]. Thus, it became important to selectively detect DA with high sensitivity, and DA related research has attracted much interest among neuroscientists and chemists.

Due to the electroactive nature of DA and its ease of oxidation at conventional electrodes [135], electrochemical DA detection has received huge attention over the past few decades. Electrochemical methods offer advantages such as simplicity, speed, and sensitivity as well. However, there are some challenges associated with the electrochemical detection of DA.

A major problem is the interference from other biological compounds such as ascorbic acid (AA), and uric acid (UA) [136]. As a result, the accuracy of DA detection is very low in real sample analysis. The basal concentration of DA is 0.01–1 μM, whereas AA and UA concentration is 100 – 1000 times higher than that of DA. Therefore, it is crucial to develop sensitive and selective methods for determination of DA [137]. A great advancement in disease diagnosis would be a development of an electrochemical sensor that would measure DA at low levels of the characteristic of a living system i.e., 26 – 40 $nmolL^{-1}$ [138].

Different strategies have been employed to solve these problems. For example, negatively charged Nafion ion-exchange membranes have been widely used in the in vivo determination of positively charged DA under physiological pH by suppressing interference from negatively charged AA [139]. Gold nanoparticle arrays, [140] self-assembled monolayers, [141] and organic polymers [142] modified electrodes have been found to selectively detect DA. However, the aforementioned methods suffered from surface deactivation of immobilized materials due to solvent evaporation, and they decay with time, resulting in nonuniform thickness and poor reproducibility. Hence, the researchers sought methods to develop new electrode materials for the detection of DA in the presence of AA and UA [143].

One-dimensional single walled CNTs and multi-walled CNTs and two-dimensional (2D) graphene compounds (graphene nanoribbon (GNR), graphene oxide (GO), and reduced graphene oxide (RGO)) possess excellent mechanical, electrical, and optical properties, opening many new approaches for their use in biosensor applications [144–146]. Numerous studies exploring the potential usage of CNTs and graphene compounds in DA detection have been reported. For instance, Mphuthi et al., have reported the fabrication of a DA biosensor using metal oxide nanoparticles doped phthalocyanine f-MWCNT. The GCE-MWCNT/ZnO/29H,31H-Pc electrode exhibited a very low detection limit (0.75μM) towards DA detection, with a sensitivity of 1.45μA/μM, and also displayed resistance to electrode fouling, and anti-interference ability [147].

The interaction of catecholamines such as dopamine with DNA can provide a novel platform for the fabrication of neuro transmitter biosensors. DNA templated synthesis of Pt nanoparticles on single walled carbon nanotubes (SWNT) demonstrates uniform deposition of ultra-nano Pt particles over well separated ss-DNA wrapped SWNT [148]. Jyothirmayee Aravind et al., have studied bio recognition ability of different types of ss-DNA in order to make use of them in DA detection in conjunction with noble metal deposited MWNT [149]. An electrode consisting of AC/ss-DNA (15 strands) was immobilized with Pt nanoparticle decorated functionalized multiwalled carbon nanotubes (f-MWNT) and coated with nafion (NA) was tested for determination of dopamine in the presence of ascorbic acid and uric acid. They also reported and compared DA biosensor fabrication and studies using 5′ -GT(15)-3′ ss-DNA functionalized Pt/MWNT. Figure 5.15 shows the structure and morphology of the Pt/MWNT and DNA functionalized Pt/MWNT. A comparison of the cyclic voltammograms of bare electrode and different modified electrodes further validate the better electrocatalytic

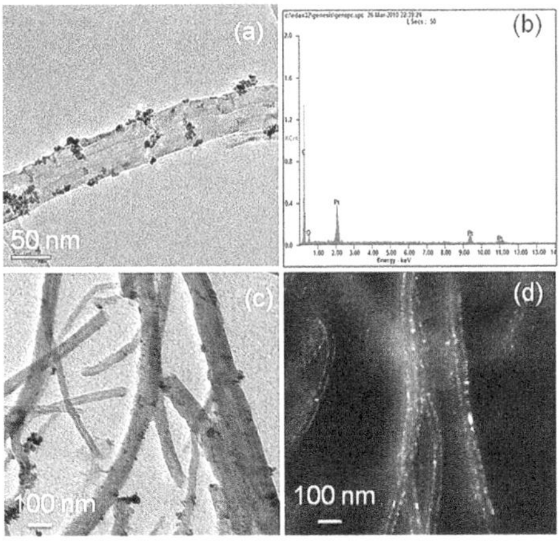

Figure 5.15 (a) TEM image (b) EDX spectrum of Pt/f-MWNT (c) bright field and (d) dark field TEM images of Pt/f-MWNT/ssDNA (AC). (Re-printed with permission from [149].)

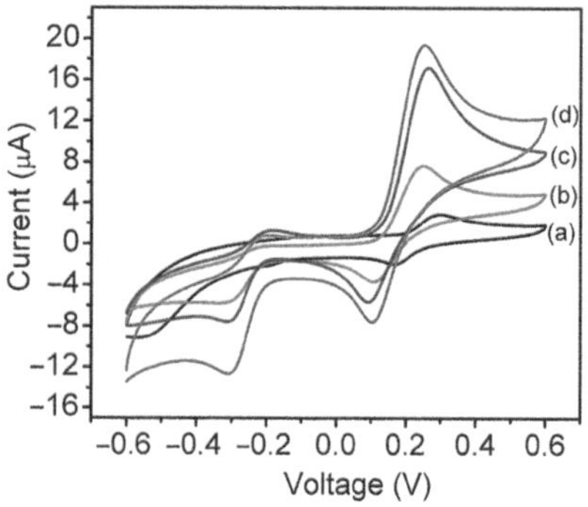

Figure 5.16 Cyclic voltammograms of (a) bare GCE (b) ssDNA(AC)/NA/GCE (c) f-MWNT/ssDNA/NA/GCE and (d) Pt/f-MWNT/ssDNA(AC)/NA/GCE in 0.5 mM DA. The supporting electrolyte is 0.1 M PB solution (pH 7). (Re-printed with permission from [149].)

activities of nanoparticles modified electrodes towards biosensing (Figure 5.16). Further, the GT-ssDNA modified electrode outperforms the AC-ssDNA based electrode, with a linearity of ~800 μM and detection limit ~0.45 μM.

Aptamer-based electrochemical biosensors are reported to have increased selectivity and good dopamine binding. Table 5.1 shows a comparison between

Table 5.1 Comparison between different DNA aptamer-nanomaterials based electrochemical biosensors

Biosensor	*Sample*	*Limit of detection (nM)*	*Linear range (μM)*	*Reference*
DNA aptamer-PEDOT/rGO	Human serum	78×10^{-6}	1×10^{-6}-0.16	[150]
DNA aptamer-Graphene/polyaniline nanocomposites film	Human serum	0.002	0.007×10^{-3} 90×10^{-3}	[151]
DNA aptamer nanowire transistor biosensor	Living PC12 cells	0.01	10^{-5}-10^{-2}	[152]
DNA aptamer-CNT/GCE	Human blood	0.22	1×10^{-3}-30×10^{-3}	[153]
DNA aptamer-Ag nanoparticle/CNT/GO composite	Human serum	0.7	0.003-0.11	[154]
DNA aptamer-GO/nile blue/GCE	Human serum	1	0.01-200	[155]
DNA aptamer-Au nanoparticle/CNTs	Human serum	2.1	5×10^{-3}-300×10^{-3}	[156]
DNA aptamer – carbon nanoparticles/Au nanoparticles	Human urine	10	0.03-3	[157]

different DNA aptamer-nanomaterials based electrochemical biosensors for the detection of dopamine in real samples.

Recently, Arumugaswamy et al., investigated the DA sensing behavior of graphene quantum dots /acid-functionalized multiwall carbon nanotubes on a glassy carbon electrode (GCE) surface. They have reported a linear range of 0.25 – 250 μM DA detection, with a detection limit of 95 nM [158].

5.2.9 Computational studies on graphene and carbon nanotube-based biosensors

Though carbon-based nanostructures are versatile biosensors owing to their physico-chemical characteristics, they suffer from structural limitations as well, such as functionalization, presence of defects, and agglomeration of particles that are often difficult to control experimentally. These attributes often hinder the device performance. Nevertheless, modeling and simulation can guide as a tool to design and predict efficient devices based on carbon nanostructures [159–170]. Computations have been extensively used to investigate the electronic properties in pristine as well as defective CNTs, to calculate the thermal, elastic, and mechanical properties [171–173]. Graphene and carbon nanotubes (CNTs) are excellent sensors for many applications, and modeling and simulation has been used to a great extent to investigate the graphene platform as well as the receptor part of the sensor unit [159]. In this context, density functional theory (DFT) based simulations are an important workhorse of materials science and widely used for such systems. In DFT calculations, CNTs are modelled as a cylindrical tube-like structure of C, with periodic boundary conditions applied along the tube axis [174]. Further, to understand the effect of adsorbates, atoms or molecules are attached to the tube surface or inside walls and the properties are compared with that of a pristine system. This will enable the comparison of electron transfer between, to, and from the surface of CNTs and the adsorbates [175].

Especially in the health sector, for various diagnoses, devices such as sensors constitute an important part for detection and quantification of biomolecules. Graphene and CNTs are used for various functionalities as their properties can be tuned to match a specific application. To achieve the socket functionality to attach the binding unit, suitable dopants such as Boron are used to tune the band structure of graphene [160], and DFT has been quite successful in predicting accurate electronic structures. Moreover, the charge transfer between the dopant and graphene layers can influence the binding efficiency of molecules. The possibility of pseudo-gap opening has also been observed in the presence of suitable dopants [160]. In general, by modifying the electron affinity and conducting properties of graphene by suitable dopants, graphene layers can be attuned to sense varied bio molecules such as urea, glucose, etc. [161]. The uniquely large surface areas of CNTs enable them to adsorb a wide range of molecules on the surface; hence, they can be used as a desirable starting material to prepare biological sensors with miniature size [162]. In DFT studies, CNTs are usually modelled with a typical diameter to length ratio of 1:1000. Further, the modification of properties can be achieved by using either metals or semiconductors. For instance, biosensors are used for detecting lethal diseases like lung cancer, and metal doped CNTs such as that with Ir and Rh have been modelled successfully that operate with change in resistivity. In such sensors, changes in conductivity occur due to metal adsorption as evidenced by electronic structure calculations. Breath analysis and biomolecular detection are two important areas CNTs and other nanomaterials find applications owing to the ease with which they can be incorporated into sensors [176]. By analyzing the exhaled breath of a person, disease conditions can be assessed. The biosensors can be successfully employed to detect exhaled air of carrier persons that may contain two probable lung cancer gases such as aniline and *o*-toluidine [163]. The CNT geometries with adsorbates and the corresponding density of states are presented in Figures 1 and 2 respectively [163]. It is evident that in the presence of the metal atom, the conducting properties of CNT enhance and facilitate the binding of adsorbate molecules. Not only precious metals, but semiconducting oxides such as ZnO also proved to be efficient in modifying properties of graphene sheets to detect different nucleotides of DNA, such as adenine, thymine, and cytosine. It is observed that the binding of nucleotides increases in the presence of ZnO compared to the pristine graphene sheets and the hetero structures possess enhanced conducting properties, and the transmission of current will be increased between the DNA and the surface of the electrodes [164].

The identification and quantification of various drugs forms an active area in the field of biosensors to avoid issues such as overdosing, illegal contents, and pollution. Nanocomposites derived from CNTS have been used to detect various drugs, such as chlorambucil, flutamide, pyrazinamide, diazepam, amphetamine, and phenylpropanolamine [165]. DFT studies have been successfully employed to model doped fullerene (C_{60}) for drug detection. The adsorption energies, electronic structure with accurate valence and conduction band positions, chemical hardness, electrophilicity, as well as dipole moments, were calculated [163]. Further, pristine and B, Al, Si, Ga, and Ge-doped C60 FLN systems were modelled using DFT to detect the presence of drugs such as amantadine [167]. Further, DFT studies have shown that, FLN-C_{60} when doped with B, Al, Ga, Si and Ge can be used to detect amphetamine efficiently as replacing carbon atoms with elements such as Si or Ge increased the reactivity and efficiency of gas sensing [168]. It is seen that the work function of these sensors decreased with doping, thereby increasing the electron emission, and another DFT study has shown that the doping of Al and Si also helps to improve the electronic sensitivity and reactivity of Al and Si doped C_{60} samples [169]. All these DFT studies have demonstrated that doping C_{60} with desirable metals helps to balance between electronic sensitivity and reactivity. Many carbon-based sensors are in use for biomedical applications, including the hybrid sensors such as FLN, GPN, and CNT/GPN hybrid sensors that can monitor nonenzymatic glucose for the diagnosis of diabetes. For instance, a bio sensor using 3-amino-capto-1,2,4-triazole functionalized over Au-C60 nanocomposite for sensing glucose has been successfully developed, and DFT studies demonstrated the effective electron transport in this sensor owing to the reduction in HOMO-LUMO gaps [170]. Experimental analysis and DFT investigations complemented each other in this study, as the large surface area of the nanocomposites helped the electrochemical oxidation of the glucose possessed by excellent selectivity and stability.

CNTs also found use in food and agricultural applications such as food ripening indicators, reducing the spoilage of food, quality assurance, and pesticide detection. Reduced size, lowered power consumption, and efficient detection capability makes CNTs ideal for tracking products, as well as the management of supply chains. CNT based sensors are successfully employed in circuits in various stages of the food industry, such as smart packages [177], fruit ripening [178], spoilage [179], detection of pesticides [180], as well as in benchmarking the flavor and smell of food products. For instance, ethylene detection using CNTs is an actively researched area computationally [181]. Ethylene is an active component in the ripening of fruits, and monitoring its concentration can help to optimize the harvesting time and resources. Since pristine CNT is inert in sensing ethylene, modifying the properties of CNT with other moieties such as SnO_2, copper complexes, etc., have been successfully achieved [182].

5.2.10 Challenges in modeling and simulation of graphene and CNTs for sensing applications

The challenges associated with the modeling and simulations studies of graphene and CNTs are as follows. In the case of CNTs the inherently large and conjugated π-system involves highly strained bonds that makes the modeling a difficult task. Furthermore, in both graphene and CNTs, small molecules may often end up adsorbed via physisorption configuration, resulting in very weak interactions with the surfaces. In the case of CNTs, the chirality also influences the electronic properties, thus making the comparison by experiment difficult. The length of the CNT included in the DFT model can also become hazardous in making direct comparison with experiments. Nevertheless, there exist a large number of computational studies in the field of graphene and CNT related to sensing applications, and the results have qualitatively reproduced most experimental findings [174, 183]. The accuracy of the computational model and the proper choice of parameters are thus important factors to consider while calculating the properties of graphene and CNTs.

5.2.11 Challenges and future trends

Compared to CNTs, graphene synthesis is cheaper, and there has been a great advancement in the preparation of graphene and related composites. However, most of the bioanalytical applications are employing multilayer graphene as single layer graphene are very costly. Hence, there is a need for dedicated research efforts to develop cost effective methods of production of single layer graphene. Further, the toxicity and biocompatibility of the graphene and carbon nanotubes also needs to be determined in case of in vivo analyte detection. Various functionalization techniques have already been demonstrated to have good biocompatibility without any apparent cytotoxicity. But there is a need for standard international guidelines for the determination of cytotoxicity of nanomaterials. The cost-effectiveness and scalability of carbon nanomaterials-based biosensors for mass production still need to be established and developed. Considering the ongoing research efforts, it is envisaged that the future endeavors will confront the existing challenges in the field, thereby lead to robust, precise and analytically superior electrochemical biosensors for bioanalytical needs (Figures 5.17 and 5.18).

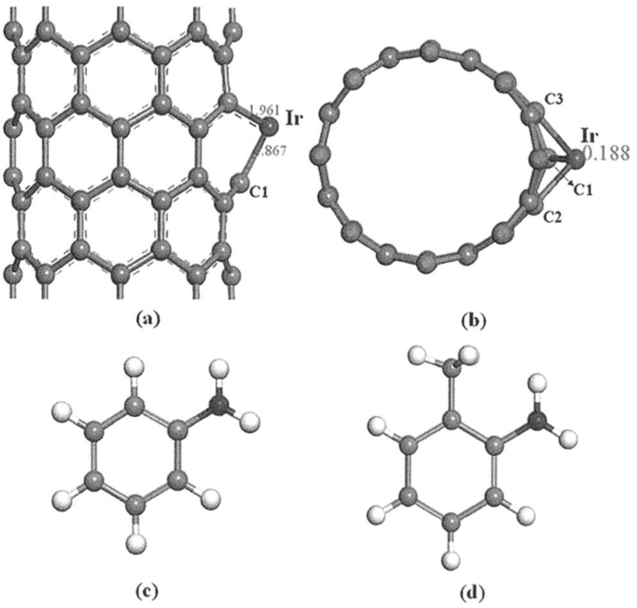

Figure 5.17 (a) and (b) represent the side and top views of CNT geometries after adsorption of Ir. (c) and (d) represent (c) Ir/C6H7N and (d) Ir/C7H9N respectively. The bond distances are also shown in the figure in red. (Re-printed with permission from [163].)

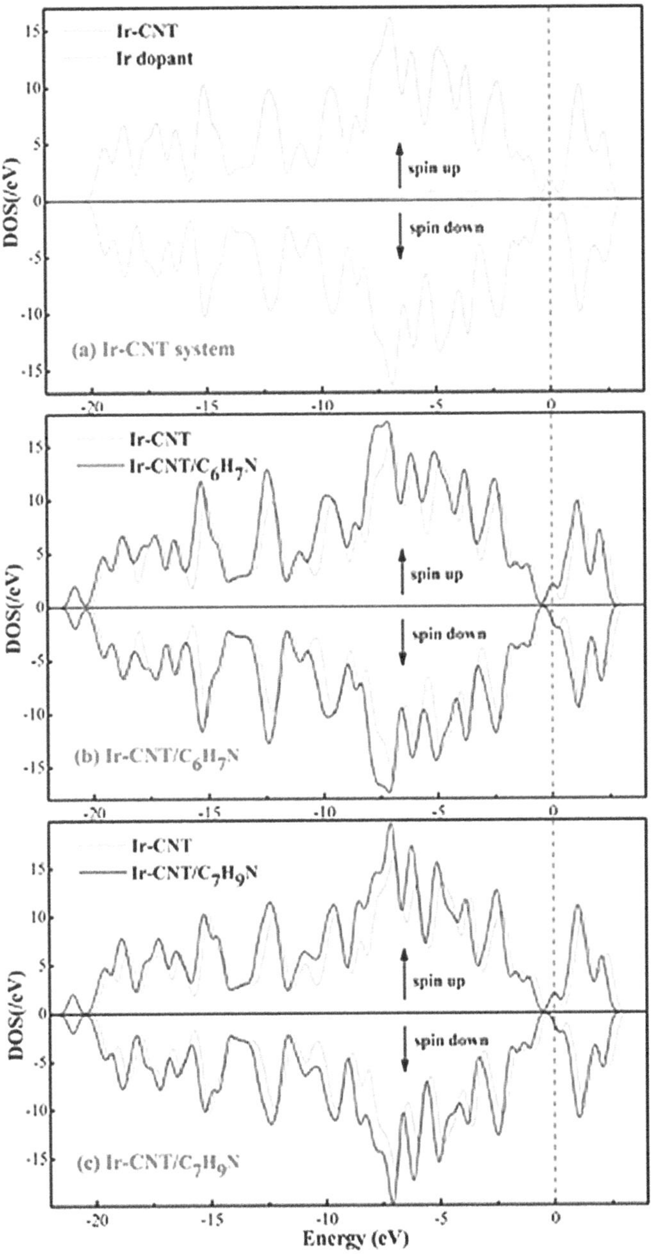

Figure 5.18 The density of states corresponding to pristine CNT as well as that in presence of adsorbates. (Re-printed with permission from [163].)

REFERENCES

1. Musameh, M., Wang, J., Merkoci, A., Lin, Y., *Electrochem. Commun.*, 2002, 4, 743–746.
2. Sireesha, M., Babu, V. J., Kiran, A. S. K., Ramakrishna, S., *Nanocomposites*, 2018, 4, 36–57.
3. Geim, A. K., Novoselov, K. S., *Nat. Mater.*, 2007, 6, 183.
4. Park, S., Ruoff, R. S., *Nat. Nanotechnol.*, 2009, 4, 217.
5. Lee, X. D., Wei, J. W., Kysar, J. H., *Science*, 2008, 321, 385.
6. Service, R. F. *Science*, 2009, 324, 875.
7. Balandin, A. A., Ghosh, S., Bao, W. Z., Calizo, I., Teweldebrhan, D., Miao, F., Lau, C. N., *Nano Lett.*, 2008, 8, 902.
8. Smalley, R. E., *The Sciences*, 1991, 31, 22–28.
9. Iijima, S., *Nature*, 1991, 354, 56–58.
10. Tîlmaciu, C. M., Morris, M. C., *Front. Chem.*, 2015, 3, 1–21.
11. Dresselhaus, M., Dresselhaus, G., Eklund, P., *Science of Fullerenes and Carbon Nanotubes*, 1996, Academic Press, San Diego.
12. Bockrath, M., Cobden, D. H., McEuen, P. L., Chopra, N. G., Zettl, A., Thess, A., Smalley, R. E., *Science*, 1997, 275, 1922–1925.
13. Wilder, J. W. G., Venema, L. C., Rinzler, A. G., Smalley, R. E., Dekker, C., *Nature*, 1998, 391, 59–62.
14. Yu, M. F., Files, B. S., Arepalli, S., Ruoff, R. S., *Phys. Rev. Lett.*, 84, 2000, 5552–5555.
15. Yu, M. F., Lourie, O., Dyer, M. J., Moloni, K., Kelly, T. F., Ruoff, R. S., *Science*, 2000, 287, 637–640.
16. Thess, A., Lee, R., Nikolaev, P., Dai, H., Petit, P., Robert, J., Xu, C., Lee, Y. H., Kim, S. G., Rinzler, A. G., Colbert, D. T., Scuseria, G. E., Tománek, D., Fischer, J. E. Smalley, R. E., *Science*, 1996, 273, 483–487.
17. Iijima, S., Ichihashi, T., *Nature*, 1993, 363, 603–605.
18. Bethune, D. S., Klang, C. H., de Vries, M. S., Gorman, G., Savoy, R., Vazquez, J., Beyers, R., *Nature*, 1993, 363, 605–607.
19. Ebbesen, T. W., Ajayan, P. M., *Nature*, 1992, 358, 220–222.
20. Liu, J., Rinzler, A. G., Dai, H., Hafner, J. H., Bradley, R. K., Boul, P. J., Lu, A., Iverson, T., Shelimov, K., Huffman, C. B., Rodriguez-Macias, F., Shon, Y.-S., Lee, T. R., Colbert, D. T., Smalley, R. E., *Science*, 1998, 280, 1253–1256.
21. Chhowalla, M, Teo, K. B. K., Ducati, C., Rupesinghe, N. L., Amaratunga, G. A. J., Ferrari, A. C., Roy, D., Robertson, J., Milne, W. I., *J. Appl. Phys.*, 2001, 90, 5308.
22. Tohji, K., Takahashi, H., Shinoda, Y., Shimizu, N., Jeyadevan, B., Matsuoka, I., Saito, Y., Kasuya, A., Ito, S., Nishina, Y., *J. Phys. Chem. B*, 1997, 101, 1974–1978.
23. Huang, S., Dai, L., *J. Phys. Chem. B*, 2002, 106, 3543–3545.
24. Chiang, I. W., Brinson, B. E., Smalley, R. E., Margrave, J. L., Hauge, R. H., *J. Phys. Chem. B*, 2001, 105, 1157–1161.
25. Li, X., Colombo, L., Rodney, R., *Adv. Mater.* 2016, 28, 6247–6252.
26. Novoselov, K. S., Geim, A. K., Morozov, S. V., Jiang, D., Zhang, Y., Dubonos, S. V., Grigorieva, I. V., Firsov, A. A., *Science*, 2004, 306, 666–669.
27. Chen, J., Li, Y., Huang, L., Li, C., Shi, G., *Carbon*, 2015, 81, 826–834.
28. Marcano, D. C., Kosynkin, D. V., Berlin, J. M., Sinitskii. A., Sun, Z., Slesarev, A., Alemany, L. B., Lu, W., Tour, J. M., *ACS Nano*, 2010, 4, 4806–4814.

29. Jankovský, O., Jiříčková, A., Luxa, J., Sedmidubský, D., Pumera, M., Sofer, Z., *ChemistrySelect*, 2017, 2, 9000–9006.
30. Lee, D. W., Santos, V. L., Seo, J. W., Felix. L. L., Bustamante. D. A., Cole. J. M., Barnes. C. H. W., *J. Phys. Chem. B* 2010, 114, 17, 5723–5728.
31. McAllister, J. M., Li, J. L., Adamson, D. H., Schniepp, H. C., Abdala, A. A., Liu, J., Alonso, M. H., Milius, D. L., Car, R., Prud'homme, R. K., Aksay, I. A., *Chem Mater.*, 2007, 19, 4396.
32. Dideikin. A. T., Vul. A. Y., *Frontiers in Physics*, 2019, 6, 1–13.
33. Eswaraiah, V., Jyothirmayee Aravind, S. S., Ramaprabhu. S., *J. Mater. Chem.*, 2011, 21, 6800–6803.
34. Hofmann, U., Frenzel, A., *Kolloid-Z*, 68, 1934, 149–151.
35. Pei, S., Cheng H-M, *Carbon*, 2012, 50, 3210–3228.
36. Stankovich, S., Dikin, D. A., Piner, R. D., Kohlhaas, K. A., Kleinhammes, A., Jia, Y., et al., *Carbon*, 2007, 45, 1558–1565.
37. De Silva, K. K. H., *Carbon*, 119, 2017.
38. Huang H. H. et al., *Sci. Rep.*, 2018, 8, 6849.
39. Hwang, H. S., *Micromachines*, 2020, 11, 814.
40. Kuila, T., Bose, S., Mishra, A., Khanra, P., Kim, N., Lee, J., *Prog. Mater. Sci.* 2012, 57, 1061–1105.
41. Kang, X., Wang, J., Wu, H., Liu, J., Aksay, I., Lin, Y., *Talanta*, 2010, 81, 754–759.
42. Shao, Y., Wang, J., Wu, H., Liu, J., Aksay, I., Lin, Y., *Electroanalysis*, 2010, 22, 1027–1036.
43. Chen, L., Tang, Y., Ke, W., Liu, C., Luo, S., *Electrochem. Commun.*, 2011, 13, 133–137.
44. Wang, Y., Li, Y., Tang, L., Lu, J., Li, J., *Electrochem. Commun.*, 2009, 11, 889–892.
45. Wang, X., Chen, S. Graphene-Based Nanocomposites. In: Mikhailov, S., editor. *Physics and Applications of Graphene – Experiments*. Croatia: InTech, 2011, 135–168.
46. Xie. Q. Z. et al., *ECS J. Solid State Sci. Technol.*, 2020, 9. 115027.
47. Grieshaber D., MacKenzie R., Vörös J., Reimhult E., *Sensors*, 8, 2008, 1400.
48. Brosel-Oliu, S., Abramova, N., Uria, N., Bratov, A., *Anal. Chim. Acta*, 2019, 1088, 1.
49. Syu, Y.-C., Hsu, W.-E., Lin, C.-T., *ECS J. Solid State Sci. Technol.*, 7, 2018, Q3196.
50. Clark, L. C., Lyons, C. *Ann. N. Y. Acad. Sci.*, 1962, 102, 29–45.
51. Newman, J. D., Turner, A. P. F., *Biosens. Bioelectron.*, 2005, 20, 2435–2453.
52. Juska, V. B., Pemble, M. E., *Sensors*, 2020, 20, 6013.
53. Cass, A. E. G., Davis, G., Francis, G. D., Hill, H. A. O., Aston, W. J., Higgins, I. J., Plotkin, E. V., Scott, L. D. L., Turner, A. P. F., *Anal. Chem.* 1984, 56, 667–671.
54. Riklin, A., Katz, E., Wiliner, I., Tocker, A., Bückmann, A. F., *Angew. Chem. Int. Ed. Engl.* 1990, 29, 82–89.
55. Degani, Y., Heller, A., *J. Phys. Chem.* 1987, 91, 1285–1289.
56. Willner, B., Katz, E., Willner, I., *Curr. Opin. Biotechnol.* 2006, 17, 589–596.
57. Katz, E., Willner, I., *Angew. Chem. Int. Ed.* 2004, 43, 6042–6108.
58. Katz, E., Willner, I. *Chem Phys Chem* 2004, 5, 1084–1104.
59. Xiao, Y., Patolsky, F., Katz, E., Hainfeld, J. F., Willner, I. *Science* 2003, 299, 1877–1881.

60. Patolsky, F., Weizmann, Y., Willner, I. *Angew. Chem. Int. Ed.* 2004, 43, 2113–2117.
61. Liu, J. Q., Chou, A., Rahmat, W., Paddon-Row, M. N., Gooding, J. J. *Electroanalysis* 2005, 17, 38–46.
62. Chen, R. J., Bangsaruntip, S., Drouvalakis, K. A., Wong Shi Kam, N., Shim, M., Li, Y., Kim, W., Utz, P. J., Dai, H., *Proc. Natl. Acad. Sci.*, 2003, 100, 4984–4989.
63. Vardharajula, S., Ali, S. Z., Tiwari, P. M., et al. *Int. J. Nanomed.* 2012, 7, 5361–5374.
64. Bianco, A., Kostarelos, K., Partidos, C. D., et al. *Chem. Commun. (Camb.).* 2005, 5, 571–577.
65. Kam, N. W., Liu, Z., Dai, H., *J. Am. Chem. Soc.* 2005, 127(36), 12492–12493.
66. Colvin, V. L., *Nat. Biotechnol.* 2003, 21(10), 1166–1170.
67. Gupta, V., Saleh, T. A., edited by Stefano Bianco. London: IntechOpen, 2011. doi:10.5772/18009
68. Nepal, D., Geckeler, K. E., *Small*, 2006, 2, 406–412.
69. Matsuura, K., Saito, T., Okazaki, T., Ohshima, S., Yumura, M., Iijima, S., *Chem. Phys. Lett.*, 2006, 429, 497–502.
70. Cai, C., Chen, J. *Anal Biochem.* 2004, 332(1), 75–83.
71. Hrapovic, S., Liu, Y., Male K. B., et al. *Anal Chem.* 2004, 76(4), 1083–1088.
72. Wu, X., Zhao, B., Wu, P., et al. *J. Phys. Chem. B.* 2009, 113(40), 13365–13373.
73. Rakhi, R. B., Sethupathi, K., Ramaprabhu, S., *J. Phys. Chem. B*, 2009, 113, 10, 3190–3194.
74. Pourasl et al., *Nanoscale Res. Lett.*, 2014, 9, 33.
75. Claussen, J. C., Hengenius, J. B., Wickner, M. M., Fisher, T. S., Umulis, D. M., Marshall Porterfield, D., *J. Phys. Chem. C* 2011, 115, 43, 20896–20904.
76. Suvarnaphaet, P., Pechprasarn, S., *Sensors*, 2017, 17, 2161.
77. Alwarappan, S., Erdem, A., Liu, C., Li, C. Z., *J. Phys. Chem. C*, 2009, 113, 8853.
78. Segal, M., *Nat. Nanotechnol.* 2009, 4, 611.
79. Banks, C. E., Crossley, A., Salter, C., Wilkins, S. J., Compton, R. G., *Angew. Chem.-Int. Edit.*, 2006, 45, 2533.
80. Shao, Y., Wang, J., Wu, H., Liu, J., Aksay, I. A., Lin, Y., *Electroanalysis*, 2010, 22(10), 1027–1036.
81. Shan, C. S., Yang, H. F., Song, J. F., Han, D. X., Ivaska, A., Niu, L., *Anal. Chem.* 2009, 81, 2378.
82. Zhou, M., Zhai, Y. M., Dong, S. J., *Anal. Chem.* 2009, 81, 5603.
83. Kang, X. H., Wang, J., Wu, H., Aksay, A. I., Liu, J., Lin, Y. H., *Biosens. Bioelectron.* 2009, 25, 901.
84. Wu, H., Wang, J., Kang, X., Wang, C., Wang, D., Liu, J., et al. *Talanta* 2009, 80(1), 403–406.
85. Lu, L. M., Li, H. B., Qu, F., Zhang, X. B., Shen, G. L., *Yu RQBiosens Bioelectron.* 2011, 26(8), 3500–3504.
86. Lu, W., Luo, Y., Chang, G., Sun, X. *Biosens Bioelectron.* 2011, 26(12), 4791–4797.
87. Wang, X., Zhang, X., *Electrochimica Acta*, December 2013, 112(1), 774–782.
88. Baby, T. T., Jyothirmayee Aravind, S. S., Arockiadoss, T., Rakhi, R. B., Ramaprabhu, S., *Sens. Actuators, B*, 4 March 2010, 145(1), 71–77.
89. Monosik, R., Stredansky, M., Luspai, K., Magdolen, P., Sturdik, E. *Enzym. Microb. Technol.* 2012, 50(4–5), 227–232.

90. Zhao, Y., Chu, J., Li, S-H, Li, W-W, Liu, G., Tian, Y-C, Yu, H-Q, *Electroanalysis*, 2014, 26(3), 656–663.
91. Wang, L., Zhu, W., Lu, W., Qin, X., Xu, X., *Biosens Bioelectron.*, 2018, 111, 41–46.
92. Heba et al., *Sci. Rep*, 2019, 9, 5524.
93. Lu, L. M., Zhang, L., Qu, F. L., Lu, H. X., Zhang, X. B., Wu, Z. S., Huan, S. Y., Wang, Q. A., Shen, G. L., Yu, R. Q. *Biosens Bioelectron.*, 2009, 25(1), 218–223.
94. Luo, P., Zhang, F., Baldwin, R. P. *Anal. Chim. Acta*, 1991, 244(2), 169–178.
95. Dhara, K., Mahapatra, D. R., *Microchim. Acta*, 2017, 185(1), 49.
96. Wang, H., Liu, R., Li, Y., Lü, X., Wang, Q., Zhao, S., Yuan, K., Cui, Z., Li, X., Xin, S., Zhang, R., Lei, M., Lin, Z. *Joule*, 2018, 2(2), 337–348.
97. He, X., Luan, S. Z., Wang, L., Wang, R. Y., Du, P., Xu, Y. Y., Yang, H. J., Wang, Y. G., Huang, K., Lei, M. *Mater. Lett.* 2019, 244, 78–82.
98. Wang, C-H, Song, Y-Y, Zhao, J-W, Xia, X-H, *Surf. Sci.*, 2006, 6(4), 38–42.
99. He, C., Wang, J., Gao, N., He, H., Zou, K., Ma, M., Zhou, Y., Cai, Z., Chang, G., He, Y., *Microchimica Acta*, 2019, 186, 722.
100. Dong, X-C, Xu, H., Wang, X-W, Huang, Y-X, Chan-Park, M. B., Zhang, H., Wang, L-H, Huang, W., Chen, P., *ACS Nano*, 2012, 6(4), 3206–3213.
101. Mohammadi, F., Vesali-Naseh, M., Khodadadi, A. A., Mortazavi, Y., 2018, 30(9), 2044–2052.
102. Ikonen, E., *Nat. Rev. Mol. Cell Biol.*, 2008, 9, 125–138.
103. Nauck, M., Marz, W., Wieland, H., *Clin. Chem.*, 2000, 46, 436–437.
104. Stapleton, P. A., Goodwill, A. G., James, M. E., Brock, R. W., Frisbee, J. C., *J. Inflamm.*, 2010, 7, 54.
105. Program, N. C. E., *Arch. Intern. Med.*, 1988, 148, 36–69.
106. Bhandaru, R. R., Srinivasan, S. R., Pargaonkar, P. S., Berenson, G. S., *Lipids*, 1977, 12, 1078–1080.
107. Mizuno, K., Toyosato, M., Yabumoto, S., Tanimizu, I., Hirakawa, H., *Anal. Biochem.*, 1980, 108, 6–10.
108. Manasterski, A., Zak, B., *Microchem. J.*, 1973, 18, 18–28.
109. Obermer, E., Milton, R., *Biochem. J.*, 1933, 27, 345–350.
110. D'Orazio, P., *Clin. Chim. Acta*, 2003, 334, 41–69.
111. Arya, S. K., Solanki, P. R., Singh, R. P., Pandey, M. K., Datta, M., Malhotra, B. D., 2006. *Talanta*, 69, 918–926.
112. Kim, K.-E., Kim, T., Sung, Y.-M., *J. Nanopart. Res.*, 2012, 14, 1–9.
113. Marazuela, M. D., Cuesta, B., Moreno-Bondi, M. C., Quejido, A., *Biosens. Bioelectron.*, 1997, 12, 233–240.
114. Brahim, S., Narinesingh, D., Guiseppi-Elie, A., *Biosens. Bioelectron.* 2002, 17, 973–981
115. Maidan, R., Heller, A., *Anal. Chem.* 1992, 64, 2889–2896.
116. Willner, I., Katz, E., *Angew. Chem. Int. Ed. Engl.* 2000, 39, 1180–1218.
117. Guo, S., Dong, S., *Trends. Anal. Chem.* 2009, 28, 96–109.
118. Niemeyer, C. M., *Angew. Chem. Int. Ed. Engl.* 2003, 42, 5796–5800.
119. Vidal, J.-C., García-Ruiz, E., Castillo, J.-R. *Electroanalysis*, 2001, 13, 229–235.
120. Yang, M., Yang, Y., Liu, Y., Shen, G., Yu, R., *Biosens. Bioelectron.*, 2006, 21, 1125–1131.
121. Zhou, N., Wang, J., Chen, T., Yu, Z., Li, G., *Anal.Chem.*, 2006, 78, 5227–5230.
122. Li, J., Peng, T., Peng, Y. *Electroanalysis*, 2003, 15, 1031–1037.

123. Hrapovic, S., Majid, E., Liu, Y., Male, K., Luong, J. H. T. *Anal. Chem.*, 2006, 78, 5504–5512.
124. Kudin, K. N., Ozbas, B., Schniepp, H. C., Prud'homme, R. K., Aksay, I. A., Car, R., *Nano Lett.*, 2007, 8, 36–41.
125. Jyothirmayee Aravind, S. S., Baby, T. T., Arockiadoss Rakhi, R. B., Ramaprabhu, S., *Thin Solid Films*, 2011, 519(16), 5667–5672.
126. Jankovic, J., *J. Neurol., Neurosurg. Psychiatry*, 2008, 79(4), 368.
127. Hornykiewicz, O., *Parkinson's Disease and Related Disorders*, Springer, 2006, 9.
128. Björklund, A., Dunnett, S. B., *Trends Neurosci.*, 2007, 30(5), 194.
129. Kapur, S., Mamo, D. *Prog. Neuropsychopharmacol. Biol. Psychiatry*, 2003, 27, 1081–1090.
130. Lang, A. E., Lozano, A. M. Parkinson's disease—first of two parts. *N. Engl. J. Med.*, 1998, 339, 1044–1053.
131. Volkow, N. D., Fowler, J. S., Wang, G. J., *Behav. Pharmacol.*, 2002, 13, 355–366.
132. Le Moal, M., Simon, H. *Physiol Rev.*, 1991, 71, 155–234.
133. Spanagel, R., Weiss, F. *Trends Neurosci.*, 1999, 22, 521–527.
134. Ungless, M. A., *Trends Neurosci.*, 2004, 27, 702–706.
135. Njagi, J., Chernov, M. M., Leiter, J., Andreescu, S., *Anal. Chem.*, 2010, 82(3), 989.
136. Fang, B. et al. *Electroanalysis*, 2005, 17, 744–748.
137. Sancy, M., Silva, J. F., Pavez, J., Zagal, J. H. *J. Chil. Chem. Soc.*, 2013, 58, 4.
138. Jackowsk, K., Krysinski, P. *Anal. Bioanal. Chem.*, 2013, 405, 3753–3771.
139. Kawagoe, K. T., Wightman, R. M., *Talanta*, 1994, 41, 865–874.
140. Raj, C. R., Okajima, T., Ohsaka, T. *J. Electroanal Chem.*, 2003, 543, 127–133.
141. Raj, C. R., Tokuda, K., Ohsaka, T. *Bioelectrochemistry* 2001, 53, 183–191.
142. Ciszewski, A., Milczarek, G. *Anal Chem.* 1999, 71, 1055–1061.
143. Alwarappan, S., Butcher, K. S. A., Wong, D. K. Y., *Sens Actuators B, Chem*, 2007, 128, 299–305.
144. Ambrosi, A., Chua, C. K., Bonanni, A., Pumera, M., *Chem. Rev.*, 2014, 114(14), 7150.
145. Stankovich, S., Dikin, D. A., Dommett, G. H., Kohlhaas, K. M., Zimney, E. J., Stach, E. A., et al. *Nature*, 2006, 442(7100), 282.
146. Balasubramanian, K., Burghard, M., *Anal. Bioanal. Chem.*, 2006, 385(3), 452.
147. Mphuthi, N. G., Adekunle, A. S., Fayemi, O. E., Olasunkanmi, L. O., Ebenso, E. E., *Sci. Rep.*, 2017, 7, 43181.
148. Dong, L., *Nanotechnology*, 2009, 20, 465602.
149. Jyothirmayee Aravind, S. S., Ramaprabhu, S., *Sens. Actuators, B*, 2011, 155, 679–686.
150. Wang, W., Wang, W., Davis, J. J., Luo, X., *Microchim. Acta.*, 2015, 182, 1123.
151. Liu, S., Xing, X. R., Yu, J. H., Lian, W. J., Li, J., Cui, M., Huang, J. D., *Biosens. Bioelectron.* 2012, 36, 186.
152. Li, B. R., Hsieh, Y. J., Chen, Y. X., Chung, Y. T., Pan, C. Y., Chen, Y. T., *J. Am. Chem. Soc.*, 2013, 135, 16034.
153. Azadbakht, A., Roushani, M., Abbasi, A. R., Derikvand, Z., *Anal. Biochem.*, 2016, 507, 47.
154. Bahrami, S., Abbasi, A. R.: Roushani, M., Derikvand, Z., Azad-Bakht, A., *Talanta*, 2016, 159, 307.

155. Jin, H., Zhao, C. Q., Gui, R. J., Gao, X. H., Wang, Z. H., *Anal. Chim. Acta.*, 2018, 1025, 154.
156. Azadbakht, A., Roushani, M., Abbasi, A. R., Menati, S., Derik-Vand, Z., *Mat. Sci. Eng. C-Mater.*, 2016, 68, 585.
157. Xu, Y. Q., Hun, X., Liu, F., Wen, X. L., Luo, X. L., *Microchim. Acta.*, 2015, 182, 1797.
158. Arumugasamy, S. K., Govindaraju, S., Yun, K., *Appl. Surf. Sci.*, 2020, 508, 145294.
159. Panchakarla, L. S., Subrahmanyam, K. S., Saha, S. K., Govindaraj, A., Krishnamurthy, H. R., Waghmare, U. V., Rao, C. N. R., *Adv. Mater.*, 2009, 21, 4726–4730.
160. Cantatore, V., Panas, I., *Carbon.*, 2016, 104, 40–46.
161. Cantatore, V., Pandit, S., Mokkapati, V. R. S. S., Schindler, S., Eigler, S., Mijakovic, I., Panas. I., *Carbon*, 2018, 137, 343–348.
162. Akhmadishina, K. F., Bobrnetskii, I. I., Komarov, I. A., Malovichko, A. M., Nevolin, V. K., Petukhov, V. A., Golovin, A. V., Zalevskii, A. O., *Nanotechnol. Russ.* 2013, 8(11–12), 721–726.
163. Wan, Q., Xu, Y., Xiao, H., *AIP Adv.*, 2018, 8, 105128.
164. Mohammadi-Manesh, E., Mir-Mahdevar, M. *Synth. Met.*, 2020, 267, 116486.
165. Prasad, B. B., Singh, R., Kumar, A., *Biosens. Bioelectron.*, 2017, 94, 115–123.
166. Ahmadi, R., Jalali Sarvestani, M. R., Sadeghi, B., *Int. J. Nano Dimens.*, 2018, 9(4), 325–335.
167. Parlak, C., Alver, Ö., *Chem. Phys. Lett.*, 2017, 678, 85–90.
168. Bashiri, S., Vessally, E., Bekhradnia, A., Hosseinian, A., Edjlali, L., *Vacuum*, 2017, 136, 156–162.
169. Moradi, M., Nouraliei, M., Moradi, R., *Physica E Low Dimens. Syst. Nanostruct.*, 2017, 87, 186–191.
170. Sutradhar, S., Patnaik, A., *Sens. Actuators, B*, 2017, 241, 681–689.
171. Saito, R., Fujita, M., Dresselhaus, G., Dresselhaus, M. S., *Appl. Phys. Lett.*, 1992, 60, 2204–2206.
172. Berber, S., Kwon, Y-K Tomanek D Unusually High Thermal Conductivity of Carbon Nanotubes. *Phys. Rev. Lett* 2000, 84, 4613–4616. [PubMed: 10990753].
173. Nardelli, M., Yakobson, B., Bernholc J Brittle and Ductile Behavior in Carbon Nanotubes. *Phys. Rev. Lett.*, 1998, 81, 4656–4659.
174. Peng, S., Cho, K. Chemical Control of Nanotube Electronics. *Nanotechnology*, 2000, 11, 57–60.
175. Santucci, S., Picozzi, S., Di Gregorio, F., Lozzi, L., Cantalini, C., Valentini, L., Kenny, J. M., Delley B., *J. Chem. Phys.*, 2003, 119, 10904–10910.
176. Zhu, R., Desroches, M., Yoon, B., Swager, T. M., *ACS Sens.*, 2017, 2, 1044–1050.
177. Esser, B., Schnorr, J. M., Swager, T. M., *Angew. Chemie – Int. Ed.*, 2012, 51, 5752–5756.
178. Liu, S. F., Petty, A. R., Sazama, G. T., Swager, T. M., *Angew. Chemie Int. Ed.*, 2015, 54, 6554–6557.
179. Chen, H., Zuo, X., Su, S., Tang, Z., Wu, A., Song, S., Zhang, D., Fan, C., *Analyst*, 2008, 133, 1182–1186.
180. Li, Y., Hodak, M., Lu, W., Bernholc, J., *Nanoscale*, 2017, 9, 1687–1698.

181. Leghrib, R., Llobet, E., Pavelko, R., Vasiliev, A. A., Felten, A., Pireaux, J. J., *Procedia Chem.* 2009, 1, 168–171.
182. Son, M., Cho, D. G., Lim, J. H., Park, J., Hong, S., Ko, H. J., Park, T. H., *Biosens. Bioelectron.*, 2015, 74, 199–206.
183. Zhao, J., Buldum, A, Han, J., Lu, J. P., *Nanotechnology.*, 2002, 13, 195–200.

Chapter 6

Transition metal dichalcogenide based surface plasmon resonance for bio-sensing

Sajal Agarwal
Rajiv Gandhi Institute of Petroleum Technology, Jais, India

Yogendra Kumar Prajapati
Motilal Nehru National Institute of Information Technology Allahabad, Prayagraj, India

CONTENTS

6.1 INTRODUCTION

The surface plasmon resonance (SPR) phenomenon was first theoretically connected with the excitation of an electromagnetic (EM) surface plasmon wave (SPW) by Fano [1] in 1941. Then, in 1968, Otto experimentally demonstrated that the attenuated total reflection method (ATR) produces a drop in reflectivity, due to the excitation of the surface plasmon. In the same year, Kretschmann and Raether [2] reported the same phenomenon using a different configuration. These pioneer works associated the surface plasmon (SP) into modern optics for various applications. In 1970, SP was used for the characterization of the thin film [3] and various other purposes.

DOI: 10.1201/9781003288633-6

In 1983, Liedberg et al. [4] first demonstrated the use of SPW for sensing applications, because of the simplicity of this method. In the last four decades, remarkable progress has been made in SPR sensors, based on their structure, materials, and applications. In this chapter, the authors intend to discuss the bio-sensing application of the SPR sensor, utilizing advanced two-dimensional (2-D) materials. But first, the basics of the SPR sensor are discussed to understand the sensing procedure and the effect of material on the sensing. Then, the properties of 2-D materials are discussed to understand their use in SPR sensing, especially bio-sensing. A short review of the various methods of bio-sensing is also given to understand the advantages of optical SPR sensing over other methods. Concluding remarks focus on pointing out the advantages and disadvantages of 2-D material-based SPR sensors for bio-sensing.

6.1.1 Basics of the surface plasmon resonance sensor

Two different configurations were proposed initially by the researchers, one by Otto [5] and the other by Kretschmann [2]. However, a realization of the Otto configuration is much more difficult than that of Kretschmann, since maintaining an air gap between the coupling element and material layers is a bit difficult, as can be seen from Figure 6.1.

Thus, the Kretschmann configuration is the most popular configuration and will be discussed in this chapter. To understand the basics of surface plasmon generation, it is important to study the electromagnetic theory of optical waveguides. In this section, the authors discuss the mathematics behind surface plasmon generation and the excitation of surface plasmon using prism coupling. To explain the generation of surface plasmon, the most basic configuration is used, which has metal-dielectric material layers. To begin, semi-infinite metal and dielectric layers are considered to have complex dielectric constants, $\varepsilon_m = \varepsilon_m' + i\varepsilon_m''$ and $\varepsilon_d = \varepsilon_d' + i\varepsilon_d''$ respectively. In the permittivity equation, ε_m' and ε_d' are the real part of permittivity and ε_m''

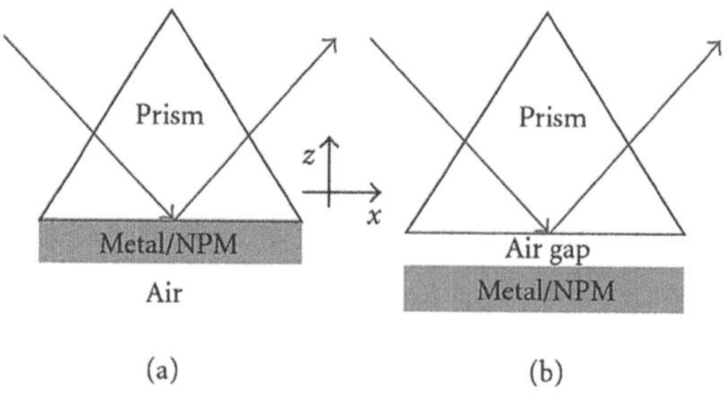

Figure 6.1 Surface Plasmon Resonance generation configuration (a) Kretschmann's (b) Otto's [6].

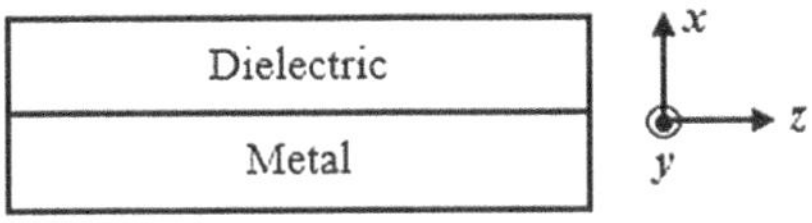

Figure 6.2 Metal dielectric waveguide configuration.

and ε_d'' are the imaginary part of permittivity of metal and dielectric respectively, as shown in Figure 6.2.

The propagation constant of the modes, which are guided and propagated along the structure, can be given for $d = 0$ for transverse electric (TE) and transverse magnetic (TM) polarized light as;

$$\gamma_m = -\gamma_d \left(\text{for TE modes}\right) \tag{6.1}$$

$$\frac{\gamma_m}{\varepsilon_m} = -\frac{\gamma_m}{\varepsilon_d}\left(\text{for TM mode}\right) \tag{6.2}$$

where $\gamma_i^2 = \beta^2 - \omega^2\mu_0\varepsilon_0\varepsilon_i$ and i is m or d. TE modes yield no solution and, thus, represent a bounded mode, whereas, TM mode eigen values can be reduced to [7];

$$\beta = \frac{\omega}{c}\sqrt{\frac{\varepsilon_d\varepsilon_m}{\varepsilon_d + \varepsilon_m}} \tag{6.3}$$

where c is the speed of light in a vacuum. For lossless material ε_m'' and ε_d'' both should be equal to zero. Equations 6.2 and 6.3 represent the guided mode provided $\varepsilon_m' < -\varepsilon_d'$, guided modes are known as Fano modes. Thus, for the Fano mode to exist it is required to have a metal of negative real permittivity. Thus, according to the free-electron theory for metals,

$$\varepsilon_m = \varepsilon_0\left(1 - \frac{\omega_p^2}{\omega^2 + i\omega\vartheta}\right) \tag{6.4}$$

where υ is collision frequency and ω_p is plasma frequency, and can be given as,

$$\omega_p = \sqrt{\frac{Ne^2}{\varepsilon_0 m_e}} \tag{6.5}$$

where N is the concentration of free electrons; e and m_e are the electron charge and mass respectively. Novel metals such as gold (Au), silver (Ag), aluminum (Al), copper (Cu), etc., exhibit the required properties and are thus used for the surface plasmon generation. On studying the generation of plasmon keenly, it is evident that plasmons are confined at the interface at the wavelengths longer than the critical wavelength, and this depends on the plasma frequency. Based on the change in the structure, the electromagnetic response will also change, and the detailed study of various interfaces can be seen through [7]. Apart from the particular material combination, coupling of light with the structure is also very important, without proper coupling, generation, and propagation of the SP and surface plasmon wave (SPW) respectively, is not possible. There are three methods available for the coupling of the light with the interface:

1. Prism coupling
2. Grating coupling
3. Waveguide coupling

Prism coupling is the most common approach for the SP generation, for this approach, where a prism with a high refractive index is used together with the metal-dielectric interface with a fixed thickness of layers, in Kretschmann's configuration.

After traveling through a prism, the light wave interacts with metal film, part of the light is reflected back, and the rest of the light propagates in the metal. This wave decays exponentially in the direction perpendicular to the interface and is thus known as an evanescent wave. For the metal thickness of less than 100 nm, the evanescent wave penetrates the metal layer and interacts with the SP at the interface of the metal dielectric. The propagation constant of SP is influenced by the dielectric and the effective refractive index of SP, and an evanescent wave depends on the interface properties. To investigate the interaction of the light and SP in the ATR method, a multilayer model can be used. The Fresnel multilayer model is proved to be the most used and simple method for ATR investigation. The amplitude of the reflection coefficient can be calculated to find the total power reflection from the structure [7].

$$R = \left| r_{pmd} \right|^2 \tag{6.6}$$

6.1.2 Interrogation approaches

While investigating the reflection from the interface, different interrogation approaches can be used based on their complexity and available equipment. The approaches are namely [8]:

1. Angular interrogation
2. Wavelength interrogation
3. Phase interrogation

In the angular interrogation, reflection is measured in regard to the incident angle of light on the prism, and the angle where minimum reflection is received is known as the resonance angle. Resonance angle is a very important parameter in SPR sensing since the difference in the resonance angle is used to calculate the sensitivity of the sensor, whereas, wavelength interrogation uses a broad wavelength source, and the reflection is measured with respect to the wavelength. The foremost drawback of wavelength interrogation is the high cost, as the broadband laser source is much more expensive than a goniometer used for angle interrogation.

However, from all the above approaches, phase interrogation is the most accurate method and uses the phase information of the reflected wave for the sensing purpose. Phase interrogation is the most complex method, as the extraction of phase information from the wave is a much more complex process than either wavelength or angle measurement.

From the above discussion, it is clear that angle interrogation is the best-optimized interrogation approach, since the main advantage of this method is that only a single wavelength source is sufficient for the sensing rather than the costly broadband wavelength source. Along with this advantage, the sensitivity of the angle interrogation method is much more than the wavelength interrogation method. However, phase interrogation offers higher sensitivity than angle interrogation. However, the dynamic detection range of this method is much narrower than the angle interrogation, which means that the refractive index can easily exceed the limit and not be detectable, which indirectly shows that phase interrogation cannot be used with a wide range of refractive indices [9]. Thus, angle interrogation is the most used method in optical SPR sensors.

6.1.3 Surface plasmon resonance sensor performance parameter

To characterize the SPR sensor, the reflection spectrum plays a very important role. Based on the parameters extracted from the reflection spectrum; sensitivity (S), detection accuracy (DA), full width at half maximum (FWHM), the figure of merit (FOM), etc., can be defined.

Sensitivity (S) is the ratio of the change in the sensing parameter (angle, wavelength, etc., based on the interrogation method) with respect to the change in the refractive index of the analyte. Since the study in this chapter is concentrated on the angle interrogation method, thus, sensitivity can be given as,

$$S = \frac{\delta\theta_{res}}{\delta n}\left({}^{\circ}/RIU\right) \tag{6.7}$$

where the numerator of Equation 6.7 shows the shift in the resonance angle and the denominator shows the change in the refractive index of the analyte. Detection accuracy (DA) is defined as the degree to which sensor output represents the true value of analyte concentration. It is mathematically denoted as;

$$DA = \frac{1}{FWHM}(/^{\circ}) \tag{6.8}$$

The figure of merit (FOM) is directly related to the resolution and can be calculated as:

$$FOM = \frac{S}{FWHM}(/RIU) \tag{6.9}$$

It is desired to have a high value of FOM, which means high resolution. For a reliable sensing procedure, sensitivity, detection accuracy, and FOM, all performance parameters should have high values, and FWHM should be as low as possible, which is practically not possible since there is a trade-off between sensitivity and detection accuracy. To optimize the sensor performance, a delicate balance has to be maintained between these parameters. In most of the studies, the above parameters are used for the characterization of the SPR sensors. In the next section, a brief introduction of various 2-D materials is given.

6.2 TWO-DIMENSIONAL MATERIALS

Since this chapter is dedicated to the application of the 2-D material in SPR sensors, especially Transition Metal Dichalcogenide (TMD) materials, a brief study is done here for all the most used 2-D materials. Since, the discovery of the 2-D materials, a large number of research groups has been attracted to the application of these materials in various fields, such as optoelectronics, electronics, etc. Sensors are an integrated part of our lives today since these are employed in most devices to collect and analyze data. Optical sensors are the most used sensors in various devices and work on the light-matter interaction [10]. This interaction can be modified/improved by utilizing the 2-D layer materials. 2-D layer materials demonstrate tunable light-matter interaction, based on the thickness and fabrication [11]. Graphene is the most used and explored 2-D layer material in the optical sensing area since its discovery in 2004 [12–14].

There are a number of different 2-D materials that have been proposed and applied in optical sensing applications up to now. Figure 6.3, shows the most used 2-D materials in the optical sensor.

Figure 6.3, depicts that 2-D materials include various materials from different families and with different properties. It can be seen that graphene and TMDs are not the only materials that lie in the 2-D material category. Many other nanomaterials are in this category, but the use of those materials is limited in optical sensors due to their stability and compatibility issues.

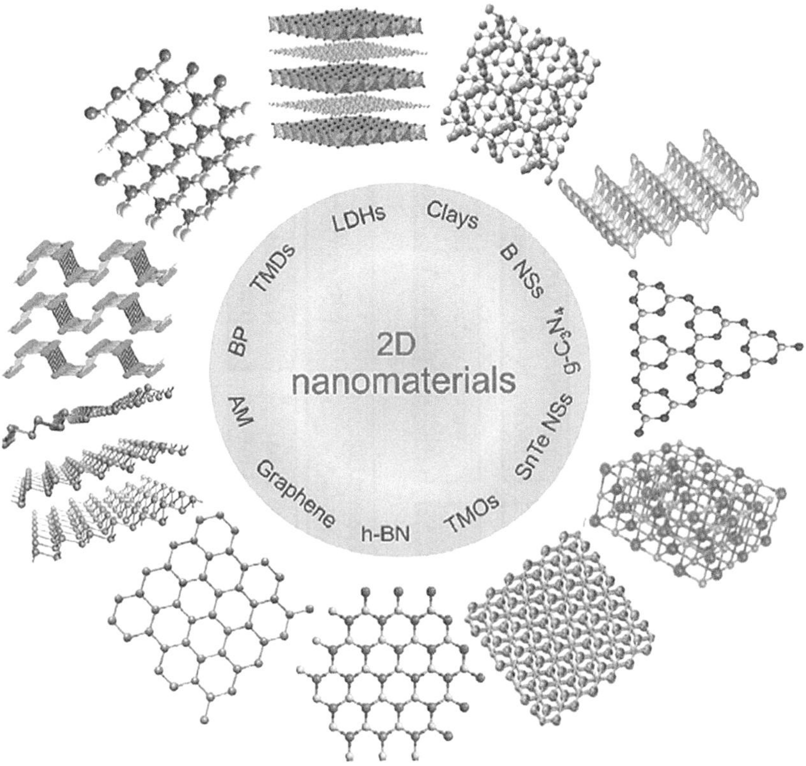

Figure 6.3 Illustration of 2-D materials including different classes [15].

Understanding the categories and their stability in the air can be perceived in Figure 6.4.

From Figure 6.4 it is observed that graphene, its compound, some 2-D oxides, and most 2-D chalcogenides are stable in the ambient temperature and environment. Also, some of the 2-D chalcogenides and oxides have the potential to be stable in ambient conditions, whereas some of the materials are not stable at all in ambient temperature. Thus, the selection of required 2-D layered material in the optical sensor is important and can be done based on different parameters, such as operating wavelength, sensing the analyte, number of layers, etc. Most used dichalcogenide materials in optical sensors are: molybdenum disulfide (MoS_2), molybdenum Di selenide ($MoSe_2$), tungsten disulfide (WS_2), tungsten Di selenide (WSe_2), MXene, and black phosphorus, etc. 2-D layered materials demonstrate that unique optical and electrical properties evolve from quantum confinement to the surface properties. The most important property that is utilized in optical sensors is their high surface-to-volume ratio, which ensures the better interaction of the analyte with the sensor surface, and in turn, better sensitivity. Tunability of the bandgap is another parameter that is absent in conventional bulk materials

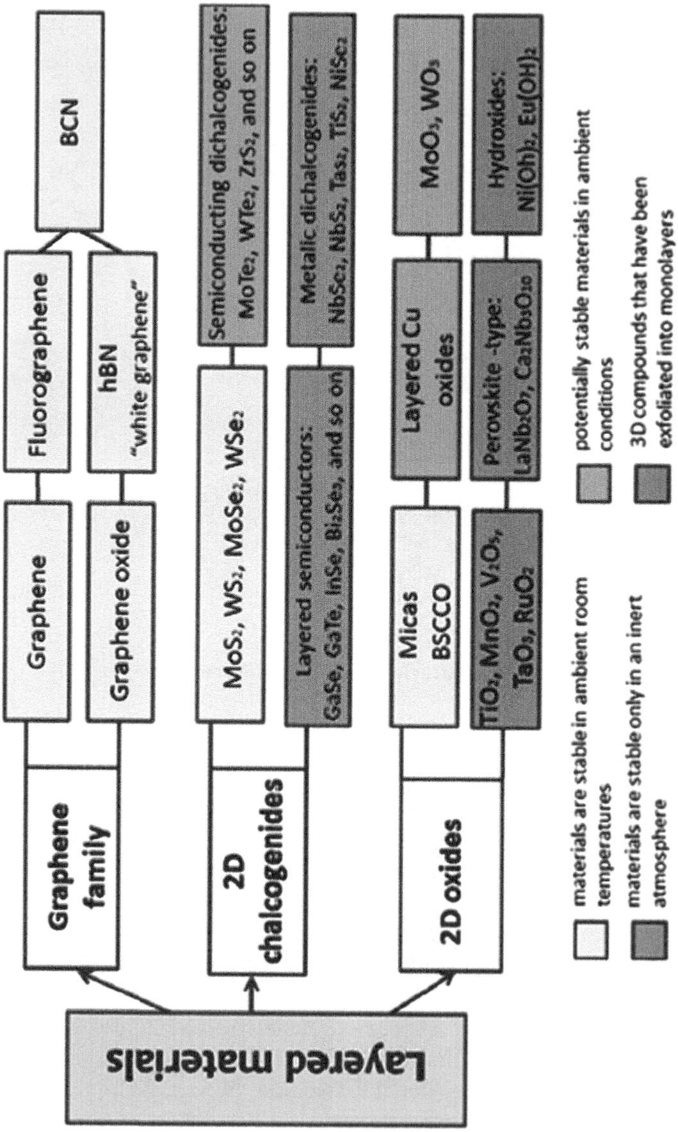

Figure 6.4 Categories of 2-D layer material by their stability in the air for different applications [16].

and strong photoluminescence. Physiochemical properties of 2-D layered materials are greatly dependent on the atomic arrangement; a few atoms' thicknesses in the vertical dimension make them extraordinary compared to the bulk materials. Graphene, an allotrope of carbon, has a hexagonal structure and has very different surface chemical properties. Due to its extraordinary surface and chemical properties, graphene is extensively used for optical sensing applications [17, 18]. TMDs are the second most used 2-D layered material category; these are three-layer atomic structures with a transition metal layer sandwiched between two chalcogen layers. The most significant property of TMD that is utilized in the optical sensors is its high surface-to-volume ratio and bandgap tunability. Properties of different TMDs vary based on the constituent metal and chalcogenide atom. In recent years, TMDs gain considerable attention due to their ultrathin feature and distinctive chemical and biological properties. Figure 6.5 shows the year-wise data of the article published for different materials.

It is evident from Figure 6.5 that, after the proposal of different TMDs, their implementation in optoelectronics applications is not limited to some materials. The use of these materials in the optical sensor is much needed due to the undeniable requirement of superior performance, small area, durability, etc. TMDs have unique electronic, photonic, magnetic, and other tunable properties different from their bulk counterpart. Diverse TMDs are also used by stacking with other 2-D materials to generate novel structures and properties, which are known as heterostructures. Heterostructures are more unique than the individual TMDs since these structures can provide entirely different properties than their constituent TMD layers. Heterostructures are

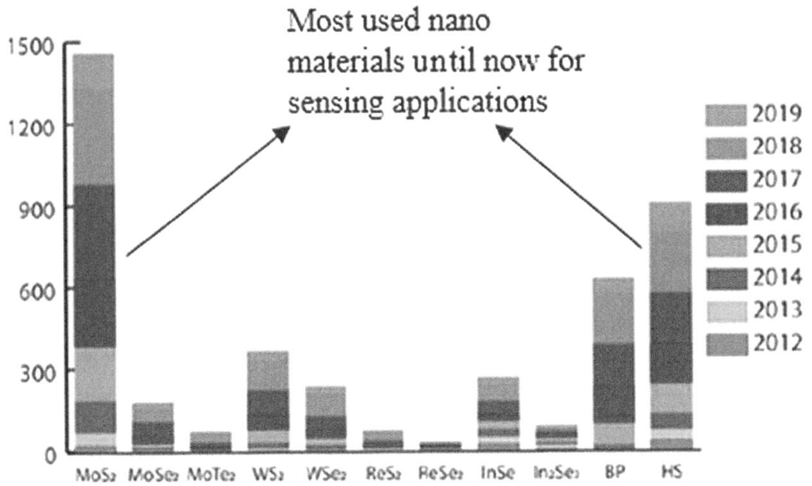

Figure 6.5 Number of articles published in a different year for different TMD materials [19].

very efficient for optical applications due to their superfast current, suitable bandgap, broad bandgap range [16], etc. Thus, in recent years, the use of TMDs heterostructure is increased for optical sensing applications. From the above discussion, it is evident that TMDs are a very attractive candidate for a high-performance optical sensor. However, before the discussion of TMDs-based optical biosensor, conventional optical biosensors are discussed in the next section.

6.3 BIOSENSOR

The first biosensor was invented by L.L. Clark in 1950 to gauge oxygen in blood [20]. The biosensor comprises the elements that can detect the targeted biomolecule. The most important feature of biosensors other than existing diagnostic methods is their selectivity. Biosensors are not only limited to sensing biomolecules, however. Ecological, pathogens, and other molecules can also be sensed by using biosensors. Since biosensors are very important in the field of medical and clinical analysis, the following features are very important for a biosensor:

1. Small size
2. Real-time sensing
3. Low cost
4. Easy to use

Based on the physics used to sense the molecule (working of biosensor), biosensors can be of various types,

1. Electrochemical
2. Piezoelectric
3. Optical
4. Magnetic
5. Thermometric, etc.

Electrochemical biosensors are the most used biosensors in which biological molecules are to be coated on the probing surface. When the probe is in place, molecules react with the sensor surface to detect and produce the appropriate electrical signal, which is proportional to the quantity measured. Electrochemical sensors can be divided into different subtypes: amperometric, potentiometric voltametric, etc., based on the parameter that is measured in accordance with the analyte.

Piezoelectric sensors are a subtype of mass-based biosensors. These are also known as acoustic biosensors since they are based on the sensing of sound vibrations. On applying pressure these sensors produce an electrical signal. It converts mass to frequency or mechanical vibrations of sensing molecules into proportional electrical signals.

Optical biosensors obey the optical measurement principle. Optical sensors can be of labeled or direct optical detection. Optical sensors allow the detection of the analyte based on a different phenomenon, like absorption, reflection, scattering, etc. The basic concept of optical sensing is to sense the change in the refractive index of the interacting surface. Since optical sensors are non-electrical, they are interference-free and can analyze multiple elements on a single layer just by varying the operating wavelength.

Magnetic biosensors use crystal or paramagnetic particles to detect the biological quantity by analyzing the changes in the magnetic properties of the sensor, such as coil inductance, resistance, etc. Thermometric biosensors are based on the heat invention due to the biological reactions. These sensors are also known as thermal biosensors. Thermal biosensors can be used to measure serum cholesterol, urea, uric acid, etc. Bio-sensing is a major area of research nowadays, due to the daily exploration of a new threat to life and use research as well. Since we are dealing with more and more complicated viruses, pathogens, etc., biosensor structure is also getting more complicated, due to the blending of both biotechnology and microelectronics. To improve the performance of the biosensors, different approaches are to be used by the researcher, such as the use of advanced materials, multilayer structure, etc. However, increased complexity of the sensor structure is not a desired thing; it is necessary in order to deal with complex disease detection. Of all the above-discussed sensors, an optical sensor is the most reasonably sensitive and accurate sensor.

6.3.1 Optical biosensor

Optical biosensors can be of different types based on the medium by which light traveled in/through the sensor, which can be fiber-based or chip-based (light traveling through free space). SPR sensors are the most advanced optical sensors, having label-free sensing. In 1982, Liedberg et al. [21] proposed the first SPR sensor for biosensing application. SPR sensors gained popularity in medical, health science, drug discovery, agriculture monitoring, etc., due to the ability to boost surface chemistry using advanced materials and other processes. In the past decade, SPR sensor development has become a fascinating area due to its immenseness to electromagnetic interference, remote sensing, and multiplexed detection.

Biorecognition and immobilization is the most important objective in biosensing. In the last decade, numerous methods have been developed to improve the specificity and sensitivity of the sensors using these immobilization methods. Direct detection of SPR sensor is the best option from the range of optical biosensors. Indirect detection and binding of the targeted particles is done with a biomolecular recognition element (BRE) layer on the surface of the designed sensor. The detection limit of the direct detection SPR sensor can be improved by using sandwiched assay and inhibit assay [22]. The performance of the SPR sensors can also be improved by increasing the fraction of SPs in the solution or at the surface. Localized surface plasmon resonance (LSPR) is one of the methods which can be used to improve the sensitivity of the sensor. Optical sensors can also be used for multiplexed sensing of many molecules using one sensor. In order to do that, different sample amounts can be immobilized into different flowing channels. To sense those different molecules, it is important to create a gradient surface for low-molecule weight detection. Thus, from the above discussion, it is quite clear that optical biosensors are a better and more sensitive method of bio-sensing. The SPR biosensor is one of the most sensitive of all the optical sensors. However, SPR sensors have a long history of evolution from a single layer of novel metal to a more complex sensor structure. Most recently, the use of advanced materials made SPR sensors much more useful, sensitive, and accurate. Thus, in this chapter SPR sensors with advanced materials are reviewed.

6.4 ADVANCED MATERIALS-BASED SURFACE PLASMON RESONANCE BIOSENSOR

With the development of material, biosensors are ever-evolving components. The selection of particular materials is critical since their interaction with the analyte or BRE element improves the sensing performance. The emergence of advanced and novel materials drastically enhances the sensor performance for biomolecule sensing. Metamaterial, graphene is the most widely used advanced material for biosensors, however, with the proposal of more advanced materials, (i.e., TMDs and other materials) performance of the biosensors can be further improved. In this chapter, a detailed review of the advanced material-based biosensors is done starting from graphene-based biosensors to TMD-based biosensors.

6.4.1 Graphene-based sensors

Graphene possesses unique physical, optical, and electrochemical properties that proved to be very useful for sensing applications. Surface engineering of graphene is a key property that is very suitable for biosensing, since analyte interaction takes place at the surface of the sensor. In the biosensing applications, not only graphene but graphene-based materials, such as graphene oxide, reduced graphene oxide, etc., also play an important role. Graphene is a single layer 2-D nanomaterial defined as a single atom thick sheet of carbon atoms, having a single layer thickness of 3.35 Å. The most basic property of the graphene layer is the large surface area (2630 m^2/g) [23], which made graphene able to interact with a wide bio-molecular range. Based on the various properties of graphene it can be said that an optical as well as electrochemical graphene-based sensor secures a special position in bio-sensing applications.

On observing both types of these sensors, it is evident that optical graphene-based sensors are more popular than electrochemical sensors.

However, the light absorption rate for single-layer graphene is too low, i.e., 2.3% only; thus, the current generated is also too low and in turn, affects the optical detection [24]. Among all the available graphene-based optical sensors, SPR sensors are the most widely used optical sensor due to their high sensitivity, label-free nature, real-time processing, etc. [25]. The graphene layer helps the sensor to absorb the biomolecules better due to π-π stacking and increases the system affinity [26]. The basic principle for the SPR sensing using a graphene layer is the same as the conventional SPR sensor.

In 2010, Wu *et al.* [24] proposed a graphene based SPR sensor having multiple layers over the gold layer, as shown in Figure 6.6 (a). It is observed in this study that the graphene layer works as the BRE layer to bind the biomolecules over the sensor surface. Figure 6.6 (b) has the reflectance spectrum for different configurations, without graphene layer and with graphene layer, and it is observed that shift in angle is increased and thus sensitivity is also increased. It is also depicted that the introduction of multiple layers of graphene improves the sensitivity of the sensor; however, it broadens the reflection spectrum, which reduces the accuracy of the sensor.

Graphene is also employed in fiber optics-based SPR sensors to improve the performance of fiber-based sensors. In 2013, Kim *et al.* [27] proposed the first graphene-based optic SPR sensor on a D-type plastic optical sensor. The conventional gold metal layer is replaced by the graphene layer for the detection of double-stranded DNA and protein. Figure 6.7 has the setup used for the sensing and schematic diagram of the proposed graphene based SPR fiber sensor.

After some time in 2015, Patnaik et al. [28] proposed a graphene-based sensor using a metal-oxide layer. In this sensor, the indium tin oxide layer is

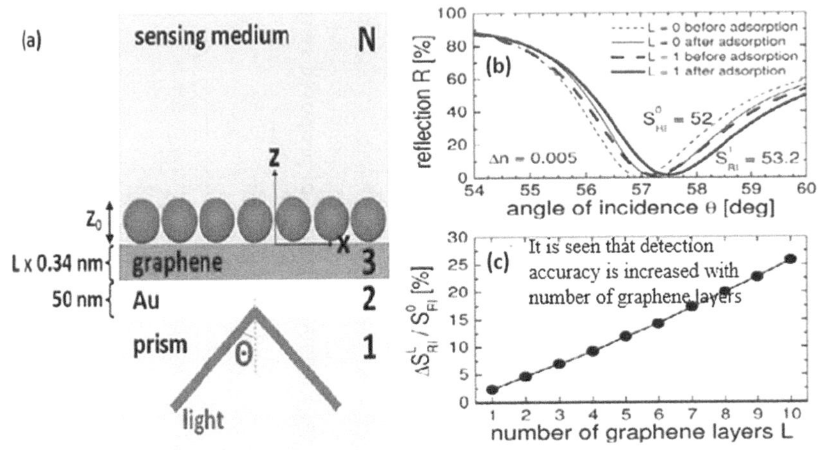

Figure 6.6 (a) Schematic diagram of proposed sensor (b) reflection spectrum of SPR sensor for angular interrogation (c) sensitivity enhancement for multiple layers of graphene [24].

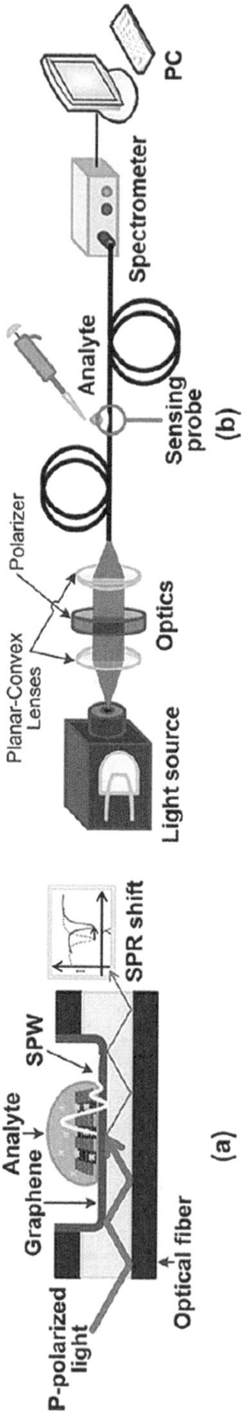

Figure 6.7 (a) Schematic diagram of proposed sensor (b) sensing set-up [27].

used to coat the D-shaped fiber, and the wavelength interrogation method is used for the sensing. It is observed that the resolution of 1.754×10^{-5} RIU is achieved with the wavelength sensitivity of 5700 nm/RIU. A number of graphene layers are also varied to observe the effect on resonance wavelength. Figure 6.8 has the proposed sensor structure with the shift in the resonance wavelength for the change in the analyte (Figure 6.9).

Gao et al. [29] proposed a tapered fiber sensor with a single graphene layer for glucose detection, with the concept of evanescent wave interaction with the analyte due to tapering. It is observed that an increased concentration of glucose will, in turn, increase the refractive index; thus, the output light intensity is reduced from 29500 to 27500. In 2017, Balaji et al. [30] proposed two different types of biosensors; graphene and graphene oxide based. Through the proposed sensor, tumor cells have been detected. This displayed that graphene and graphene-based materials are proved to be very effective for the detection of fatal disease cells such as cancer because of their biocompatibility and highly sensitive nature. Recently in 2020, Abinash et al. [31] proposed a combined sensor for blood glucose and gas detection. This sensor is prism coupled SPR sensor compatible with human blood sample ranges 25–175 mg/dl and gas detection for the refractive index variation from 1.0000 to 1.007 at 589 nm wavelength. It is observed that the proposed blood glucose sensor provides 275.15 °/RIU sensitivity and 1.41/° detection accuracy. The quality factor for the glucose sensor is also calculated, which is 76.2. When comparing the proposed sensor with the existing work, it is depicted that the proposed sensor displays much better performance. The performance of the proposed sensor as the gas sensor is also compared with the existing work, and again proved to be better. Sensitivity for the gas sensing is calculated as 92.1 °/RIU and detection accuracy is 2.55/°, whereas the quality factor is 230.2. This improvement is achieved due to high field enhancement at the interface of graphene and the analyte layer.

From the above discussion, it is clear that the use of graphene layer/s in the optical sensor greatly enhances the sensor performance due to the better interaction of the layer with the biomolecules. Along with this, it is also proved that the stacking of graphene layers further affects the sensing performance. However, sensing performance can further be improved by the use of much advanced TMD materials and hybrid structures of graphene and TMDs. These sensors are discussed further in the next section.

6.4.2 TMDs based sensors

TMDs are a very important part of the 2-D material family. Basic details, their structure, properties and advantages are already discussed in Section 1.2. TMDs are very attractive for various applications, due to their mechanical/chemical/optical properties, biocompatibility, etc. Optical sensing is one of the most used applications of TMDs. Since 2-D materials have the highest specific surface area due to their thinness, TMDs have large anchoring

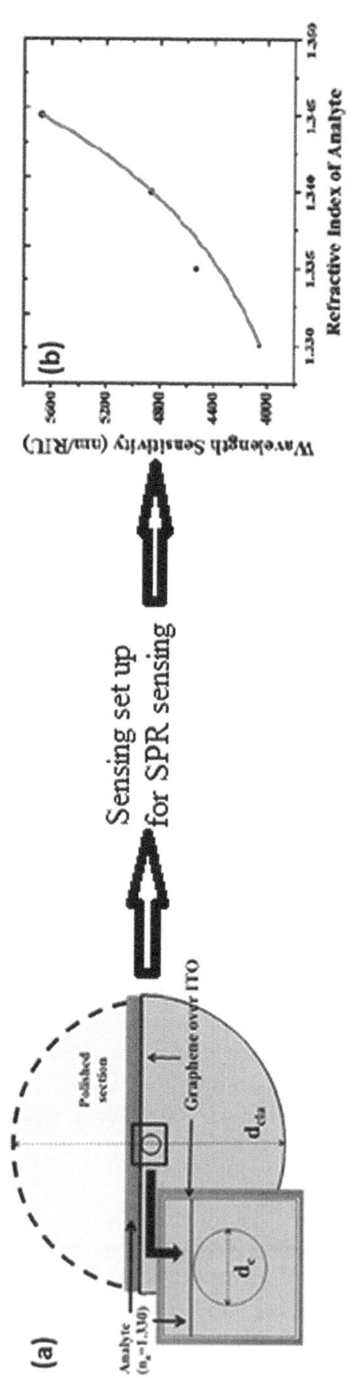

Figure 6.8 (a) Schematic diagram of proposed sensor (b) shift in the wavelength with the change in the analyte refractive index [28].

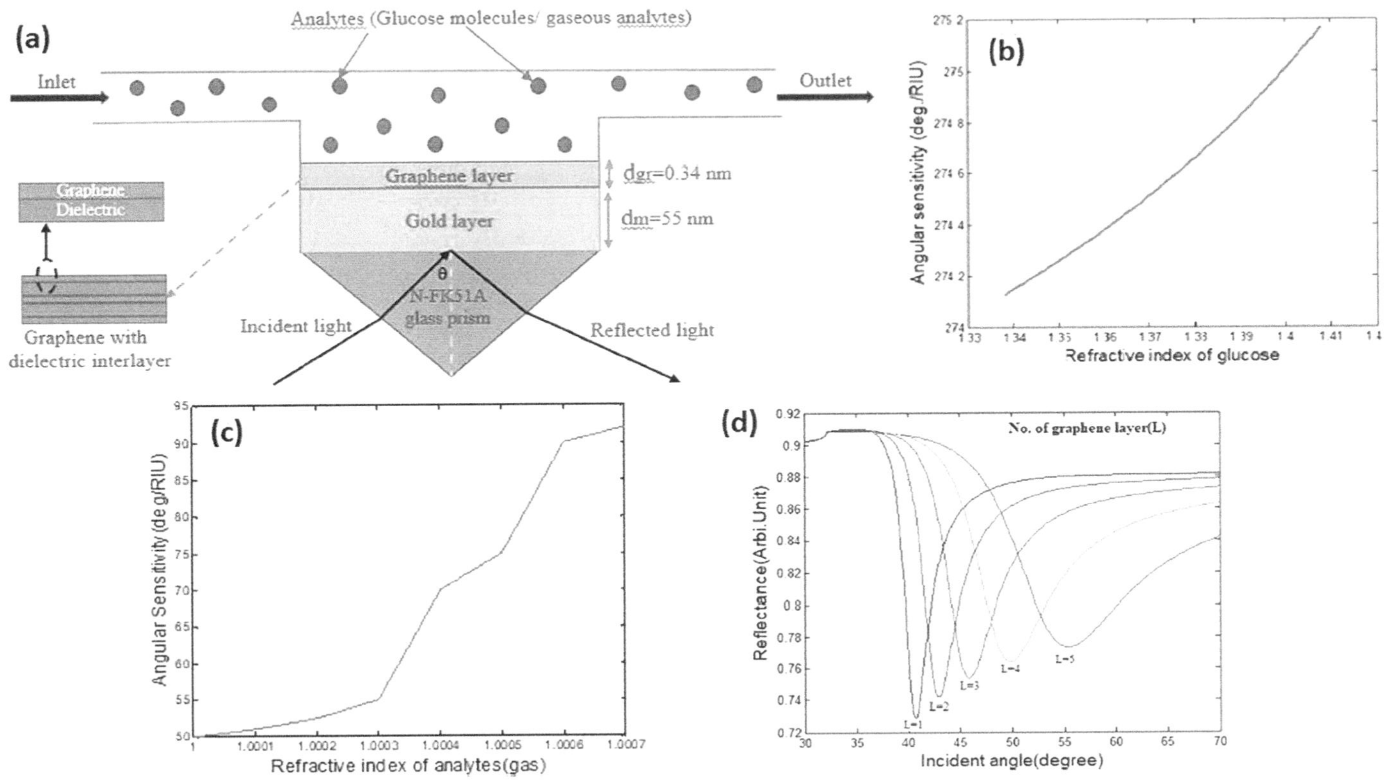

Figure 6.9 (a) Schematic diagram of a proposed prism based SPR sensor (b) angle interrogation for blood glucose sensing (c) angle interrogation for gas sensing (d) reflectance curve dependency on the number of graphene layers [31].

sites for the efficient interaction of biomolecules with the sensor. In 2017, Eugene et al. [32] proposed a hydrogen adsorption sensor using a MoS_2 monolayer. The investigation is done based on the first-principal calculation; it is observed that hydrogen atoms diffused through the monolayer because of the energy barrier. Diffusion is enhanced by strain and showed that the MoS_2 monolayer is a very effective hydrogen sensor. The proposed study is not an exact sensing application, but this indicates that MoS_2 can prove to be a very effective sensor for bio-sensing. In 2018, Nan and Ting [33] proposed an SPR sensor using a functional MoS_2 sheet. Monochloroacetic acid is used to modify and functionalize the carboxyl MoS_2. It is observed from the study that MoS_2-COOH based SPR sensor displays high sensitivity and affinity for biomolecule detection for a very low volume of 20 µl, which is not possible with conventional bulk materials. A comparison is done between the functionalized TMD layer, simple TMD layer, and traditional SPR sensor. It is depicted that the functionalized SPR sensor responds by 3.1-fold and enhances the association rate by 212 folds compared to the traditional SPR chip. This excellent performance is given due to the high binding affinity and biocompatibility of the TMD layer. Figure 6.10 has the

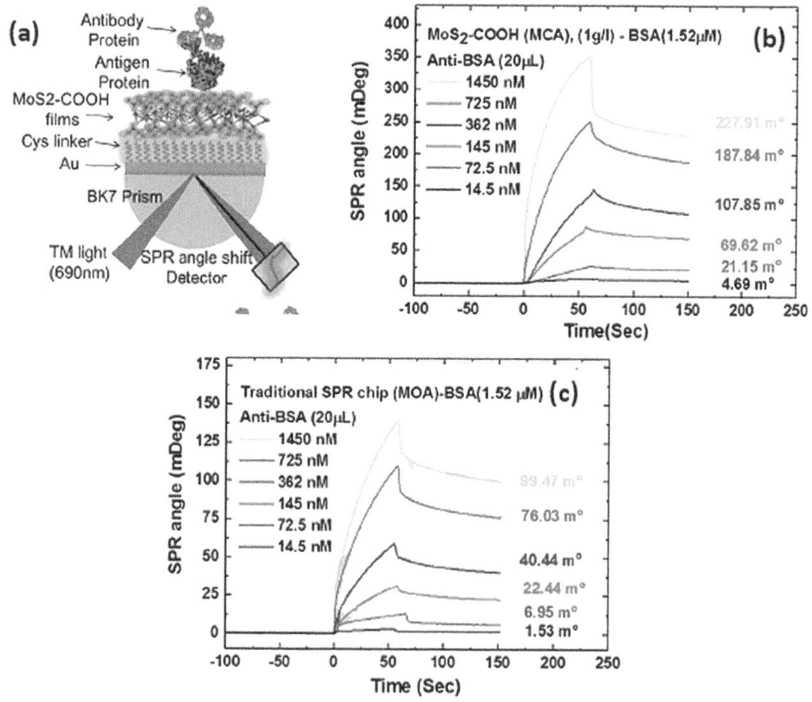

Figure 6.10 (a) Proposed sensor design, (b) sensorgram of the proposed sensor, (c) sensorgram of the traditional sensor [33].

proposed sensor structure with the sensorgram of the proposed and traditional SPR sensor.

It is believed that the proposed sensor can be proved beneficial for the detection of clinical diseases in blood samples, etc. Sharma and Pandey [34] proposed a heterostructure SPR sensor with enhanced sensitivity for bio and gas sensing applications. Figure 6.11 has the proposed sensor structure having blue phosphorene/MoS_2 heterostructure as the active layer. Various combinations of layers and stacking (L = 1 to 4) have been analyzed for their effect on the reflectance spectrum. It is observed that for repeated layers, sensitivity is increased, however, detection accuracy is reduced. This variation is achieved due to the variation in the effective refractive index of the sensor. On analyzing the stacking effect, it is seen that for $L > 4$, performance is not acceptable (very large spectrum width). It is also demonstrated that the use of a graphene layer in place of the proposed heterostructure reduces the sensitivity of the sensor. Along with this, the effect of the wavelength of incident light is also observed by varying the wavelength from 660 nm to 910 nm, i.e., visible to near-infrared wavelength region. From the simulation results, it is depicted that a trade-off has to be maintained between the sensitivity and detection accuracy since, at lower wavelengths, sensitivity is high and detection accuracy is low, whereas, at higher wavelengths, detection accuracy is high, and sensitivity is low. After the detailed analysis, it is determined that at 662 nm wavelength an adequately optimized value of sensitivity and detection accuracy can be achieved for a single heterostructure layer sensor.

In 2018, Aksimsek *et al.* [35] proposed a heterostructure having graphene–MoS_2 – metal layers for SPR sensors. Figure 6.12 has the finding of the study, including the reflectance curve for a stacked layer of graphene or MoS_2. The effect of metal layer thickness is also studied, with the result that if the metal thickness is reduced from 55 to 32 nm, sensor performance is improved by 87% to 13% with the graphene and MoS_2 layers respectively. It is observed that the TMD-based sensor gives better performance than the graphene-based sensor.

In 2019, Yi Xu *et al.* [36] proposed a hybrid sensor having MXene 2-D layer with TMD layer in combination for SPR sensor. This study is the theoretical study, which analyzed that sensitivity of 198°/RIU can be achieved through this hybrid structure where WS_2 is used as the TMD layer. For the study, four TMD layers are used, namely: MoS_2, $MoSe_2$, WS_2, and WSe_2. Figure 6.13 has the comparative graph for the sensitivity, having different TMD layers in the hybrid structure.

It can be concluded from the study that the refractive index sensitivity of 174°/RIU, 176°/RIU, 198°/RIU, and 192°/RIU is achieved for monolayer MXene and four-layer MoS_2, five-layer $MoSe_2$, five-layer WS_2, and six-layer WSe_2 respectively at 633 nm incident light wavelength. It is seen that not only MoS_2 but other TMD materials are also eligible for the sensing application and provide better performance than the overrated MoS_2 layer. The reason

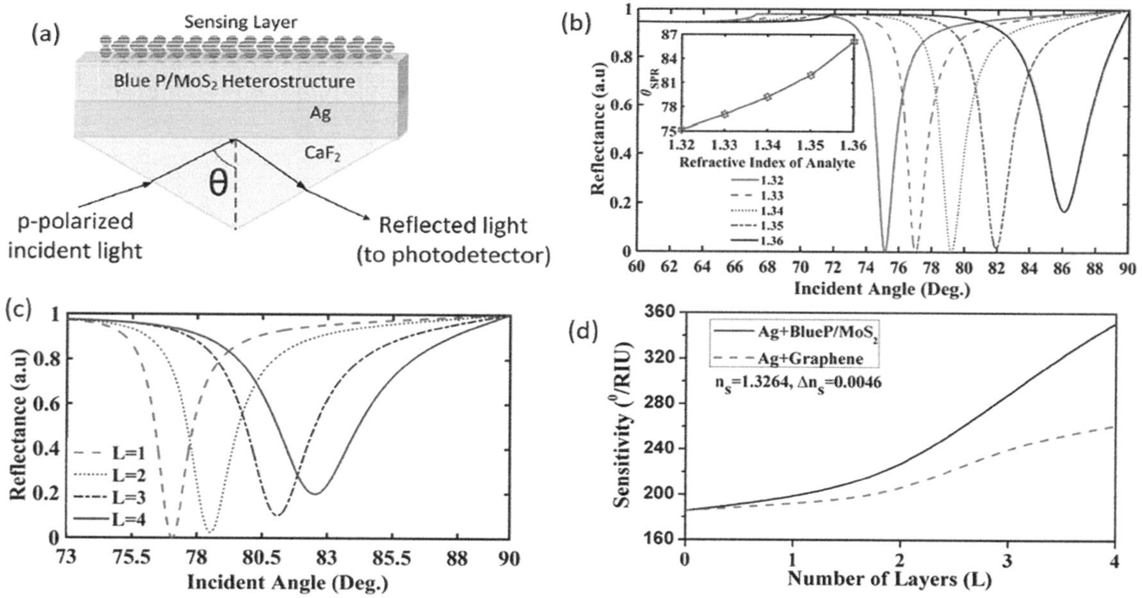

Figure 6.11 (a) Schematic diagram of the proposed sensor (b) SPR curve for monolayer blue phosphorene/MoS_2 (c) SPR curve for multilayer heterostructure (d) sensitivity variation for proposed heterostructure and graphene layer [34].

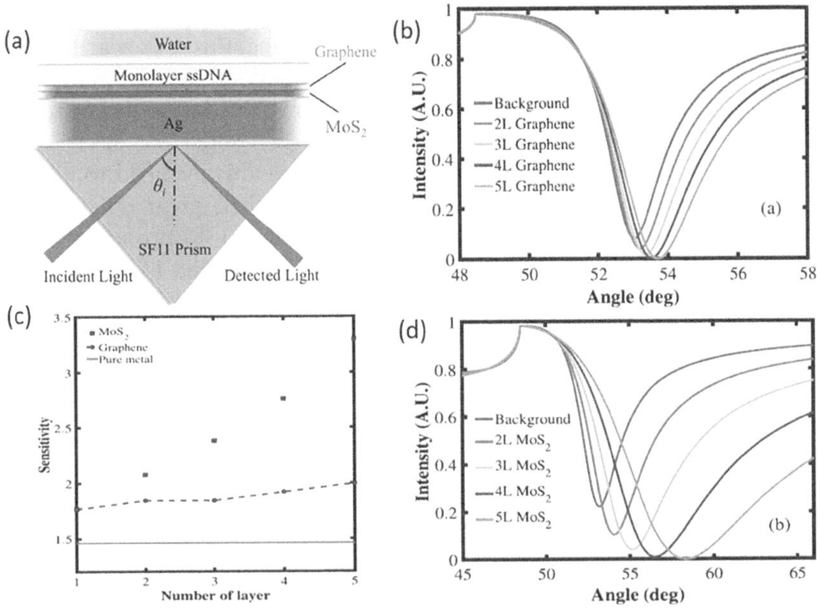

Figure 6.12 (a) Proposed sensor structure (b) reflectance for stacked graphene layers (c) effect of the number of layers on the sensitivity of the sensor (d) reflectance for stacked MoS_2 layer [35].

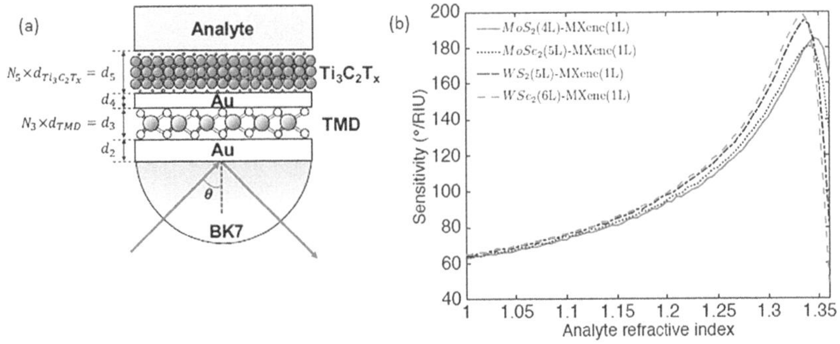

Figure 6.13 (a) Proposed sensor structure (b) sensitivity variation for optimized MXene-TMD-based SPR sensor for different analyte refractive indices [36].

behind the better sensitivity of WS_2 is its high surface-to-volume ratio. Huawen Hu et al. reviewed the use of 2-D TMD materials for biological sensing in detail [37]. In this study, nanolayer TMDs are discussed for sensing applications through different physical phenomena, such as optical, electrical, electrochemical, etc. Since we are interested in optical sensing in this chapter, this study is very useful. The different optical platform is discussed, like fluorescence and chemiluminescence, label-free, absorption-based, colorimetric sensor, etc. Along with the concept of sensing, this review article also talked about the fabrication/exfoliation process in detail for different TMD nanosheets. In conclusion, TMDs based systems are being developed to improve the overall efficiency of systems, whether optical or electrical. A small discussion on TMD-based-nanocomposite is also included to address the shortcoming of existing pristine TMDs, and their great potential for novel, high-performance devices. In 2016, Rou et al. [38] proposed a WS_2 based biosensor for protein detection. In this study, metallic 1T phase TMD, i.e., 1T WS_2 is proposed for immobilization of hemoglobin (Hb). The proposed sensor not only displays the reduction of hydrogen peroxide (H_2O_2) but also provides a high surface-to-volume ratio and conductivity. 1T-phase WS_2 based sensor is also compared with the 1T-phase MoS_2, $MoSe_2$, and WSe_2 to find out that 1T-WS_2 provides high analytical performance with a wide range, sensitivity, repeatability, and stability, etc., compared to others. However, this is an electrochemical sensor but does exhibit promising results for optical sensing also, due to its easy structure (Figure 6.14 (a)) and good optical properties.

In 2016, Q. Ouyang et al. [39] did a simulation comparative study for TMD-based SPR biosensors. This study has a silicon nanosheet and TMD layer over it for the enhanced sensitivity. The sensor structure has triangular prism/gold film/silicon nanosheet/ on any one of the TMD layers (MoS_2, $MoSe_2$, WS_2, and WSe_2)/BRE/sensing medium. To characterize the performance of the sensor, various parameters are analyzed, such as sensitivity, full width at half maximum, reflectivity, etc (Figure 6.15).

To investigate the performance, the influence of incident light is also studied by varying the wavelength, namely, 600, 633, 660, 784, and 1024 nm. The study showed that each aspect of the sensor affects the sensor performance,

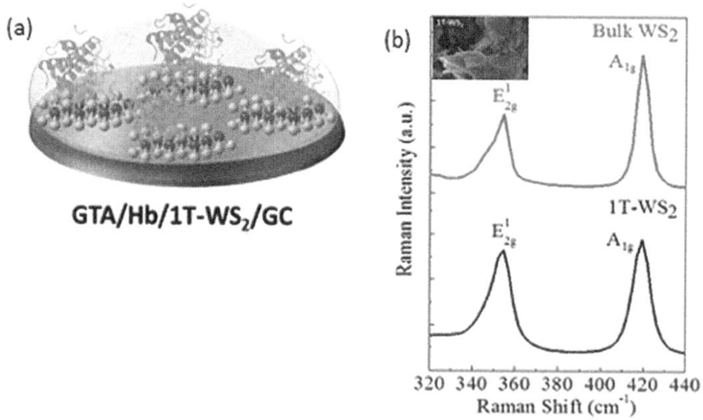

Figure 6.14 (a) Schematic diagram of proposed sensor (b) Raman spectrum of bulk WS_2 and 1T WS_2 [38].

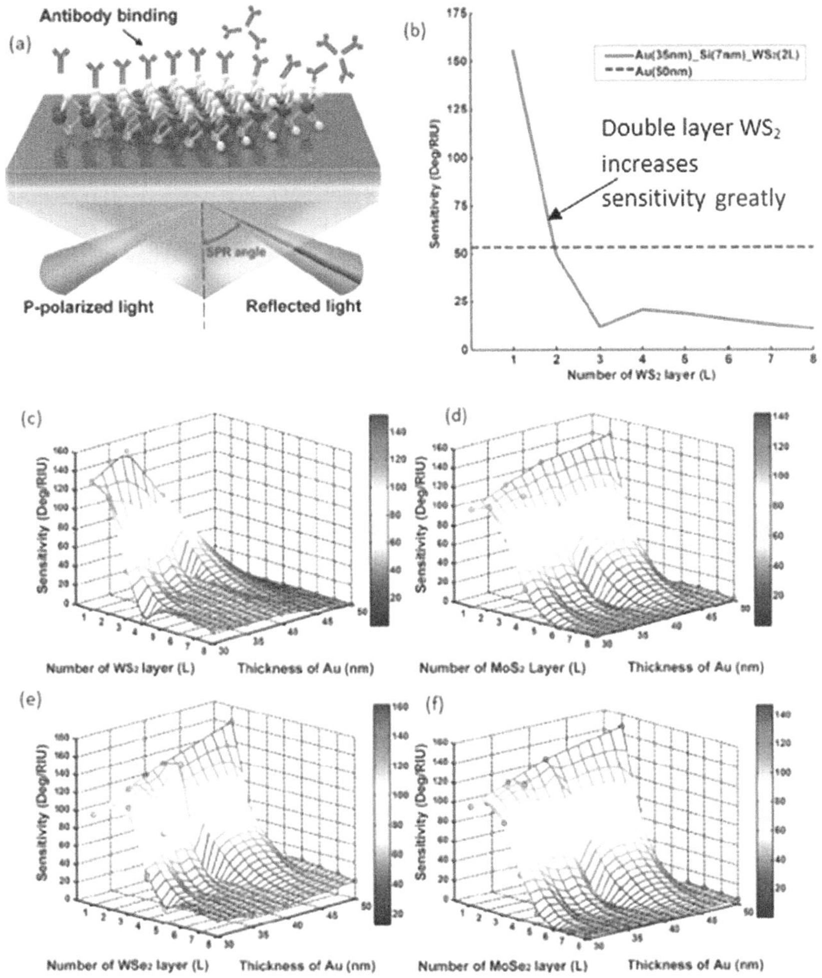

Figure 6.15 (a) Proposed sensor structure and set-up (b) comparison of sensitivity for conventional and WS_2 based sensor (c-f) sensitivity variation as a function of gold film thickness and number of TMD layers WS_2, MoS_2, WSe_2, and $MoSe_2$ respectively [39].

whether it is film thickness or the number of TMD layers or incident light wavelength. With optimized parameters, the highest sensitivity of 155.68°/RIU is achieved by having 35 nm thick gold film, 7 nm thick silicon sheet, and WS_2 monolayer at 600 nm excitation wavelength.

Recently in 2020, P. Zhao et al. [40] proposed a very complex SPR sensor consisting of MoS_2 nanoflower and gold nanoparticles to sense immunoassay of mouse IgG experimentally. Sensor structure has a gold film over a glass slide; over that, MoS_2 nanoflowers are deposited through a hydrothermal method. After that, gold nanoparticles are prepared using an aqueous

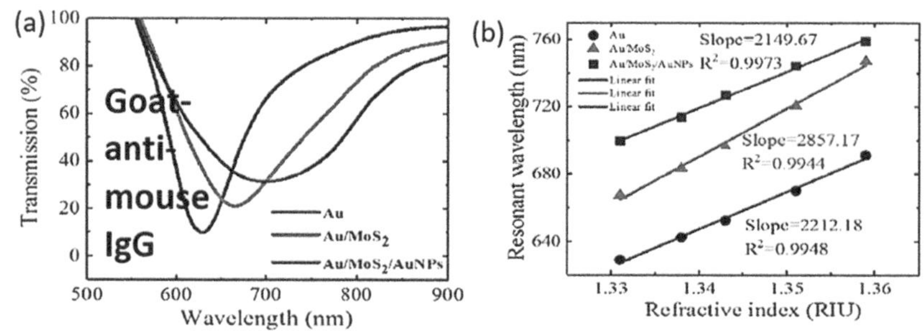

Figure 6.16 (a) Transmission spectrum for 1.33 refractive index value (b) plot between resonant wavelength and sensing refractive index for different kinds of sensors [40].

solution and then deposited over MoS_2 nanoflowers. Over the sensor, the immobilization layer and BRE layers are deposited for effective sensing of mouse IgG. This study used the wavelength interrogation method for SPR sensing, and it is observed that the use of MoS_2 improved the sensitivity from 2212.2 to 2857.2 nm/RIU, however, deposition of gold nanoparticles reduced the sensitivity to 2149.7 nm/RIU. Even so, the sensitivity is approximately three times higher than the unmodified sensor. A comparative study is done to show the difference between the conventional and proposed SPR sensors as shown in Figure 6.16.

It is also observed that the proposed sensor displays the best performance for the mouse IgG having a limit of detection of 0.06 μg/ml, which is much higher than the other existing work. In addition to improvement in sensitivity, it also demonstrates the simple fabrication of sensor structure, which was not proposed till now. Again in 2020, Ashish Bijalwan et al. [41] proposed an SPR sensor using nanoribbon of graphene and WSe_2. This study is a numerical study that used a rigorous coupled-wave analysis method for calculations. Two combinations are studied: one having a continuous graphene/WSe_2 layer and the second having graphene/WSe_2 nanoribbon. It is observed that the maximum achieved quality factor is 184.97/RIU, 181.11/RIU and detection accuracy is 0.92, 0.90 for graphene and WSe_2 nanoribbon, respectively. On analyzing the results critically, it is observed that, identical to the previous studies, an increased number of layers of graphene and TMD improved the sensitivity and broadened the spectrum, which in turn reduced the detection accuracy and dip strength. The point to be noted here is that the proposed sensor can be used optimally for the sensing refractive index that lies in the range of 1.33.

Akash Srivastava and Y. K. Prajapati [42] proposed a more complex sensor structure recently for biosensing applications using the TMD layer.

This study is a detailed study that analyzed various combinations of materials and their order in the sensor structure, comparison is also made with the traditional sensor having only a single metal layer based on the performance

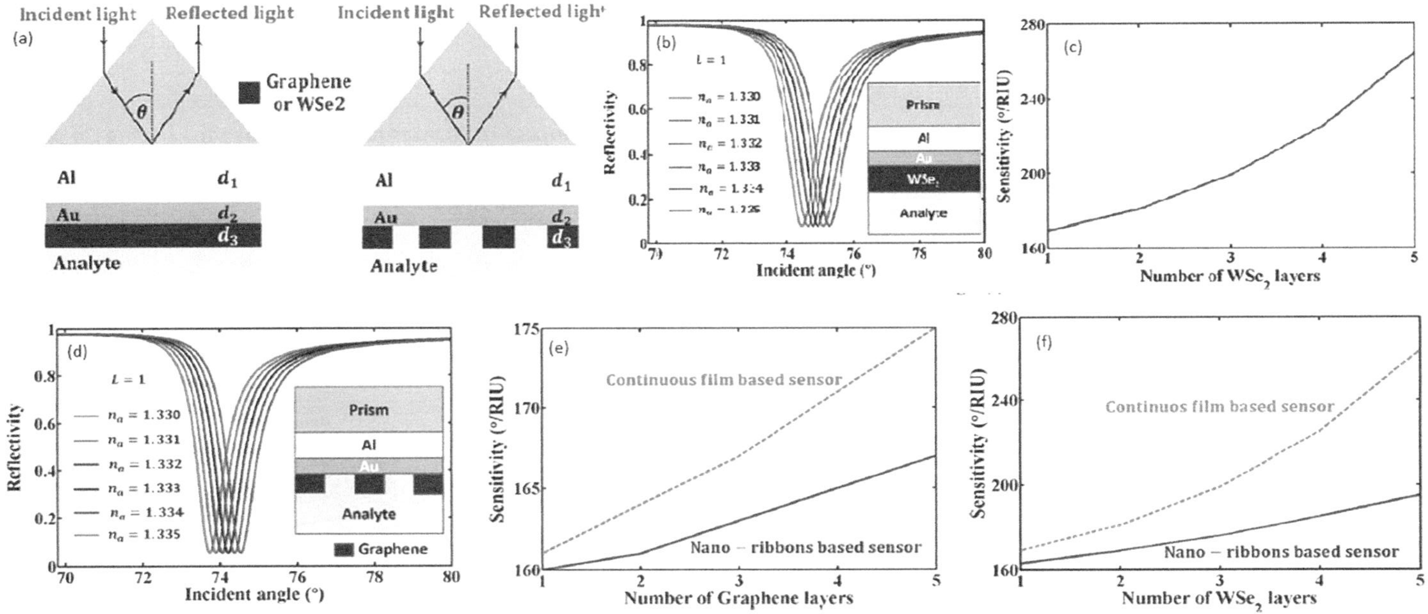

Figure 6.17 (a) schematic diagram of the proposed sensor for continuous and nanoribbon structure (b) reflection spectrum for continuous graphene layer for L=1 (c) change in sensitivity for different WSe_2 layers (d) reflection spectrum for nanoribbon-based sensor for L=1 (e) variation in sensitivity for stacked graphene layers (f) variation in sensitivity for stacked WSe_2 layers [41].

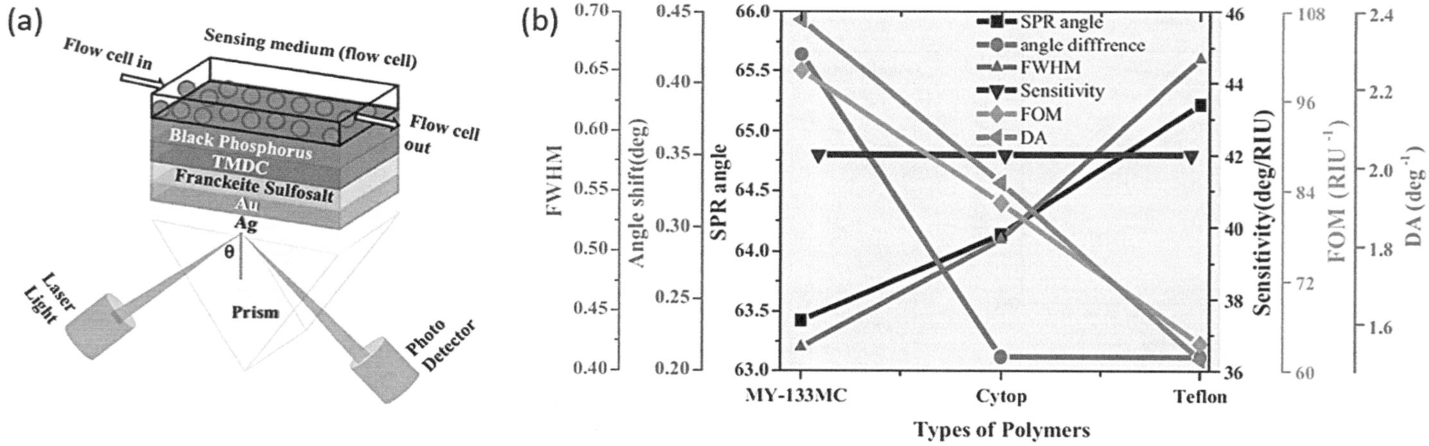

Figure 6.18 (a) Proposed sensor structure (b) performance of the sensor for different polymer materials over glass substrate [42].

parameters. Proposed sensor structure consists of prism/bimetallic layer/Franckeite/TMD layer/black phosphorus. It is observed that the proposed sensor showed two times better sensitivity as compared to the traditional SPR sensor, at 633 nm excitation wavelength. The maximum sensitivity achieved through the sensor is 208°/RIU. On comparing this sensitivity with traditional sensors, it is observed that there is an increase of 46.47% in sensitivity. The best-suited polymer detected through the study is fluoropolymer (MY-133MC) due to its high sensitivity, the figure of merit, and detection accuracy (Figures 6.17 and 6.18).

From all the discussion done in this chapter, it is realized that bio-sensing is a very critical field nowadays, since we come face to face daily with a new virus or disease, and life is very valuable. Thus, advancement in bio-sensing is also very important. With the progression in material research, senses become more and more accurate. However, the introduction of advanced 2-D materials made biosensing more reliable and precise. There are a number of studies available for 2-D materials based optical biosensors. In this chapter, authors tried to compile appropriate and adequate studies to give insight to readers about the use of 2-D materials in SPR biosensors.

6.5 CONCLUSION

This chapter is dedicated to the biosensing application of TMD materials. The main focus is on the optical SPR sensor. In this chapter, a detailed study is done on the basics of SPR sensors as well as 2-D materials. First, the basics about the SPR sensors with the well-known mathematical method for modeling (i.e., transfer matrix method) is discussed, then interrogation methods for optical sensor and performance parameters of SPR sensors are discussed. After that, a detailed study of the 2-D materials is done, materials are categorized based on their constituent materials, use and basics of biosensors is discussed, along with a discussion of optical biosensors. After discussing all the basic terminologies, 2-D material-based SPR biosensors are reviewed in detail, based on graphene as well as TMD materials. On analyzing the SPR biosensors based on 2-D materials, it is shown that the unique properties of 2-D materials proved to be very beneficial for sensors, since the large surface area and exclusive optical properties of these materials provide better interaction of biomolecules with the sensor and a large dynamic range. It is also seen that among various TMD nanomaterials, tungsten-based nanomaterial (WS_2 and WSe_2) provides better performance than the most popular MoS_2. However, widely proposed fabrication methods and compatibility with an extensive range of biomolecules made MoS_2 much more popular than tungsten-based nanomaterials. Although, there is a large scope for further improvement in this field, both of material and the technology, the development and sensing performance of optical biosensors became much more accurate by using TMD nanomaterials.

REFERENCES

1. Fano, Ugo. "The theory of anomalous diffraction gratings and quasi-stationary waves on metallic surfaces (Sommerfeld's waves)." *JOSA* 31, no. 3 (1941): 213–222.
2. Kretschmann, Erwin, and Heinz Raether. "Radiative decay of non-radiative surface plasmons excited by light." *Zeitschrift für Naturforschung A* 23, no. 12 (1968): 2135–2136.
3. Pockrand, I., J. D. Swalen, J. G. Gordon Ii, and M. R. Philpott. "Surface plasmon spectroscopy of organic monolayer assemblies." *Surface Science* 74, no. 1 (1978): 237–244.
4. Liedberg, Bo, Claes Nylander, and Ingemar Lunström. "Surface plasmon resonance for gas detection and biosensing." *Sensors and Actuators* 4 (1983): 299–304.
5. Otto, Andreas. "Excitation of nonradiative surface plasma waves in silver by the method of frustrated total reflection." *Zeitschrift für Physik A Hadrons and Nuclei* 216, no. 4 (1968): 398–410.
6. Estroff, Andrew, and Bruce W. Smith. "Tuning metamaterials for applications at DUV wavelengths." *International Journal of Optics* 2012 (2012).
7. Homola, Jirí, Marek Piliarik, and Róbert Horváth. Springer Series on Chemical Sensors and Biosensors Methods and Applications. Series Ed.: Wolfbeis, OS (2006): 1612–7617.
8. Homola, Jiří. "Present and future of surface plasmon resonance biosensors." *Analytical and Bioanalytical Chemistry* 377, no. 3 (2003): 528–539.
9. Zeng, Youjun, Xueliang Wang, Jie Zhou, Ruibiao Miyan, Junle Qu, Ho-Pui Ho, Kaiming Zhou, Bruce Zhi Gao, and Yonghong Shao. "Phase interrogation SPR sensing based on white light polarized interference for wide dynamic detection range." *Optics Express* 28, no. 3 (2020): 3442–3450.
10. Vasa, Parinda, and Christoph Lienau. "Strong light–matter interaction in quantum emitter/metal hybrid nanostructures." *ACS Photonics* 5, no. 1 (2018): 2–23.
11. Ashraf, Naveed, Abdul Majid, Muhammad Rafique, and Muhammad Bilal Tahir. "A review of the interfacial properties of 2-D materials for energy storage and sensor applications." *Chinese Journal of Physics* (2020).
12. Allen, Matthew J., Vincent C. Tung, and Richard B. Kaner. "Honeycomb carbon: a review of graphene." *Chemical Reviews* 110, no. 1 (2010): 132–145.
13. Xing, Fei, Zhi-Bo Liu, Zhi-Chao Deng, Xiang-Tian Kong, Xiao-Qing Yan, Xu-Dong Chen, Qing Ye, Chun-Ping Zhang, Yong-Sheng Chen, and Jian-Guo Tian. "Sensitive real-time monitoring of refractive indexes using a novel graphene-based optical sensor." *Scientific Reports* 2, no. 1 (2012): 1–7.
14. Zhao, Yong, Xue-gang Li, Xue Zhou, and Ya-nan Zhang. "Review on the graphene based optical fiber chemical and biological sensors." *Sensors and Actuators B: Chemical* 231 (2016): 324–340.
15. Hu, Tingting, Xuan Mei, Yingjie Wang, Xisheng Weng, Ruizheng Liang, and Min Wei. "Two-dimensional nanomaterials: fascinating materials in biomedical field." *Science Bulletin* 64, no. 22 (2019): 1707–1727.
16. Khan, Karim, Ayesha Khan Tareen, Muhammad Aslam, Renheng Wang, Yupeng Zhang, Asif Mahmood, Zhengbiao Ouyang, Han Zhang, and Zhongyi Guo. "Recent developments in emerging two-dimensional materials and their applications." *Journal of Materials Chemistry C* 8, no. 2 (2020): 387–440.

17. Xing, Fei, Zhi-Bo Liu, Zhi-Chao Deng, Xiang-Tian Kong, Xiao-Qing Yan, Xu-Dong Chen, Qing Ye, Chun-Ping Zhang, Yong-Sheng Chen, and Jian-Guo Tian. "Sensitive real-time monitoring of refractive indexes using a novel graphene-based optical sensor." *Scientific Reports* 2, no. 1 (2012): 1–7.
18. Patil, Pravin O., Gaurav R. Pandey, Ashwini G. Patil, Vivek B. Borse, Prashant K. Deshmukh, Dilip R. Patil, Rahul S. Tade et al. "Graphene-based nanocomposites for sensitivity enhancement of surface plasmon resonance sensor for biological and chemical sensing: a review." *Biosensors and Bioelectronics* 139 (2019): 111324.
19. Thakar, Kartikey, and Saurabh Lodha. "Optoelectronic and photonic devices based on transition metal dichalcogenides." *Materials Research Express* 7, no. 1 (2020): 014002.
20. Clark, J.R., C. Leland, Richard Wolf, Donald Granger, and Zena Taylor. "Continuous recording of blood oxygen tensions by polarography." *Journal of Applied Physiology* 6, no. 3 (1953): 189–193.
21. Liedberg, Bo, Claes Nylander, and Ingemar Lundström. "Biosensing with surface plasmon resonance—how it all started." *Biosensors and Bioelectronics* 10, no. 8 (1995): i–ix.
22. Guo, Xiaowei. "Surface plasmon resonance-based biosensor technique: a review." *Journal of Biophotonics* 5, no. 7 (2012): 483–501.
23. Stoller, Meryl D., Sungjin Park, Yanwu Zhu, Jinho An, and Rodney S. Ruoff. "Graphene-based ultracapacitors." *Nano Letters* 8, no. 10 (2008): 3498–3502.
24. Wu, L., H. S. Chu, W. S. Koh, and E. P. Li. "Highly sensitive graphene biosensors based on surface plasmon resonance." *Optics Express* 18, no. 14 (2010): 14395–14400.
25. Rifat, Ahmmed A., G. Amouzad Mahdiraji, Desmond M. Chow, Yu Gang Shee, Rajib Ahmed, and Faisal Rafiq Mahamd Adikan. "Photonic crystal fiber-based surface plasmon resonance sensor with selective analyte channels and graphene-silver deposited core." *Sensors* 15, no. 5 (2015): 11499–11510.
26. Li, Zongwen, Wenfei Zhang, and Fei Xing. "Graphene optical biosensors." *International Journal of Molecular Sciences* 20, no. 10 (2019): 2461.
27. Kim, Jang Ah, Taehyun Hwang, Sreekantha Reddy Dugasani, Rashid Amin, Atul Kulkarni, Sung Ha Park, and Taesung Kim. "Graphene based fiber optic surface plasmon resonance for bio-chemical sensor applications." *Sensors and Actuators B: Chemical* 187 (2013): 426–433.
28. Patnaik, A., Senthilnathan, K. and Jha, R. "Graphene-based conducting metal oxide coated D-shaped optical fiber SPR sensor." *IEEE Photonics Technology Letters* 27, no. 23 (2015): 2437–2440.
29. Gao, S. S., H. W. Qiu, C. Zhang, S. Z. Jiang, Z. Li, X. Y. Liu, W. W. Yue et al. "Absorbance response of a graphene oxide coated U-bent optical fiber sensor for aqueous ethanol detection." *RSC Advances* 6, no. 19 (2016): 15808–15815.
30. Balaji, Aditya, and Jin Zhang. "Electrochemical and optical biosensors for early-stage cancer diagnosis by using graphene and graphene oxide." *Cancer Nanotechnology* 8, no. 1 (2017): 1–12.
31. Panda, Abinash, Puspa Devi Pukhrambam, and Gerd Keiser. "Performance analysis of graphene-based surface plasmon resonance biosensor for blood glucose and gas detection." *Applied Physics A* 126, no. 3 (2020): 1–12.
32. Koh, Eugene Wai Keong, Cheng Hsin Chiu, Yao Kun Lim, Yong-Wei Zhang, and Hui Pan. "Hydrogen adsorption on and diffusion through MoS_2 monolayer: First-principles study." *International Journal of Hydrogen Energy* 37, no. 19 (2012): 14323–14328.

33. Chiu, Nan-Fu, and Ting-Li Lin. "Affinity capture surface carboxyl-functionalized MoS_2 sheets to enhance the sensitivity of surface plasmon resonance immunosensors." *Talanta* 185 (2018): 174–181.
34. Sharma, Anuj K., and Ankit Kumar Pandey. "Blue phosphorene/MoS_2 heterostructure based SPR sensor with enhanced sensitivity." *IEEE Photonics Technology Letters* 30, no. 7 (2018): 595–598.
35. Aksimsek, Sinan, Henri Jussila, and Zhipei Sun. "Graphene–MoS_2–metal hybrid structures for plasmonic biosensors." *Optics Communications* 428 (2018): 233–239.
36. Xu, Yi, Yee Sin Ang, Lin Wu, and Lay Kee Ang. "High sensitivity surface plasmon resonance sensor based on two-dimensional MXene and transition metal dichalcogenide: a theoretical study." *Nanomaterials* 9, no. 2 (2019): 165.
37. Hu, Huawen, Ali Zavabeti, Haiyan Quan, Wuqing Zhu, Hongyang Wei, Dongchu Chen, and Jian Zhen Ou. "Recent advances in two-dimensional transition metal dichalcogenides for biological sensing." *Biosensors and Bioelectronics* 142 (2019): 111573.
38. Toh, Rou Jun, Carmen C. Mayorga-Martinez, Zdenek Sofer, and Martin Pumera. "1T-phase WS_2 protein-based biosensor." *Advanced Functional Materials* 27, no. 5 (2017): 1604923.
39. Ouyang, Qingling, Shuwen Zeng, Li Jiang, Liying Hong, Gaixia Xu, Xuan-Quyen Dinh, Jun Qian et al. "Sensitivity enhancement of transition metal dichalcogenides/silicon nanostructure-based surface plasmon resonance biosensor." *Scientific Reports* 6, no. 1 (2016): 1–13.
40. Zhao, Peili, Yaofei Chen, Yu Chen, Shiqi Hu, Hui Chen, Wei Xiao, Guishi Liu et al. "A MoS_2 nanoflower and gold nanoparticle-modified surface plasmon resonance biosensor for a sensitivity-improved immunoassay." *Journal of Materials Chemistry* C 8, no. 20 (2020): 6861–6868.
41. Bijalwan, Ashish, Bipin Kumar Singh, and Vipul Rastogi. "Surface plasmon resonance-based sensors using nano-ribbons of graphene and WSe_2." *Plasmonics* (2020): 1–9.
42. Srivastava, Akash, and Yogendra Kumar Prajapati. "Effect of sulfosalt and polymers on performance parameter of SPR biosensor." *Optical and Quantum Electronics* 52, no. 10 (2020): 1–14.

Chapter 7

Graphene and carbon nanotube-based sensors

Neeraj Kumar
Indian Institute of Science Bangalore, Bengaluru, India

Prakash Chander Thapliyal
CSIR–Central Building Research Institute, Roorkee, India

CONTENTS

7.1 INTRODUCTION

Carbon is a primary element that exists in all known forms of life on earth. It is present in diverse allotropic forms that have different chemical and physical behaviors. Carbon allotropes have been widely applied in science and engineering for several decades due their novel properties, which makes them suitable for various applications. The well–known allotropes of carbon

DOI: 10.1201/9781003288633-7

are graphite and diamond. In 1985, the discovery of fullerenes opened the door to a new group of carbon allotropes in nano–scale levels [1]. However, after the discovery of carbon nanotubes and graphene, they continuously attracted the scientific community to develop the new materials and applications. Carbon based nanomaterials have gained high impact over the recent few decades in development and applications in various fields. The scientific community has applied several attempts to grow the field in terms of inventions, and unique applications for these extraordinary materials [2]. Several numbers of carbon–based materials have developed, e.g., fullerene, nano–onions, nano–diamonds, carbon nanotubes (CNTs), graphene, graphene nanoribbons, graphene quantum dots, etc., among which the sp^2 hybridized carbon materials have been the focus of the research community for the last three decades due to their excellent physiochemical properties. Among them, the graphene and CNTs are the most prominent sp^2 hybrid carbons and have gained enormous attention in the scientific community [1, 3, 4]. These have identical composition with sp^2 hybridized carbon but are different in structure. With identical structure, the properties would seem to be comparable, but this is not always true, and variations in their structure unlock new ways for advanced development. Carbon nanotubes have cylindrical structure that are rolled–up sheets of single–layer graphene. These are defined as single or multiple walled CNTs [3, 4]. Graphene is a single carbon atom thick layer organized like honeycomb, with six membered rings. The graphene is considered as a two–dimensional (2D) material [5, 6].

Graphene and CNT based materials have various applications, including biomedical, electronics, field emission, batteries, supercapacitors, and sensors and biosensors [9–15, 18, 19]. Carbon based materials are frequently applied for sensors due to their excellent electron transfer kinetics, high electrical conductivity, larger surface area, and biocompatibility [7, 8, 10, 18, 19]. These materials have excellent carrier mobility with lower electrical and thermal noise [9, 16, 17]. Graphene has superior mechanical adhesion to CNTs, as CNTs have tubular structure and minimal contact area, while the contact area is greater in graphene [20, 21]. Additionally, a small deviation in the charge circumstances due to adsorption of biomolecules can provide quantifiable variations in its properties. Based on excellent optical and electronic properties, CNTs and graphene have been frequently applied for sensing of biomolecules and drugs [22, 23]. Graphene based materials are novel and have significantly different properties than one dimensional carbon nanotubes and spherical nanoparticles [41, 42]. Thus, graphene and CNT were used to develop the sensors that will be helpful for real time monitoring of patient's health. In this scenario, a number of sensors/biosensors have been developed to monitor the biomolecules and drugs.

A sensor is a device that can detect or measure any change or reaction in chemical or physical stimulus and responds to various types of inputs or signals from the physical environment. A sensor contains two main fundamental components; one is a receptor that can interact with a specific substance/

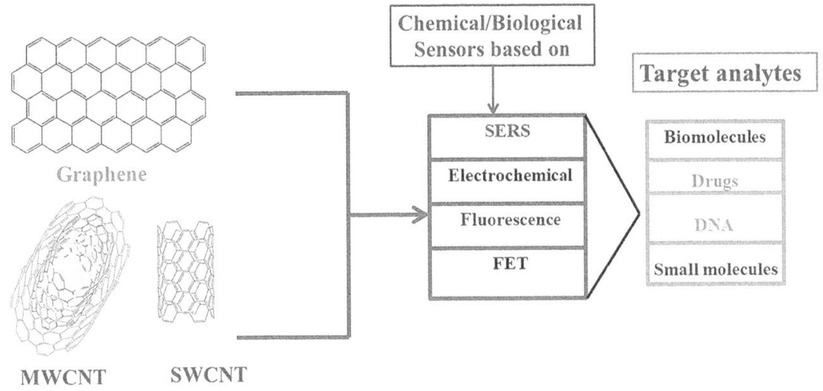

Figure 7.1 Representation of graphene and carbon nanotubes-based sensors based on different approaches, such as SERS, electrochemical, fluorescence, and FET based sensors, for detection of target analytes.

analytes, and the other is a transducer that can detect the observed event or changes (chemical or physical) in between the analyte [28]. Clark and Lyons proposed the first biosensor for the detection of glucose by using enzyme glucose oxidase [24]. The commercial and clinical application of the sensor for diabetes monitoring opened a new way for the research community and inspired many researchers to develop more similar systems to benefit mankind [27]. The detection of small molecules, protein, DNA, biomolecules and other disease related organisms or biomolecules and biologically significant drugs in real samples have become a crucial and fundamental part of human therapy and diagnostics. Therefore, in the last two or three decades, several attempts have been made to develop more sensitive and selective techniques/methods for analysis of biomolecules and drugs in real samples [25–27, 29]. A thousand tactics have been reported previously to develop the different types of sensors by using different types of modifiers to reach the trace level detection of clinically important biomolecules and drugs. This chapter deals with the graphene and CNT based electrochemical sensors, which are macroscale sensors, then fluorescence sensors, SERS based sensors, and finally FET–based sensors, which are nanoscale sensors as represented in the flow chart (Figure 7.1). This chapter also briefly explains the graphene-based materials and types of CNTs used for making the different types of sensors.

7.2 CLASSIFICATION OF GRAPHENE AND ITS DERIVATIVES

Graphene and related derivatives got a great deal of attention in all domains, like physics, chemistry, materials science, and others. Though they are promising candidates for sensors and biosensors application due of their exceptional properties, they easily form agglomerations, thereby limiting their

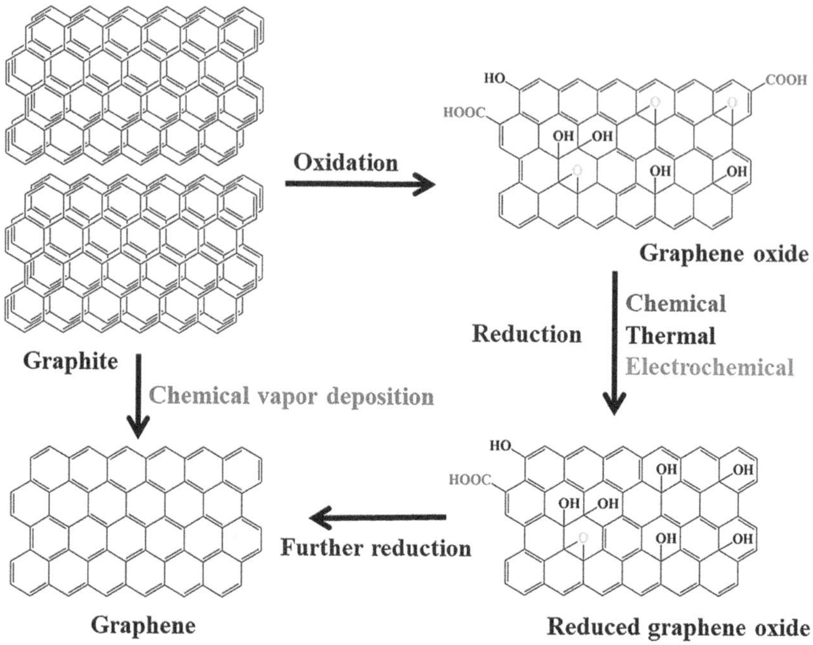

Figure 7.2 A scheme for synthesis of Graphene, GO, and RGO.

activity as sensors in terms of sensitivity and selectivity [44, 45]. To reduce agglomerating, various conducting polymers and metal nanoparticles are used [9]. Further, they have been employed for the determination of numerous biomolecules and drugs, like Deoxyribonucleic Acid (DNA), miRNA, small molecules, Antibodies, enzymes, protein, biomarkers, etc. [43–47]. Generally, graphene-based biosensors are developed based on two factors; i) electrostatic forces, ii) charge biomolecule interaction at π–π systems, and charge interchange leading to electrical changes in graphene. Another is based on the effect of distortion and chemical modifications to immobilization of the molecules on graphene oxide (GO), graphene quantum dots (GQDs), and reduced GO (RGO) [43]. Each sensor has different properties. There are several methods to synthesize graphene related materials. These methods are capable and are broadly classified into thermal, chemical, or electrochemical reduction and chemical vapor deposition (Figure 7.2). First, the basics of graphene and its derivatives are discussed, followed by their properties and synthesis methods.

7.2.1 Graphene

Graphene was developed by A. Geim and K. Novoselov in 2004. Later in 2010, both researchers were awarded the Nobel Prize in Physics [5, 6]. They used a "Scotch tape" process to grow the graphene. But the yield from this method was extremely low. To increase the production of graphene for their applications, new methods and surface treatment schemes have been developed [30]. Graphene is a 2D nanomaterial with high mechanical strength, structural, thermal, and optical properties [30]. The application of graphene and related derivatives demonstrates great promise in the field of sensors and biosensors due to its extraordinary conductivity, high specific surface area, biocompatibility, and exceptional ability to bind with biomolecules. It is capable of interaction with other materials, including metals, π–conjugate systems, and so on. In addition to pristine graphene, its other derivatives also have attractive properties. One main challenge for graphene is to attain the decent dispersion of the nano–scale filler in the composite materials. Good dispersion is important for attaining the desired improvement in the final chemical and physical properties of the composite materials. Graphene has a strong tendency to reconvert or agglomerate to graphite, due to intrinsic van der Waals Forces [31]. Numerous approaches have been proposed for synthesizes of graphene; however, chemical exfoliation [33, 34], mechanical cleaving (exfoliation) [34], chemical vapor deposition (CVD) [37] and chemical synthesis [38] are the frequently applied methods [32–35]. Other methods were also developed, like microwave synthesis [31] and unzipping of carbon nanotubes [35, 36].

7.2.2 Graphene oxide

GO is the oxidized form of graphene. It can be simply dispersed in physiological surroundings and water due to the presence of hydrophilic groups like carboxylic, epoxide and hydroxyl on its surface, which makes it hydrophilic [31]. GO has received significant attention and has been frequently applied for electrochemical sensors due to its extraordinary properties like larger surface area, water dispensability, and versatility of the surface modification. It is widely prepared by the oxidative treatment of graphite by using the Brodie method, Hummer's method, the Improved Hummers method, or the Staudenmeir method [39, 40, 43]. However, GO usually suffers from serious aggregation due to intermolecular interaction. The conductivity of the graphene oxide can be affected by the irreversible aggregation and poor dispersion after reduction [9]. To avoid the aggregation problems, other materials can be applied, such as conducting polymers, metal nanoparticles, etc. [9]. However, due to the presence of oxygenated groups, it becomes thermally unstable. GO has unique properties like larger surface area and dispersion ability in an aqueous medium. Thus, it was widely applied for the fabrication of sensors to detect different types of drugs.

7.2.3 Reduced graphene oxide

RGO is a reduced form of GO that is treated by using thermal, chemical, and other methods to reduce the oxygen percentage. To synthesize GO from graphite, strong oxidants are used, like $KMnO_4$. So, GO contains major amount of imperfections (or defects) in its structure. Hence, conductivity of GO is comparably lower than graphene although it's mechanical and optical properties have a lesser impact. GO can recover properties such as graphene by reductive exfoliation treatments that convert GO into RGO. At this stage, RGO has graphene–like properties, including good conductivity, and becomes a better compromise between graphene and GO. It can be easily prepared in desired amounts from GO by using various methods such as microwave, thermal, photo–assisted, and electrochemical methods. The diversity of the reduction methods leads to distinct qualities of RGO with (extremely) transformed properties depending on the amount of reduction [48–50].

7.3 GRAPHENE BASED SENSORS

7.3.1 Graphene based SERS sensor

In 1973, the SERS based approach was proposed by Martin Fleischmann. Then, the interest of the scientific community in Raman spectroscopy as spectroscopic and analytical techniques has been reenergized. The important theoretical and experimental efforts have been directed to understanding the SERS effect and showing its impact on numerous types of highly sensitive sensing applications in various realms. SERS is an effective method that enhanced Raman scattering by absorbing the substance/molecules on the surface of irregular metal. It is an important method for the investigation of specific molecules with extreme sensitivity and selectivity [47, 51, 52, 65]. Graphene has strong interaction with target molecules and provides an ideal flat surface and applied as Raman enhancement substrates. SERS sensors developed based on the SERS have been considered as promising systems for selective determination of trace amounts of biomolecules [47, 53]. The Raman enhancement effect of graphene is primarily attributed to charge transfer among target molecules and graphene [55]. Graphene may be employed as a substrate to improve the Raman signals of adsorbed molecules [56]. It is a fast, sensitive, and non–destructive spectroscopic technique for several analyses, has received huge attention, and displays wide application in biomedicine, diagnosis, catalysis monitoring, food safety, environmental analysis and bio–chemical sensing [56]. SERS is a highly effective platform for the determination of a very small amount of target analytes, such as microRNA, DNA, bacteria, blood, and proteins; bio–imaging; single–cell identification and detection; and disease diagnosis, providing the various important structural information for biological analytes [66]. Several biomolecules and biomarkers, such as nucleosides, DNA, protein bacteria, and fungi, have been detected by using graphene and its derivative

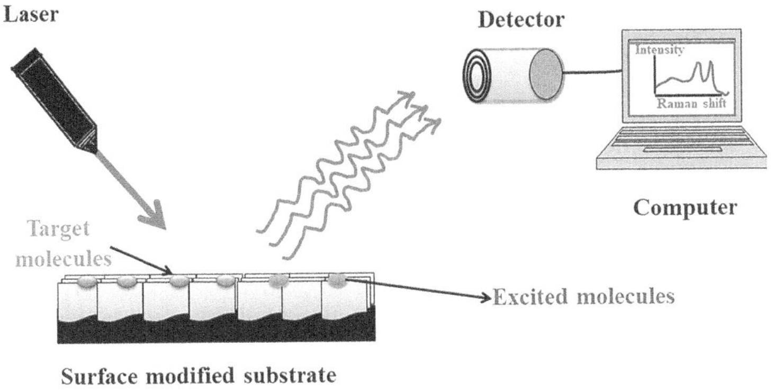

Figure 7.3 The sensing scheme of target molecules on the SERS enhancement substrate.

substrates [56, 60–64]. This technique is also suitable for practical application because of its low detection limit. Carbon based nanomaterials are widely applied for developing SERS based biosensors/sensors [66]. A schematic sensing representation has been shown in Figure 7.3.

Muntean et al. [62] developed the SERS platform for biomedical applications at low pH. Wang et al. reported the ultrasensitive SERB based sensor for the analysis of dopamine and serotonin by using graphene–gold nanopryamid [70]. Neri et al. develop the nitrogen doped graphene/gold nanocomposite-based SERS sensors for determination of dopamine [69]. Zhao et al. [57] developed hybrid materials to improve the Raman response up to 137–fold of graphene, and a limit of detection was found 0.1 pM for rhodamine 6G molecules. Moreover, the RGO–SnO_2 showed 50% enhancement in sensitivity for carbon dioxide gas investigation as compared to a commercial SnO_2 related gas sensor [54]. Lin et al. [58] proposed the silver nanoparticles (AgNPs) and the GO (AgNPs–GO) based composite SERS sensing system in the existence of two complementary DNAs. The limit of detection was found on the pM level for target DNAs at self–assembly of larger sized AgNPs onto AgNPs–GO. He et al. [59] proposed gold nanoparticles (AuNPs) modified CVD growth graphene-based system, which is applied for the determination of DNA. The proposed system shows exceptional specificity and sensitivity for the investigation of DNA. The limit of detection was observed as 10 pM. The two simultaneous DNA targets can be detected by using the same protocol while applying one light source.

Zhu et el. [67] proposed the SERS based glucose biosensor by using GO–decorated silver–coated gold nano bones. Luo et al. [68] applied the RGO with silver nanotriangle composite sol to develop the SERS sensor for dopamine. The limit of detection was found to be 1.2 μM, on the label–free probe (Figure 7.4).

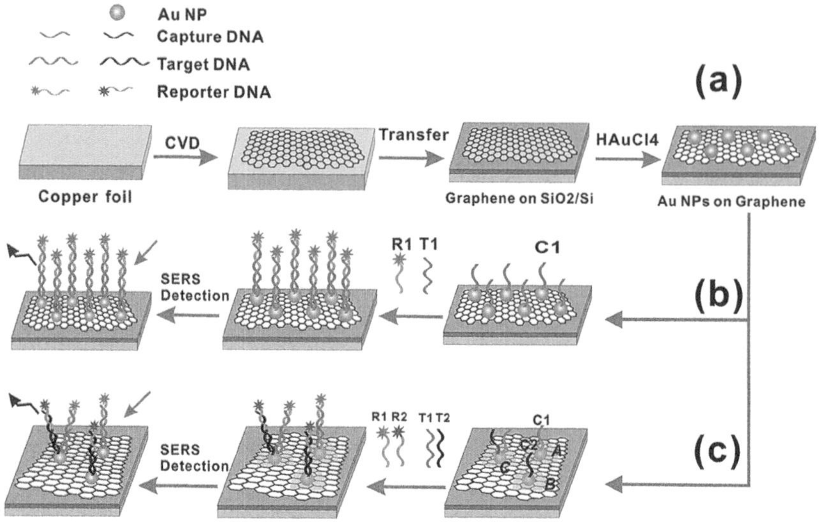

Figure 7.4 A flow diagram of SERS–Active Substrate analysis of DNA. a) Graphene films grown at copper foil substrate by CVD and transferred onto SiO2/Si substrate and deposition of AuNPs. (b) SERS based sensor for the determination of DNA on Au–G–SiO2/Si. (c) The analysis of multiplexing of two different DNAs at developed sensor; A, B, and C show different locations for multiplex DNA analysis. Reproduced from [59] with permission (2012, American Chemical Society).

7.3.2 Graphene based electrochemical sensor

The development in the electrochemical sensors has received huge attention in the research community in the last few decades. The researchers are actively looking for impactful and cost-effective materials to develop the sensitive and selective sensors. In recent years, nanotechnology and nanomaterials have opened a new prospective to design and fabricate the nanomaterials modified sensors for electrochemical applications [71–75]. To investigate the diverse types of biomolecules and chemicals at the electrode, the main factors should be to investigate first such issues as electron transfer rate, electrochemical potential window, redox potential, etc. [72]. Most significant, the sensors or biosensors should be more specific and sensitive towards specific analytes. A variety of effective nanomaterials with well–controlled structural properties have been used to modify the electrode to improve the electrochemical response [74]. Due to increasing the demand for highly selective and sensitive sensors in clinical, pharmaceutical, and industrial sectors, graphene related materials have gained much attention worldwide. Graphene has novel structural and extraordinary chemical and physical properties. These properties makes it suitable for sensor and biosensor applications [75–77, 81, 83, 88, 89]. Graphene supports the fast electron transfers that assist selective and accurate investigation of biomolecules and drugs [77, 78, 84, 88]. RGO and GO are frequently applied in various applications, such environmental monitoring, food industry, and biomedical fields, as well as being used to biosensor device fabrication [90]. The richness of oxygenated groups in the hybrid materials act as a catalysis for the fabrication of sensors, which helps in surface functionalization, recognition of species, and friendliness with the micro– and nano– bio–environment [82]. In the literature, several strategies have been applied to development of growing the RGO, but the electrochemical method was considered as a simple and green methodology to reduce the GO to RGO [86, 87, 91, 95]. Graphene based sensors have also been realized in the immobilization of enzymes, like horseradish peroxidase, glucose oxidase, and hemoglobin, etc., and selective analysis of various biomolecules [92–93]. A schematic diagram is represented for surface modification of bare electrodes and their electrochemical response for the target analyte (Figure 7.5).

Sun et al. [79] proposed a graphene-based sensor for detection of caffeine. The Nafion–graphene modified GCE demonstrated excellent electro–catalytic activity towards caffeine. The electrochemical sensor demonstrates satisfactory performance for detecting caffeine with a 0.12 μM and shows the recovery greater than 98%. Adhikari et al. [80] reported a sensor for determination of acetaminophen. The electrochemical method has been used to reduce the GO to ERGO at GCE surface (Figure 7.6). The ERGO was applied to detection of acetaminophen by using chrono–amperometry, CV, and DPV. The interference studies were also executed with commonly present interfering biomolecules, such as uric acid, adenine, glucose, and sucrose, and with combination of all these biomolecules with 40 μM, only 3% deviation was observed. The results showed that the ERGO/GCE sensor has good catalytic activity electro–oxidation of acetaminophen in the rages of 5.0 nM – 800 mM. The lower limit of detection was observed, 2.13 nM and greater than 96% recovery in human serum sample.

Peng et al. [85] proposed graphene decorated GCE for the analysis of vanillin. A drop casting method has been used to modify the electrode surface. First, graphene (1 mg) suspension was prepared in 1 mL N–N–dimethylformamide

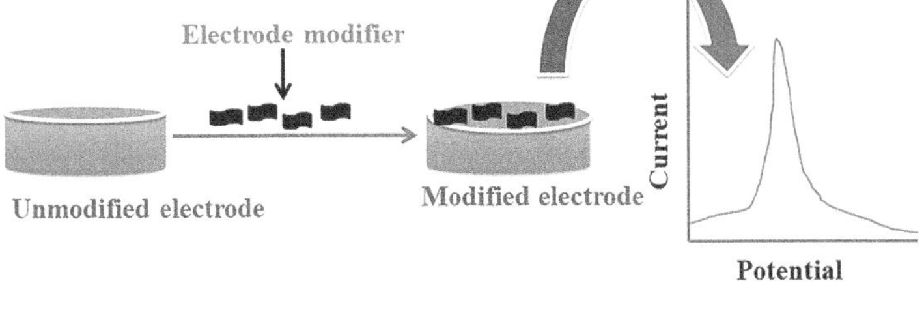

Figure 7.5 A schematic representation of the surface modification of the bare electrode.

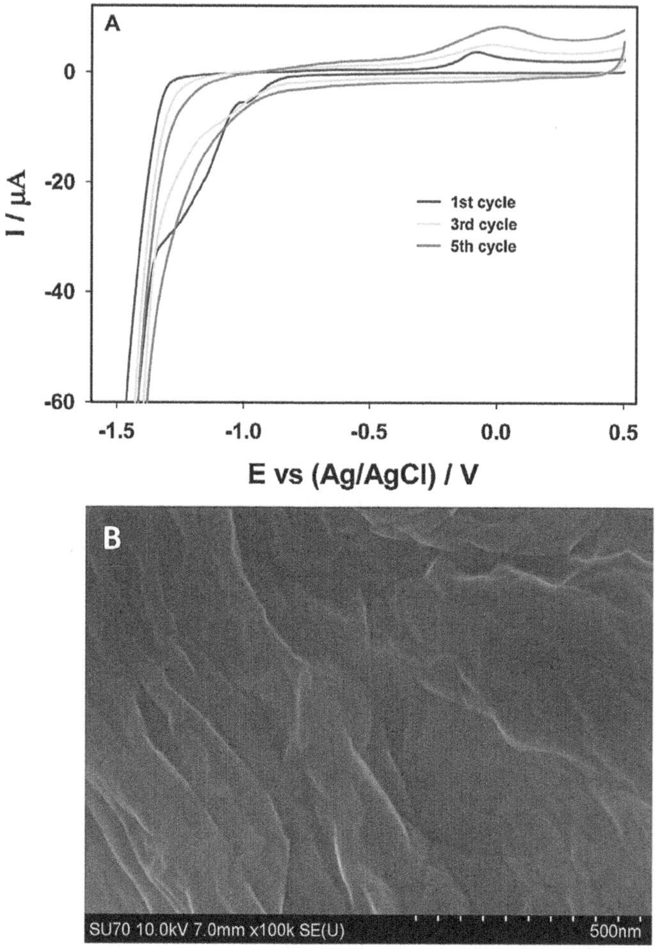

Figure 7.6 (A) Electrochemical reduction of GO (0.3 mg/mL at 7.2 pH in PBS) into electrochemical reduced graphene (ERGO) at 10 m Vs^{-1} sweep rate. (B) SEM image of ERGO at glassy carbon surface. Regenerated with consent from Ref. [80] Copyright @Elsevier.

(DMF) and drop casted on GCE. The oxidation current was found proportional to the concentration 0.6–0.48 µM range with 0.056 µM LOD. The graphene-based sensor was employed for selective and sensitive analysis of hydroquinone at GCE. Fritea et al. [94] proposed a system to develop the RGO decorated with electrochemically generated AuNPs for selective and sensitive determination of nitrazepam. The synergic effect of both materials provides the excellent catalytic activity, high conductivity, and surface area. The sensor showed better response in the range of 0.5–400 µM with 0.166 µM LOD. The nitrazepam was determined in serum sample and a pharmaceutical sample with excellent recovery rate (99%–102.4%). Kumar et al. [95] applied the electrochemical method to develop the Au–PdNPs decorated graphene oxide at GCE for detection of two important antibiotic drugs. Both lomefloxacin and amoxicillin were detected by using AuNP–PdNP modified ErGO (AuNP–PdNP–ErGO). The linear calibration curves for amoxicillin and lomefloxacin were observed in the range of 30 – 350 μM and 4 – 500 μM, respectively. The LOD were found to be 9 μM and 81 nM for amoxicillin and lomefloxacin, respectively. The proposed platform was also used for analysis of both drugs in the presence of a complex matrix such as urine samples, which contains a high concentration of interfering molecules like uric acid, hypoxanthine, ascorbic acid, etc. The practicable applicability of the sensor was also investigated in pharmaceutical samples. Wang et al. [96] proposed ionic liquid functionalized graphene biosensors to sense the uric acid and dopamine content. The GCE was modified with 1–butyl–3–methylimidazolium 2–amino–3–mercaptopropionic acid salt modified graphene used for estimation of uric acid (UA) and dopamine (DA) with ascorbic acid (AA). The peak–to–peak separations of oxidation potential of AA–DA, AA–UA and DA–UA were found as 147, 292, and 145 mV, respectively. The electrochemical biosensor used to detect the dopamine and uric acid in urine samples by applying the standard addition plot method with more than 97.0% recovery. Li et al. [97] used graphene and AuNPs composite film to modify the GCE and found significant improvements in the oxidation peak current response and oxidation peak potential difference between DA and AA as compared to unmodified electrode. Yang et al. [98] applied the ERGO to modify the GCE for simultaneous investigation of ascorbic acid, uric acid, and dopamine. The ERGO decorated electrode has high electro–catalytic activity compared to unmodified GCE toward the oxidation of AA, DA, and UA. Zhang et al. [99] proposed nitrogen–doped graphene materials for the determination of DA, AA, and UA. The developed composites were characterized by using Raman, SEM, and Infrared (IR) spectroscopy. The linear ranges to develop sensors for DA, AA, and UA were 1.0 – 60 μM, 0.1 – 4 mM, and 1.0 – 30 μM, respectively. The detection limit was observed for DA, AA, and UA as 0.1 μM, 9.6 μM, and 0.2 μM, respectively.

7.3.3 Graphene based fluorescence sensor

The fluorescence of GO is due to electronic energy transitions. It has been seen that each "fingerprint" fluorescence band of GO demonstrates because of specific electron transitions among the bonding and anti–bonding molecular orbitals, like π^*–n, σ^*–n, and π^*– π. GO can have higher oxygenated functional groups, such as carboxyl group (COOH), hydroxyl groups (COH), epoxy groups (C–O–C), carbonyl (C=O), and aromatic rings (C=C). Therefore, the several fluorescence peaks were found related to different electronic transitions and excited simultaneously. However, a broad fluorescence peak is observed due to overlapping to each other of fluorescence peaks. GO and graphene quantum dots are pretty fluorophores, which are photo–stable, non–toxic, water soluble, biocompatible, environmentally friendly, and cost effective. Hence their wide applications as fluorescent chemi–sensors and biosensors, in which they either serve as quencher or fluorophores. In recent years, various GO related fluorescent sensing methods have been described for the investigation of small substances, metal ions, DNA, proteins, glucose, and other chemicals [100, 105, 106, 111, 116–118, 120, 121]. As quenchers, GO demonstrated an extraordinary quenching efficiency through its Forster Resonance Energy Transfer (FRET) or electron transfer process. As fluorophores, they showed the "Giant Red–Edge Effect" and tunable photoluminescence emission. The process made excited state fluorophores (donor) and adjacent ground state fluorophores (acceptor). In the presence of external light excitation, emission of the donor will excite the acceptor via fluorescence energy transfer [100, 104]. GO showed a novel platform to develop the DNA related biosensors for two main properties: (i) GO absorbed DNA with appropriate affinity, (ii) its fluorescence quenching properties [109, 113]. Moreover, GO can attain adsorption of biomolecules via hydrogen bonding, electrostatic interaction, or π–π interactions, which can also offer feasible parameters for interaction between biomolecules [112–115]. Thus, these properties make graphene and its related materials feasible for fluorescence sensors. Tan et al. [113] reported a fluorescent apta–sensor for the estimation of leukemia by using GO–aptamer complex (GO–apt).

The fluorescence determination has become an effective technique, since the development of fluorescent nanomaterials such as graphene oxide, quantum dots, dye–loaded silica–NPs, etc., due to its excellent sensitivity, low background noise, easy surface manipulation, and low fabrication cost. Moreover, the attractive properties of graphene–based materials, like larger surface area, cyclic organic network, and high percentage of surface atom number, make them suitable for other organic molecules through π–π interactions, hydrogen bonding, or other probable hydrophobic interactions. Graphene and their composites were extensively applied to the investigation of various small biomolecules, enzymes, protein, DNA, and many more small molecules [101]. Graphene quantum dots also have attractive properties and are applied in a fluorescence platform for biological labeling, drug, and biomolecule delivery, diagnostics, and electronic devices for environmental monitoring [103]. Cheng et al. [102] reported a fluorescence sensor for analysis of glutamate in aqueous solution and serums. The GO was modified with 5–aminofluorescein for analysis of glutamate. The good linear relation was found from 1 to 45 mg/L for glutamate. In serum, the linear range of glutamate detection was 6 – 30 mg/L. The study demonstrated the specificity and selectivity of the protocol for the determination of glutamate in aqueous solutions and bovine serums. In other study, Tomita et al. [105] develop a signature–based protein sensor by using DNA/GO conjugate. The signature–based protein sensing has been considered as an effective substitute to standard lock–and–key methods. ssDNA strongly adsorbs on GO due to its π–stacking and hydrophobicity between the cyclic organic network of GO and the DNA nucleobases. In another study, Kitamura et al. [107] develop the GO amplified fluorescence sensing protocol for the determination of nucleic acid with the method of DNA circuit. The fluorescence signal amplification of a DNA sensor was amplified using a catalytic hairpin assembly. The LOD was observed to be 0.1 nM. The fluorescence signal was improved by 50 to 100–fold at sub–nanomolar level at 35^0C under isothermal parameter. In addition to this, Weng et al. [108] used self–assembled multilayers of graphene to fabricate a surface–enhanced fluorescence platform. The specificity for detection of vitamin B_{12} with graphene–based platform was found higher than 50–times compared to the case when fluorophores are not attached to the surface of graphene. Kushwaha et al. [110] explored GO based fluorescent biosensor for selective and sensitive analysis of the estriol. The fluorescence and quenching properties of GO was used as an energy receptor in the FRET based sensor. The linear range was found between 0.0 to 10.0 nM, with LOD of 1.3 nM. Chen et al. [112] reported fluorescent aptasensor for the determination of mercury(II), silver(I), cysteine, thrombin, and sequence–specific DNA. The linear range for Ag^+, Hg^+, cysteine, sequence–specific DNA, thrombin were found as 0.5 – 30 µM, 0.5 – 50 µM, 0.5 – 6.0 µM, 0.005 – 1.0 µM, and 0.01 – 20 µg/milliliter and the limit of detection were observed as 200, 400, 300, 3.0 nM, and 0.5 µg/milliliter, respectively. The probe–GO platform was versatile and excellent for determination of multiple target analytes. Xiang et al. [119] reported a fluorescent approach for determination of sinapine. The GO related fluorescent platform was found fast and selective for detection of sinapine (Figure 7.7). The investigation of sinapine (0.04 – 1.5 µg/mL) was conducted at optimized concentration of fluorescein amidite decorated single–stranded DNA (FAMssDNA) probe and GO. The selectivity of the proposed protocol was analyzed in the presence of the commonly observed analogues in plants, such as sinapine, cinnamic acid, sinapic acid, caffeic

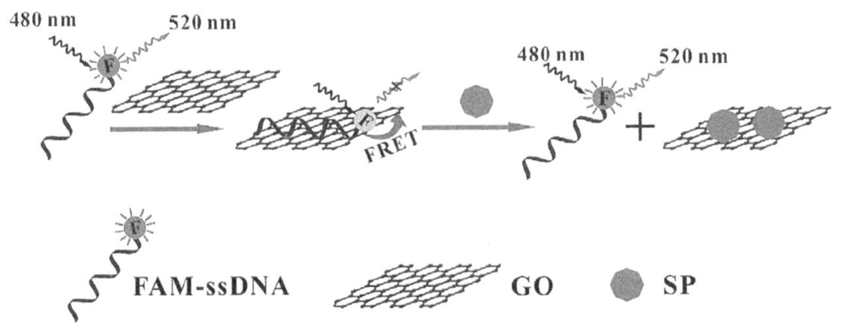

Figure 7.7 Schematic representation of GO related fluorescent platform for determination of sinapine. FAM–ssDNA, SP, sinapine. Regenerated with permission [119] Copyright @Elsevier.

acid, protocatechuic acid, ferulic acid, p–coumaric acid, and syringic acid. The recovery studies were carried out in real samples and higher than 93.4% recovery found in meal samples. The present sensor may be used for detection of the plant derived small molecules in real samples.

In another approaches, a sensing system for detection of pathogenic bacteria, simultaneous investigation of multiple relevant RNAs and hemin by using DNAzyme/graphene hybrid material, GO and DNA, and RGO as nano quencher were applied [121–123].

7.3.4 Graphene based FET sensor

A FET contains a gate electrode, channel section connecting source electrode and a drain electrode, and a barrier isolating the gate from the channel [130]. A semiconducting material is applied to make the channel between the source electrode and drain electrode, although an insulating material is required to separate the channel from the gate electrode. Thus, the current flow through the channel between source and drain (source–drain current, I_{SD}) is regulated by voltage applied between source–gate (gate voltage, V_G) at a constant voltage applied between source and drain (source–drain voltage, V_{SD}) [130, 133, 134]. In the last two decades, the FET related systems have also been receiving much attention for analysis of biomolecules due to their high sensitivity, cost reduction, label–free operation, and miniaturization. A FET related sensor is also known as a biosensor FET (Bio–FET) or field–effect biosensor. A FET sensor is gated by variations in the surface potential encouraged by the adsorption or binding of molecules. When target substances, like analytes/biomolecules, adsorbed on the FET gate electrode can change the charge distribution of the semiconducting material or channel leads to a change in conductance of FET channel. A FET system has two major sections; one is the FET, and another is the biological recognition element [124–128, 139]. However, the bulk channels, applied in planner devices, reduced the interaction between channel and target analytes only on the channel surface, demonstrating the low sensitivity of the FET sensors [124, 125]. The high transporter mobility and novel band structure of graphene make it an ultimate material to be applied in integrated circuits. The holes and electrons can be continuously tuned to change the Field–effect. The electronic investigation of target analytes is based on the conductance change of FET semiconducting channels upon adsorption of molecules [129]. In recent years, the 2D–materials, graphene, and related materials have been widely used in electronics for various purposes such as active material for energy harvesting, electronic applications in transparent electrodes and storage, and channel stuffs for FETs [125]. The graphene-based FETs have stabilized quickly and are now considered an alternative for post–silicon electronics [130]. Besides the excellent electronic properties, graphene demonstrates the highly flexibility and biocompatibility, simple chemical functionalization, and larger surface area compared to CNTs, making it suitable for sensors. More prominently, all atoms of carbon interact directly with the target molecule due to one atomic thickness of graphene. It is the most effective feature of graphene to apply in transistors, and mobility is the most useful property of graphene for application in nano–electronics [125, 130]. A graphene-based FET biosensor was reported for studying the binding affinities and kinetics between protein and drugs study. The single crystal graphene sheets were synthesized and used to fabricate the FET based biosensors. It is capable of converting the molecular interactions into electrical signals [136, 137].

An RGO–based FETs platform was proposed for the sensing of dopamine. The electrochemical pulse deposition technique is used for deposition and reduction of GO among two carbon nano–electrodes to form the channel of the RGO related FET. The fabrication procedure was estimated and optimized to develop the RGO related FETs with excellent performance. The developed device was found sensitive for sensing the dopamine in the range 1.0 nM to 1.0 µM and shows the excellent performance with 500 µM ascorbic acid [131]. An organic electrochemical transistor (OECTs) for the investigation of dopamine was reported by Liao et al. [132]. The improvement in selectivity of the proposed sensor was achieved by using the biocompatible polymer chitosan or Nafion on the gate electrode surface. The interference of the AA and UA was removed by using the Nafion. The sensitivity of the developed device enhanced by using the graphene co–modify on the gate electrode. Graphene flakes were applied to enhance the detection limit and response of the fabricated protocol, due to the larger surface to volume ratio and excellent charge transport properties of graphene. A linear calibration was found 5 nM – 1 µM and limit of detection was observed lower to 5 nM at fabricated device. The diagram of OECTs–based sensor for the estimation of dopamine by using Nafion and graphene is represented in Figure 7.8.

Fenoy et al. [135] describe a strategy for detection of Acetylcholine by using graphene–based FETs. The acetylcholinesterease was immobilized on graphene FETs to develop the Acetylcholine sensor. The constructed sensor demonstrated a limit of detection 2.3 µM and the concentration of acetylcholine was investigated in the range of 5 – 1000 µM. The developed biosensor showed the good device–to–device reproducibility, excellent selectivity with interfering molecules, and greater than 97% recovery. You et al. [138] proposed a FET enzymatic sensor based on silk fibroin–encapsulated graphene for analysis of glucose. The developed FET biosensor exhibited ambipolar transfer properties. The fabricated biosensor can detect the glucose amount through detecting the Dirac point shift and the differential drain–source current. The proposed biosensor demonstrated larger linear range from 0.1 – 10 mM, which is suitable for diabetes analysis, and the detection limit was ~ 0.1 mM. Shin et al. [140] reported a palladium nanoflower (PdNFs) modified graphene by glucose oxidase (GOx) and a Nafion based system for detection of glucose concentration in human biological samples.

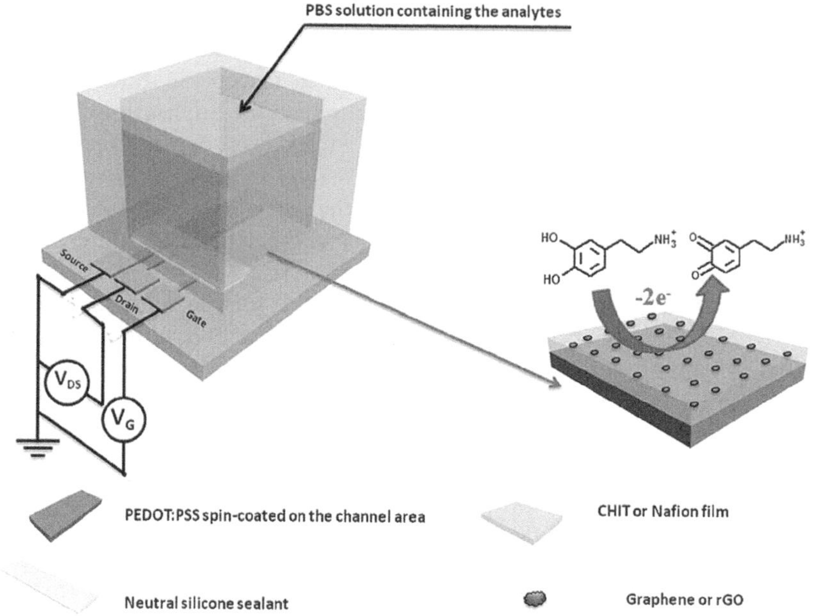

Figure 7.8 Schematic representation of an OECT based sensor for dopamine detection. Restored with permission [132] Copyright 2014 @The Royal Society of Chemistry.

The graphene was applied as substrate for electrodeposition and p–type channel for FET based electrode. The shape–controlled palladium nanoflowers were synthesized by adjusted the H_2SO_4 amount in the electrolyte in the electrodeposition. The GOx and Nafion were electrodeposited on the shape controlled PdNFs modified graphene by using spin coating method. The proposed FET sensor demonstrates the high sensitivity towards the glucose and shows the high selectivity for AA and UA with lower LOD for glucose (1 nM). Schuck et al. [141] reported a biosensor for determination of L–lactic acid concentration in different samples (plasma and buffer solution). The linear relation was found from 0 to 7.5 mM at the biosensor device in the different fluids. The selectivity of the developed biosensor was also performed in the presence of uric acid, ascorbic acid, and glucose, and the biosensor device operation was found stable for 12 days.

7.4 CLASSIFICATION OF CARBON NANOTUBES

Carbon nanotubes (CNTs) were also important allotropes of the carbon and can be considered as members of the same group due to the presence of sp^2 hybrid carbon atoms arranged in a hexagonal network. These have a common structure; however, they have significant differences in their

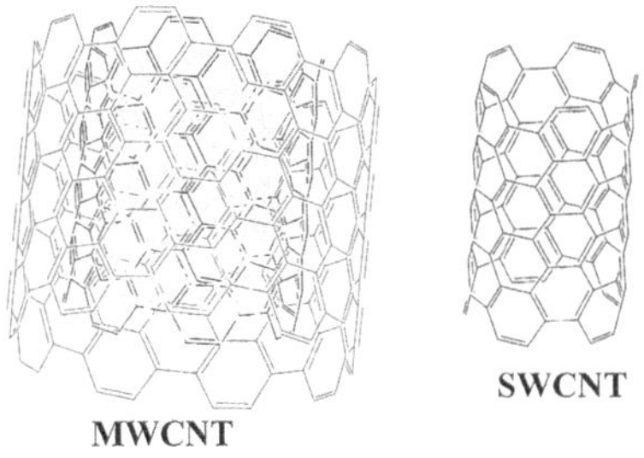

Figure 7.9 Structure representation of MWCNT and SWCNT

shapes and sizes. CNTs also have similar electrical, mechanical, optical, and chemical properties [149]. In 1991, Sumio Iijima was the discoverer of the microtubules of graphitic carbon with 4–30 nm outer diameters and up to 1 mm length. These tubes contained two or more unified graphene cylinders concentrically assembled and considered as multi–walled carbon nanotubes (MWCNTs) due to the tubes containing diameter in nanometers. The field of nanoscience and nanotechnology explores their investigation to develop CNTs with suitable features for future applications. Several approaches have been reported for synthesis of the CNTs, including arch–discharge, chemical vapor deposition (CVD), flame pyrolysis, bottom–up organic approach, laser ablation, etc. [144]. However, CVD, laser ablation, and arc–discharge are applied frequently to grow the CNTs. Apart from these methods, electrolysis, the hydro–thermal method, and Ball milling methods have also been applied to synthesize the CNTs [148–150]. Some other approaches were also applied to produce the diverse types of CNTs, like as SWCNTs, MWCNTs, double walled carbon nanotubes (DWCNTs), and open ended carbon nanotubes [144, 148, 150]. In this chapter, we will only consider the application of the SWCNTs and MWCNTs in sensors. Carbon nanotubes (CNTs) are divided into two groups, the SWCNTs and MWCNTs, based on their structural features and properties (Figure 7.9).

7.4.1 Multi–walled carbon nanotubes

In 1991, Sumio Iijima prepared MWCNTs with inner diameter ~4 nm [142, 143]. The discovery of CNTs is credited to Iijima, as the first scientist who develop the MWCNTs after a random event during the experiment of a new arc evaporation method for C_{60} synthesis [144]. In the arc discharge method, usually a high temperature (above 1700 °C) is used for CNT synthesis and applied to grow the CNTs with less structural defects in comparison to other approaches. The CNTs have since attracted huge attention in the research community, due to their exceptional properties and unique structure. The MWCNTs have concentric graphene tubules with low van der Waals bonds among the shells and powerful C–C bonds with the graphene shells. These can be synthesized in larger quantities by using the arc–discharge method or by CVD methods. The MWCNTs usually have 5–50 nm outer diameters and can be one micron long [145].

7.4.2 Single–walled carbon nanotubes

In 1993, Sumio Iijima NEC and D.S. Bethune of IBM, corporation independently prepared the SWCNTs. SWCNTs are made of a graphene sheet, which contains uniform cylinders. SWCNTs are considered as a "rolled up" form of one–atom–thick sheets of graphene. The diameters vary from 0.4 to 2–3 nm, and length is generally in micrometers. These are hexagonally placed in bundles to form a crystal–like structure [142, 143]. The SWCNTs were initially started as a smaller side product in the arc–discharge method. The high purity was observed in the catalysis–assisted arc–discharge method and laser ablation. These are typically 1–10 micron in length and 1–2 nm in diameter. They were found close–packed bundles by van der Waals inter–nanotube bounding [145–148].

7.5 APPLICATION OF CARBON NANOTUBES–BASED SENSORS

7.5.1 Carbon nanotubes-based SERS sensors

Since the development in the SERS method, it has been used in various realms like single–molecule analysis, biomolecules, drugs, and environmental monitoring, etc. [151, 152]. MWCNTs have nano–scale hollow tubular morphology, strong adsorption capacity, larger specific surface area, durability, and stability. The carbon atoms in CNTs make π–bonds, and π–electrons can easily combine with molecules having π–electrons, structures (like benzene rings) via π–π bonding [153]. CNTs seem biocompatible and crucial for better SERS sensing of biomolecules or biological substances. These properties of CNTs are dynamic for a SERS biosensor compound with a high usual enhancement effect [154, 155]. Zhang et al. [153] reported a method for the investigation of deltamethrin by using MWCNTs as dispersive solid–phase extraction sorbents coupled with SERS. Chen et al. [151] developed a pH sensor based of p–aminothiophenol functionalized Au@Ag/MWCNTs by using the SERS method. The p–aminothiophenol modified Au@Ag NPs/MWCNTs demonstrated high biocompatibility for intracellular pH detection and can be applied for cell physiological processes.

7.5.2 Carbon nanotubes-based electrochemical sensors

CNTs have developed enormous attention in electrochemical analysis and bio–sensing because of their exceptional electronic, chemical, and morphological properties, for illustration, novel tube–like nanostructure, excellent conductivity, larger surface area, good biocompatibility, adjustable side wall, high reactivity, and so on. These are commonly applied in optoelectronic and electronic, biomedical, energy, pharmaceutical, catalytic, material, and analytical fields. These properties make them ideal materials for electrochemical sensors. CNTs can efficiently enhance the redox currents of the target analytes (organic compounds, inorganic molecules, macro–biomolecules, or even biological cells, and help to minimize the redox over potential. Along with the chemical stability, well–defined structural properties, and electrocatalytic activity with minimal surface fouling towards many substance, CNTs are widely used as carrier platforms for developing the sensors. Several electrochemical sensors and biosensors were reported by using the CNTs. Electrochemical sensors based on CNTs were found simple, reproducible, sensitive, and selective, with ease of miniaturization. The unique properties of CNTs have smoothed the fabrication of a wide range of sensors and biosensors [156–159].

Savk et al. [160] reported an MWCNTs based sensor for determination of UA, AA, and dopamine. The sensor was fabricated by using ZnNi bimetallic nanoalloy modified MWCNT (ZnNiNPs@f–MWCNT) for simultaneous investigation of UA, dopamine, and AA. The ZnNiNPs@f–MWCNT was found by the microwave irradiation method. The developed sensor demonstrated high electrochemical activity towards the target analytes, and linear responses were found for AA, dopamine, and UA in the range 0.3 to 1.1 mM, 0.2 to 1.2 mM, and 0.2 to1.1 mM, respectively. The limit of detection was observed as 0.0655, 0.511, and 0.0882 μM for dopamine, AA, and UA, respectively. In another study, Anirudhan et al. [161] used the MWCNTs based molecularly imprinted polymer (MWCNTs–MIP) for the analysis of dopamine in urine and blood serum samples. The acrylamide graphite MWCNTs with the vinyl group was synthesized by the free radical grafted co–polymerization method. Molecularly imprinted polymer was fabricated by polymerization of MWCNTs–acrylamide graphite MWCNTs with the vinyl group with itaconic acid as the monomer with target molecules dopamine by α,α′–azobisisobutyronitrile as a initiator and ethylene glycol dimethacrylate as a cross–linker. The modified sensor can be applied to investigation of dopamine in the linear range of 10^{-9} to 10^{-5} M. The response time of the sensor was ~2 minutes, and the limit of detection was 1.0 nM. The lifetime of the developed dopamine sensor was found longer than two months and was tested in real samples. In addition to this, Huang et al. [162] reported an electrode based on composite of graphene quantum dots/MWCNTs for ultrasensitive analysis of dopamine. The graphene quantum dots enhanced the conductivity of the electrochemical sensor due larger surface area, and MWCNTs were considered as excellent electrode materials. The excellent selectivity of the sensor was expected for dopamine among common interfering biomolecules. The sensor showed the broad linear range from 0.0005 to 100.0 μM with the LOD of 0.87 nM. Moreover, the developed sensor was employed for estimation of dopamine in human serum. Kan et al. [163]

proposed a molecularly imprinted polymer system for analysis of dopamine. The electrode was fabricated by electro–polymerization of pyrrole with dopamine after electro–polymerization carboxyl functionalized MWCNTs (MWCNTs–COOH) on GCE. The DPV response was found to be linear in the range from 6.25×10^{-7} to 1×10^{-4} M, with 6×10^{-8} M limit of detection. Ardakani et al. [164] developed a carbon paste electrode modified by CNTs and a synthesized hydroquinone derivative. The electrochemical sensor was used for simultaneous detection of acetaminophen, norepinephrine, and tryptophan. The oxidation potential of norepinephrine decreased about 220 mV at the modified electrode in comparison to a bare electrode, due to electrocatalytic oxidation of norepinephrine through the electrochemical mechanism at a fabricated electrode. The DPV was applied to the investigation of norepinephrine, and two linear ranges of 0.2 to 20.0 µM and 20.0 to 1500.0 µM were found, with a detection limit of 40.0 nM. Also, simultaneous investigation of norepinephrine, tryptophan, and acetaminophen was performed by using modified sensors, and in the range of 20.0 to 800.0 µM was observed for acetaminophen and tryptophan. Kumar et al. [165] proposed the palladium NPs decorated MWCNTs (PdNPs: MWCNTs) for electrochemical sensing of 5–hydroxytryptophan. The PdNPs:MWCNTs was used to decorate the GCE and applied for investigation of 5–hydroxytryptophan. The square wave voltammetry (SWV) was used for quantitative analysis of 5–hydroxytryptophan and in linear range from 2 – 400 µM. The limit of detection was observed 77 nM with 0.2122 µA/µM sensitivity. The fabricated electrochemical sensor demonstrated the excellent selectivity in the existence of commonly present inferring biomolecules in the urine samples. Yapasan et al. [166] explored the application of MWCNTs–COOH modified graphite electrode for electrochemical investigation of bio–molecular interactions of platinum derivatives.

7.5.3 Carbon nanotubes-based fluorescence sensors

The SWCNTs have been considered as ideal candidate for optics and optoelectronics because of direct band gap, well–defined band, and sub–band structure [167]. The outstanding fluorescence quenching efficiency of carbon-based nanocomposite was observed for organic dye and quantum dots. Carbon based materials frequently applied as fluorescence quenchers and energy acceptors. Hence, the homogeneous FRET suspension tests based on quantum dots and carbon nanomaterials are substantial for the DNA monitoring [171]. The CNTs fluoresce is in a region of the Near Infrared (NIR) [168]. The inherent NIR fluorescence is a newly developed optical behaviour of the single SWCNT. The fluorescence of SWCNT opens the new area for highly effective investigation in biological systems and complex environmental areas, due to photo–stability, high sensitivity, and the minimum auto fluorescent background at NIR wavelengths in biological media [174]. But, the fluorescent approach has inherent merits, like the possibility for investigation of a broad range of substances, including those that are redox inactive, the significance for fabrication for nano–sensors functioned in confined environments like realizing in vivo imaging, living cells. The CNTs based fluorescence sensors usually depend on either the NIR photoluminescence (PL) of semiconducting SWCNTs or on quenching ability of CNTs over conventional organic quenchers [167]. Thus, fluorescence-based sensor has become an interesting topic between researchers. Fluorescence based sensors can also be applied for detection of metals ions. A DNA/SWCNTs related fluorescence sensor was used for estimation of mercury (Hg^{+2}) [170]. A pH sensor was developed by using the fluorescence quenching of CNTs [173]. Hu et al. [169] proposed a CNTs based fluorescent method for investigation of adenosine deaminase and inhibitors screening by adenosine as the substrate. The developed sensor can identify adenosine deaminase effectively from 0.005 – 0.2 U/mL with 0.002 U/mL limit of detection. In another investigation, Li et al. [171] reported a FRET based protocol by using effective self–assembly between the quantum dots and MWCNTs@GO nanoribbons to form the fluorescent "on–off–on" switching for simultaneous detection of dual target DNAs of terminator nopaline synthase (TNOS) and cauliflower mosaic virus 35s (P35s). The developed sensor showed the lower LOD 0.35 nM for P35s and 0.5 nM for the terminator nopaline synthase. In other study, Qian et al. [172] proposed a sensitive sensor by using the FRET platform between graphene quantum dots and CNTs for analysis of DNA. The constructed system has a larger linear range, up to 133.0 nM and lower LOD 0.4 nM.

7.5.4 Carbon nanotubes-based FET sensor

Carbon materials have been considered favorable potential materials for unique nano–electronic biosensors for the analysis of biomolecules. The applications of 1D nanomaterial like nano rods/wires, into electrical sensors/devices have considerable benefits for the determination of biologically significant substances. The nanowires have two main advantages in electronic circuitry: one is related to size compatibility, and the other to electrostatic interactions and charge transfer. The diameters of nanowires can be similar to biomolecules like nucleic acids and proteins; these binding effects can be sensitively determined by using the nanowire FETs sensors [175–177]. The size compatibility and biocompatibility of CNTs is seen with biomolecules. In SWCNTs, each atom is on the surface and visible to the environment and, therefore, even minor deviation in the charge environment can cause drastic changes to its electrical properties. Furthermore, the SWCNTs diameter is similar to the size of single molecules (e.g., DNA is 1 nm in size), SWCNTs are in micrometer in length, thus offering a convenient interface with a micrometer–scale system. Additionally, all carbons of SWCNTs provided a match to organic molecules [176, 180]. Hence, FET based devices were found suitable for detection and sensing of targeted chemical and biomolecules, including proteins, nucleic acids, viruses, and small molecules [178, 179, 182]. Sorgenfrei et al. [181] reported a CNTs based FET system for the analysis of DNA hybridization at single–molecule scale. In another study, Dastagi et al. [183] proposed the determination of a sequence of Hepatitis C virus at pM level range by using SWCNTs FET modified with peptide nucleic acid. The unambiguous, real–time, and label–free investigation of an RNA sequence of the Hepatitis C virus, with a LOD down to 0.5 pM was observed by using peptide nucleic acid modified SWCNT devices.

7.6 CONCLUSION

The sensitive and selective determination of the biomolecules released from living cells in real time is highly impactful in both major investigations and monitoring the drugs for the dealing of section associated infections. In this chapter, we have discussed graphene and CNTs based sensors for determination of biomolecules and drugs by using SERS, electrochemicals, fluorescence, and FET based sensors/biosensors. The sensor and biosensor based on graphene and CNTs are capable of sensing or monitoring health status and offer a significant enhancement in the management and initial analysis of related diseases. The amazing size comparability and surface compatibility of CNTs and graphene were found with biomolecules. In recent years, several methods/techniques have been proposed to prepare the defect free graphene and CNTs. The pristine graphene and CNTs and their nanocomposites have been used to fabricate the chemical and biosensors. The selective and sensitive sensors/biosensors have been reported by using different approaches, such as SERS, electrochemical, fluorescence, and FET based sensors. The studies of graphene and CNTs based sensor showed the immense importance for further development in monitoring the biologically significant molecules in real samples. The graphene and CNTs based nanomaterials play a significant role in developing the sensors and biosensors for the determination of biomolecules and drugs.

ACKNOWLEDGEMENTS

Author (Neeraj Kumar) is grateful to SERB, DST, Govt. of India, for the award of National Post–Doctoral Fellowship (File No. **PDF/2020/001201**).

REFERENCES

1. Li Z, Liu Z, Sun H, Gao C (2015). Superstructured assembly of nanocarbons: Fullerenes, nanotubes, and graphene. *Chem Rev* 115:7046–7117.

2. Kinloch IA, Suhr J, Lou J, Young RJ, Ajayan PM (2018) Composites with carbon nanotubes and graphene: An outlook. *Science* 362:547–553.
3. Hirsch A (2010) The era of carbon allotropes. *Nat Mater* 9:868–871.
4. Geim A K (2009) Graphene: Status and prospects. *Science* 324:1530–1534.
5. Georgakilas V, Tiwari JN, Kemp KC, Perman JA, Bourlinos AB, Kim KS, Zboril R (2016) Noncovalent functionalization of graphene and graphene oxide for energy materials, biosensing, catalytic, and biomedical applications. *Chem Rev* 116:5464–5519.
6. Clancy AJ, Bayazit MK, Hodge SA, Skipper NT, Howard CA, Shaffer MSP (2018) Charged Carbon Nanomaterials: Redox chemistries of fullerenes, carbon nanotubes, and graphenes. *Chem Rev* 118:7363–7408.
7. Eatemadi A, Daraee H, Karimkhanloo H, Kouhi M, Zarghami N, Akbarzadeh A, Abasi M, Hanifehpour Y, Joo SW (2014) Carbon nanotubes: properties, synthesis, purification, and medical applications. *Nanoscale Res Lett* 9:393.
8. Jacobs CB, Peairs MJ, Venton BJ (2010) Review: Carbon nanotube based electrochemical sensors for biomolecules. *Anal Chim Acta* 662:105–127.
9. Misra A (2014) Carbon nanotubes and graphene–based chemical sensors. *Curr Sci* 107:419–429.
10. Zhu Z (2017) An overview of carbon nanotubes and graphene for biosensing applications. *Nano–Micro Lett* 9:1–25.
11. Fang R, Chen K, Yin L, Sun Z, Li F, Cheng H-M (2019) The regulating role of carbon nanotubes and graphene in lithium–ion and lithium–sulfur batteries. *Adv Mater* 31:1800863.
12. Yuan W, Zhang Y, Cheng L, Wu H, Zheng L, Zhao D (2016) The applications of carbon nanotubes and graphene in advanced rechargeable lithium batteries. *J Mater Chem A* 4:8932.
13. Sainz-Urruela C, Vera-Lopez S, San Andres MP, Diez-Pascual AM (2021) Graphene–based sensors for the detection of bioactive compounds: A review. *Int J Mol Sci* 22(7):3316.
14. Aval LF, Ghoranneviss M, Pour GB (2018) High–performance supercapacitors based on the carbon nanotubes, graphene and graphite nanoparticles electrodes. *Heliyon* 4:e00862.
15. Zhang R, Palumbo A, Kim JC, Ding J, Yang E-H (2019) Flexible graphene–, graphene–oxide–, and carbon–nanotube–based supercapacitors and batteries. *Ann Phys (Berlin)* 531:1800507.
16. Geim AK, Novoselov KS (2007) The rise of graphene. *Nat Mater* 6:183–191.
17. Stolyarov MA, Liu G, Rumyantsev SL, Shur M, Balandin AA (2015) Suppression of 1/*f* Noise in Near–Ballistic *h*–BN–Graphene–*h*–BN Heterostructure field–effect transistors. *Appl Phys Lett* 107:023106.
18. Jesion I, Skibniewski M, Skibniewska E, Strupinski W, Szulc-Dąbrowska L, Krajewska A, Pasternak I, Kowalczyk P, Pinkowski R (2015) Graphene and carbon nanocompounds: biofunctionalization and applications in tissue engineering. *Biotechnol Biotechno Equip* 29:415–422.
19. Ryoo S-R, Kim Y-K, Kim M-H, Min D-H (2010) Behaviors of NIH–3T3 fibroblasts on graphene/carbon nanotubes: proliferation, focal adhesion, and gene transfection studies. *ACS Nano* 4:6587–6598.
20. Perez-Lopez B, Merkoçi A (2012) Carbon nanotubes and graphene in analytical sciences. *Microchim Acta* 179:1–16.

21. Biswas C, Lee YH (2011) Graphene versus carbon nanotubes in electronic devices. *Adv Funct Mater* 31:3806–3826.
22. Jariwala D, Sangwan VK, Lauhon LJ, Marks TJ, Hersam MC (2013) Carbon nanomaterials for electronics, optoelectronics, photovoltaics, and sensing. *Chem Soc Rev* 42:2824.
23. Yang W, Ratinac KR, Ringer SP, Thordarson P, Gooding JJ, Braet F (2010) Carbon nanomaterials in biosensors: Should you use nanotubes or graphene?. *Angew Chem Int Ed* 49:2114–2138.
24. Shahdeo D, Roberts A, Abbineni N, Gandhi S (2020) Graphene based sensors. *Compre Anal Chem*. https://doi.org/10.1016/bs.coac.2020.08.007
25. Wang Y, Xu H, Zhang J, Li G, (2008) Electrochemical sensors for clinic analysis. *Sensors* 8:2043–2081.
26. Baptista FR, Belhout SA, Giordani S, Quinn SJ (2015) Recent developments in carbon nanomaterial sensors. *Chem Soc Rev* 44:4433.
27. Zanfrognini B, Pigani L, Zanardi C (2020) Recent advances in the direct electrochemical detection of drugs of abuse. *J Solid State Electr* 24:2603–2616.
28. Eggins BR (2002) *Chemical sensors and biosensors*. David J. Ando (Ed.), John Wiley & Sons Inc., England.
29. Teymourian H, Parrilla M, Sempionatto JR, Montiel NF, Barfidokht A, Echelpoel RV, Wael KD, Wang J (2020) Wearable electrochemical sensors for the monitoring and screening of drugs. *ACS Sens* 5:2679–2700.
30. Loh KP, Bao Q, Ang PK, Yang J (2010) The chemistry of graphene. *J Mater Chem* 20:2277–2289.
31. Aldosari MA, Othman AA, Alsharaeh EH (2013) Synthesis and characterization of the *in situ* bulk polymerization of PMMA containing graphene sheets using microwave irradiation. *Molecules* 18:3152–3167.
32. Bhuyan MSA, Uddin MN, Islam MM, Bipasha FA, Hossain SS (2016) Synthesis of graphene. *Int Nano Lett* 6:65–83.
33. Allen MJ, Tung VC, Kaner RB (2010) Honeycomb carbon: A review of graphene. *Chem Rev* 110:132–145.
34. Viculis LM, Mack JJ, Kaner RB (2003) A Chemical Route to Carbon Nanoscrolls. *Science* 299:1361.
35. Jiao L, Wang X, Diankov G, Wang H, Dai H (2010) Facile synthesis of high-quality graphene nanoribbons. *Nat Nanotechnol* 5:321–325.
36. Kosynkin DV, Higginbotham AL, Sinitskii A, Lomeda JR, Dimiev A, Price BK, Tour JM (2009) Longitudinal unzipping of carbon nanotubes to form graphene nanoribbons. *Nature* 458:872–877.
37. Reina A, Jia X, Ho J, Nezich D, Son H, Bulovic V, Dresselhaus MS, Kong J (2009) Large area, few-layer graphene films on arbitrary substrates by chemical vapor deposition. *Nano Lett* 9:30–35.
38. Park S, Ruoff RS (2009) Chemical methods for the production of graphenes. *Nat Nanotechnol* 4:217–224.
39. Marcano DC, Kosynkin DV, Berlin JM, Sinitskii A, Sun Z, Slesarev A, Alemany LB, Lu W, Tour JM (2010) Improved synthesis of graphene oxide. *ACS Nano* 4:4806–4814.
40. Paulchamy B, Arthi G, Lignesh BD (2015) A simple approach to stepwise synthesis of graphene oxide nanomaterial. *J Nanomed Nanotechnol* 6:253.
41. Chung C, Kim Y-K, Shin D, Ryoo S-R, Hong BH, Min D-H (2013) Biomedical applications of graphene and graphene oxide. *Acc Chem Res* 46:2211–2224.

42. Seabra AB, Paula AJ, de Lima R, Alves OL, Duran N (2014) Nanotoxicity of graphene and graphene oxide. *Chem Res Toxicol* 27:159–168.
43. Suvarnaphaet P, Pechprasarn S (2017) Graphene-based materials for biosensors: A review. *Sensors* 17:2161.
44. Zhao Y, Zheng X, Wang Q, Zhe T, Bai Y, Bu T, Zhang M, Wang L (2020) Electrochemical behavior of reduced graphene oxide/cyclodextrins sensors for ultrasensitive detection of imidacloprid in brown rice. *Food Chem* 333:127495.
45. Wang L, Zhang Y, Wu A, Wei G (2017) Designed graphene–peptide nanocomposites for biosensor applications: A review. *Anal Chim Acta* 985:24–40.
46. Gaoa N, Gaoa T, Yanga X, Daia X, Zhoua W, Zhanga A, Lieber CM (2016) Specific detection of biomolecules in physiological solutions using graphene transistor biosensors. *PNAS* 113:14633–14638.
47. Bai Y, Xu T, Zhang X (2020) Graphene-based biosensors for detection of biomarkers. *Micromachines* 11:60.
48. Tarcan R, Todor-Boer O, Petrovai I, Leordean C, Astilean S, Botiz I (2020) Reduced graphene oxide today. *J Mater Chem C* 1:1198.
49. Habte AT, Ayele DW (2019) Synthesis and characterization of reduced graphene oxide (rGO) started from graphene oxide (GO) using the tour method with different parameters. *Adv Mater Sci Eng* 2019:1–9.
50. Azizighannad S, Mitra S (2018) Stepwise reduction of graphene oxide (GO) and its effects on chemical and colloidal properties. *Sci Rep* 8:2–7.
51. Langer J, de Aberasturi DJ, Aizpurua J et al. (2020) Present and future of surface–enhanced Raman scattering. *ACS Nano* 14:28–117.
52. Li J-F, Zhang Y-J, Ding S-Y, Panneerselvam R, Tian Z-Q (2017) Core–Shell Nanoparticle–Enhanced Raman Spectroscopy. *Chem Rev* 117:5002–5069.
53. Wang L, Wu A, Wei G (2018) Graphene–based aptasensors: from molecule–interface interactions to sensor design and biomedical diagnostics. *Analyst* 143:1526.
54. Singh NS, Mayanglambam F, Nemade HB, Giri PK (2021) Facile synthetic route to exfoliate high quality and super–large lateral size graphene–based sheets and their applications in SERS and CO2 gas sensing. *RSC Adv* 11:9488.
55. Gao X-G, Cheng L-X, Jiang W-S, Li X-K, Xing F (2021) Graphene and its derivatives–based optical sensors. *Front Chem* 9:615164.
56. Lai H, Xu F, Zhang Y, Wang L (2018) Recent progress on graphene–based substrates for surface–enhanced Raman scattering applications. *J Mater Chem B* 6(24):4008–4028.
57. Zhao Y, Li X, Liu Y, Zhang L, Wang F, Lu Y (2017) High performance surface–enhanced Raman scattering sensing based on Au nanoparticle–monolayer graphene–Ag nanostar array hybrid system. *Sens Actuators B Chem* 247: 850–857.
58. Lin TW, Wu HY, Tasi TT, Lai YH, Shen HH (2015) Surface–enhanced Raman spectroscopy for DNA detection by the self–assembly of Ag nanoparticles onto Ag nanoparticle–graphene oxide nanocomposites. *Phys Chem Chem Phys* 17(28):18443–18448.
59. He S, Liu KK, Su S, Yan J, Mao X, Wang D, He Y, Li LJ, Song S, Fan C (2012) Graphene–based high–efficiency surface–enhanced Raman scattering–active platform for sensitive and multiplex DNA detection. *Anal Chem* 84(10): 4622–4627.

60. Pyrak E, Krajczewski J, Kowalik A, Kudelski A, Jaworska A (2019) Surface enhanced Raman spectroscopy for DNA biosensors—How far are we?. *Molecules* 24(24):4423.
61. Cialla D, Pollok S, Steinbrücker C, Weber K, Popp J (2014) SERS–based detection of biomolecules. *Nanophotonics* 3(6):383–411.
62. Muntean CM, Dina NE, Coroş M, Toşa N, Turza AI, Dan M, (2019) Graphene/silver nanoparticles–based surface–enhanced Raman spectroscopy detection platforms: Application in the study of DNA molecules at low pH. *J Raman Spectrosc* 50(12):1849–1860.
63. Silver A, Kitadai, H, Liu H, Granzier-Nakajima T, Terrones M, Ling X, Huang S, (2019) Chemical and bio sensing using graphene–enhanced Raman spectroscopy. *Nanomaterials* 9(4):516.
64. Feliu N, Hassan M, Rico EG, Cui D, Parak W, Alvarez-Puebla R (2017) SERS quantification and characterization of proteins and other biomolecules. *Langmuir* 33(38):9711–9730.
65. Chen M, Liu D, Du X, Lo KH, Wang S, Zhou B, Pan H, (2020) 2D materials: Excellent substrates for surface–enhanced Raman scattering (SERS) in chemical sensing and biosensing. *Trends Anal Chem* 130:115983.
66. Liang X, Li N, Zhang R, Yin P, Zhang C, Yang N, Liang K, Kong B (2021) Carbon–based SERS biosensor: from substrate design to sensing and bioapplication. *NPG Asia Mater* 13(1):1–36.
67. Zhu J, Du HF, Zhang Q, Zhao J, Weng GJ, Li JJ, Zhao JW (2019) SERS detection of glucose using graphene–oxide–wrapped gold nanobones with silver coating. *J Mater Chem C* 7(11):3322–3334.
68. Luo Y, Ma L, Zhang X, Liang A, Jiang Z (2015) SERS detection of dopamine using label–free acridine red as molecular probe in reduced graphene oxide/silver nanotriangle sol substrate. *Nanoscale Res Lett* 10(1):1–9.
69. Neri G, Fazio E, Mineo PG, Scala A, Piperno A (2019) SERS sensing properties of new graphene/gold nanocomposite. *Nanomaterials* 9(9):1236.
70. Wang P, Xia M, Liang O, Sun K, Cipriano AF, Schroeder T, Liu H, Xie YH (2015) Label–free SERS selective detection of dopamine and serotonin using graphene–Au nanopyramid heterostructure. *Analy Chem* 87(20):10255–10261.
71. Wu S, He Q, Tan C, Wang Y, Zhang H (2013) Graphene–based electrochemical sensors. *Small* 9(8):1160–1172.
72. Shao Y, Wang J, Wu H, Liu J, Aksay IA, Lin Y (2010) Graphene based electrochemical sensors and biosensors: A review. *Electroanalysis* 22(10):1027–1036.
73. Kang X, Wang J, Wu H, Liu J, Aksay IA, Lin Y, (2010) A graphene–based electrochemical sensor for sensitive detection of paracetamol. *Talanta* 81(3): 754–759.
74. Coros M, Pruneanu S, Stefan-van Staden RI (2019) Recent progress in the graphene–based electrochemical sensors and biosensors. *J Electrochem Soc* 167(3):037528.
75. Kumar S, Bukkitgar SD, Singh S, Singh V, Reddy KR, Shetti NP, Reddy CV, Sadhu V, Naveen S (2019) Electrochemical sensors and biosensors based on graphene functionalized with metal oxide nanostructures for healthcare applications. *ChemistrySelect* 4(18):5322–5337.
76. Martin A, Hernandez-Ferrer J, Martínez MT, Escarpa A (2015) Graphene nanoribbon–based electrochemical sensors on screen–printed platforms. *Electrochim Acta* 172:2–6.
77. Lawal AT (2015) Synthesis and utilisation of graphene for fabrication of electrochemical sensors. *Talanta* 131:424–443.

78. Karuwan C, Wisitsoraat A, Chaisuwan P, Nacapricha D, Tuantranont A (2017) Screen-printed graphene-based electrochemical sensors for a microfluidic device. *Anal Methods* 9(24):3689–3695.
79. Sun JY, Huang KJ, Wei SY, Wu ZW, Ren FP (2011) A graphene-based electrochemical sensor for sensitive determination of caffeine. *Colloids Surf B: Biointerfaces* 84(2):421–426.
80. Adhikari BR, Govindhan M, Chen A (2015) Sensitive detection of acetaminophen with graphene-based electrochemical sensor. *Electrochim Acta* 162:198–204.
81. Zhang C, Zhang Z, Yang Q, Chen W (2018) Graphene-based Electrochemical Glucose Sensors: Fabrication and Sensing Properties. *Electroanalysis* 30(11):2504–2524.
82. Taniselass S, Arshad MM, Gopinath SC (2019) Graphene-based electrochemical biosensors for monitoring noncommunicable disease biomarkers. *Biosens Bioelectron* 130:276–292.
83. Pandikumar A, How GTS, See TP, Omar FS, Jayabal S, Kamali KZ, Yusoff N, Jamil A, Ramaraj R, John SA, Lim HN (2014) Graphene and its nanocomposite material based electrochemical sensor platform for dopamine. *RSC Adv* 4(108):63296–63323.
84. Gan T, Hu S (2011) Electrochemical sensors based on graphene materials. *Microchim Acta* 175(1):1–19.
85. Peng J, Hou C, Hu X (2012) A graphene-based electrochemical sensor for sensitive detection of vanillin. *Int J Electrochem Sci* 7(2):1724–1733.
86. Yang C, Denno ME, Pyakurel P, Venton BJ (2015) Recent trends in carbon nanomaterial-based electrochemical sensors for biomolecules: A review. *Anal Chimica Acta* 887:17–37.
87. Li SJ, Xing Y, Wang GF (2012) A graphene-based electrochemical sensor for sensitive and selective determination of hydroquinone. *Microchim Acta* 176(1–2):163–168.
88. Tiwari JN, Vij V, Kemp KC, Kim KS (2016) Engineered carbon-nanomaterial-based electrochemical sensors for biomolecules. *ACS Nano* 10(1):46–80.
89. Song Y, Luo Y, Zhu C, Li H, Du D, Lin Y (2016) Recent advances in electrochemical biosensors based on graphene two-dimensional nanomaterials. *Biosens Bioelectron* 76:195–212.
90. Chen A, Chatterjee S (2013) Nanomaterials based electrochemical sensors for biomedical applications. *Chem Soc Rev* 42(12):5425–5438.
91. Kumar N, Goyal RN (2016) Nanopalladium grained polymer nanocomposite based sensor for the sensitive determination of melatonin. *Electrochim Acta* 211:18–26.
92. Zhang Y, Shen J, Li H, Wang L, Cao D, Feng X, Liu Y, Ma Y, Wang L (2016) Recent progress on graphene-based electrochemical biosensors. *Chem Rec* 16(1):273–294.
93. Song H, Zhang X, Liu Y, Su Z (2019) Developing graphene-based nanohybrids for electrochemical sensing. *Chem Rec* 19(2–3):534–549.
94. Fritea L, Banica F, Costea TO, Moldovan L, Iovan C, Cavalu S (2018) A gold nanoparticles-Graphene based electrochemical sensor for sensitive determination of nitrazepam. *J Electroanal Chem* 830:63–71.
95. Kumar N, Rosy S, Goyal RN (2017) Gold-palladium nanoparticles aided electrochemically reduced graphene oxide sensor for the simultaneous estimation of lomefloxacin and amoxicillin. *Sens Actuators B Chem* 243:658–668.

96. Wang C, Xu P, Zhuo K (2014) Ionic liquid functionalized graphene–Based electrochemical biosensor for simultaneous determination of dopamine and uric acid in the presence of ascorbic acid. *Electroanalysis* 26(1):191–198.
97. Li J, Yang J, Yang Z, Li Y, Yu S, Xu Q, Hu X (2012) Graphene–Au nanoparticles nanocomposite film for selective electrochemical determination of dopamine. *Anal Meth* 4(6):1725–1728.
98. Yang L, Liu D, Huang J, You T (2014) Simultaneous determination of dopamine, ascorbic acid and uric acid at electrochemically reduced graphene oxide modified electrode. *Sens Actuators B Chem* 193:166–172.
99. Zhang H, Liu S (2020) Electrochemical sensors based on nitrogen–doped reduced graphene oxide for the simultaneous detection of ascorbic acid, dopamine and uric acid. *J Alloy Compd* 842:155873.
100. Zheng P, Wu N (2017) Fluorescence and sensing applications of graphene oxide and graphene quantum dots: A review. *Chem Asian J* 12(18):2343–2353.
101. Mitra R, Saha A (2017) Reduced graphene oxide based "turn–on" fluorescence sensor for highly reproducible and sensitive detection of small organic pollutants. *ACS Sustain Chem Eng* 5(1):604–615.
102. Cheng R, Cheng L, Ou S (2019) A graphene oxide–based fluorescent sensor for recognition of glutamate in aqueous solutions and bovine serum. *Spectrochim Acta A Mol Biomol Spectrosc* 221:117204.
103. Ryu J, Lee E, Lee K, Jang J (2015) A graphene quantum dots based fluorescent sensor for anthrax biomarker detection and its size dependence. *J Mater Chem B* 3(24):4865–4870.
104. Balaji A, Yang S, Wang J, Zhang J (2019) Graphene oxide–based nanostructured DNA sensor. *Biosensors* 9(2):74.
105. Tomita S, Ishihara S, Kurita R (2017) A multi–fluorescent DNA/graphene oxide conjugate sensor for signature–based protein discrimination. *Sensors* 17(10):2194.
106. Guo H, Li J, Li Y, Wu D, Ma H, Wei Q, Du B (2018) A turn–on fluorescent sensor for Hg 2+ detection based on graphene oxide and DNA aptamers. *New J Chem* 42(13):11147–11152.
107. Kitamura Y, Miyahata T, Matsuura H, Hatakeyama K, Taniguchi T, Koinuma M, Matsumoto Y, Ihara T (2015) Graphene oxide–based amplified fluorescence sensor for nucleic acid detection through target–catalyzed hairpin assembly. *Chem Lett* 44(10):1353–1355.
108. Weng W, Sun X, Liu B, Shen J (2018) Enhanced fluorescence based on graphene self–assembled films and highly sensitive sensing for VB 12. *J Mater Chem C* 6(16):4400–4408.
109. Liu Z, Liu B, Ding J, Liu J (2014) Fluorescent sensors using DNA–functionalized graphene oxide. *Anal Bioanal Chem* 406(27):6885–6902.
110. Kushwaha HS, Sao R, Vaish R (2014) Label free selective detection of estriol using graphene oxide–based fluorescence sensor. *J Appl Phys* 116(3):034701.
111. Ebrahim S, Shokry A, Khalil MMA, Ibrahim H, Soliman M (2020) Polyaniline/Ag nanoparticles/graphene oxide nanocomposite fluorescent sensor for recognition of chromium (VI) ions. *Sci Rep* 10(1):1–11.
112. Chen X, Zhou H, Zhai N, Liu P, Chen Q, Jin L, Zheng Q (2015) Graphene oxide–based homogeneous fluorescence sensor for multiplex determination of various targets by a multifunctional aptamer. *Anal Lett* 48(12):1892–1906.
113. Tan J, Lai Z, Zhong L, Zhang Z, Zheng R, Su J, Huang Y, Huang P, Song H, Yang N, Zhou S (2018) A graphene oxide–based fluorescent aptasensor for the turn–on detection of CCRF–CEM. *Nanoscale Res Lett* 13(1):1–8.

114. Gao X, Cheng L, Jiang W, Li XK, Xing F (2021) Graphene and its derivatives–based optical sensors. *Front Chem* 9:615164.
115. Szunerits S, Boukherroub R (2018) Graphene–based biosensors. *Interface Focus* 8(3):20160132.
116. Gao L, Lian C, Zhou Y, Yan L, Li Q, Zhang C, Chen L, Chen K (2014) Graphene oxide–DNA based sensors. *Biosens Bioelectron* 60:22–29.
117. Wen Y, Xing F, He S, Song S, Wang L, Long Y, Li D, Fan C (2010) A graphene–based fluorescent nanoprobe for silver (I) ions detection by using graphene oxide and a silver–specific oligonucleotide. *Chem Comm* 46(15):2596–2598.
118. Chang H, Tang L, Wang Y, Jiang J, Li J (2010) Graphene fluorescence resonance energy transfer aptasensor for the thrombin detection. *Anal Chem* 82(6): 2341–2346.
119. Xiang X, Han L, Zhang Z, Huang F (2017) Graphene oxide–based fluorescent sensor for sensitive turn–on detection of sinapine. *Spectrochim Acta A Mol Biomol Spectrosc* 174:75–79.
120. Bai Y, Xu T, Zhang X (2020) Graphene–based biosensors for detection of biomarkers. *Micromachines* 11(1):60.
121. Liu M, Zhang Q, Brennan JD, Li Y (2018) Graphene–DNAzyme–based fluorescent biosensor for Escherichia coli detection. *MRS Commun* 8(3):687–694.
122. Pan W, Liu B, Gao X, Yu Z, Liu X, Li N, Tang B (2018) A graphene–based fluorescent nanoprobe for simultaneous monitoring of miRNA and mRNA in living cells. *Nanoscale* 10(29):14264–14271.
123. Shi Y, Huang WT, Luo HQ, Li NB (2011) A label–free DNA reduced graphene oxide–based fluorescent sensor for highly sensitive and selective detection of hemin. *Chem Comm* 47(16):4676–4678.
124. Lowe BM, Sun K, Zeimpekis I, Skylaris CK, Green NG (2017) Field–effect sensors–from pH sensing to biosensing: Sensitivity enhancement using streptavidin–biotin as a model system. *Analyst* 142(22):4173–4200.
125. He Q, Wu S, Yin Z, Zhang H (2012) Graphene–based electronic sensors. *Chem Sci* 3(6):1764–1772.
126. Brand U, Brandes L, Koch V, Kullik T, Reinhardt B, Rüther F, Scheper T, Schügerl K, Wang S, Wu X, Ferretti R (1991) Monitoring and control of biotechnological production processes by Bio–FET–FIA–sensors. *Appl Microb Biotechnol* 36(2):167–172.
127. Lin MC, Chu CJ, Tsai LC, Lin HY, Wu CS, Wu YP, Wu YN, Shieh DB, Su YW, Chen CD (2007) Control and detection of organosilane polarization on nanowire field–effect transistors. *Nano Lett* 7(12):3656–3661.
128. Maddalena F, Kuiper MJ, Poolman B, Brouwer F, Hummelen JC, de Leeuw DM, De Boer B, Blom PW (2010) Organic field–effect transistor–based biosensors functionalized with protein receptors. *J Appl Phys* 108(12):124501.
129. Zhu Y, Murali S, Cai W, Li X, Suk JW, Potts JR, Ruoff RS (2010) Graphene and graphene oxide: synthesis, properties, and applications. *Adv Mater* 22(35):3906–3924.
130. Schwierz F (2010) Graphene transistors. *Nat Nanotechnol* 5(7):487.
131. Quast T, Mariani F, Scavetta E, Schuhmann W, Andronescu C (2020) Reduced–graphene–oxide–based needle–type field–effect transistor for dopamine sensing. *ChemElectroChem* 7(8):1922–1927.
132. Liao C, Zhang M, Niu L, Zheng Z, Yan F (2014) Organic electrochemical transistors with graphene–modified gate electrodes for highly sensitive and selective dopamine sensors. *J Mater Chem B* 2(2):191–200.

133. Qing X, Wang Y, Zhang Y, Ding X, Zhong W, Wang D, Wang W, Liu Q, Liu K, Li M, Lu Z (2019) Wearable fiber–based organic electrochemical transistors as a platform for highly sensitive dopamine monitoring. *ACS Appl Mater Interfaces* 11(14):13105–13113.
134. Andronescu C, Schuhmann W (2017) Graphene–based field effect transistors as biosensors. *Curr Opin Electrochem* 3(1):11–17.
135. Fenoy GE, Marmisollé WA, Azzaroni O, Knoll W (2020) Acetylcholine biosensor based on the electrochemical functionalization of graphene field–effect transistors. *Biosens Bioelectron* 148:111796.
136. Veeralingam S, Badhulika S (2020) Surface functionalized β–Bi2O3 nanofibers based flexible, field–effect transistor–biosensor (BioFET) for rapid, label–free detection of serotonin in biological fluids. *Sens Actuators B Chem* 321:128540.
137. Xu S, Wang T, Liu G, Cao Z, Frank LA, Jiang S, Zhang C, Li Z, Krasitskaya VV, Li Q, Sha Y (2021) Analysis of interactions between proteins and small-molecule drugs by a biosensor based on a graphene field–effect transistor. *Sens Actuators B Chem* 326:128991.
138. You X, Pak JJ (2014) Graphene–based field effect transistor enzymatic glucose biosensor using silk protein for enzyme immobilization and device substrate. *Actuators B Chem* 202:1357–1365.
139. Park CS, Yoon H, Kwon OS (2016) Graphene–based nanoelectronic biosensors. *J Ind d Eng Chem* 38:13–22.
140. Shin DH, Kim W, Jun J, Lee JS, Kim JH, Jang J (2018) Highly selective FET–type glucose sensor based on shape–controlled palladium nanoflower–decorated graphene. *Sens Actuators B Chem* 264:216–223.
141. Schuck A, Kim HE, Moreira JK, Lora PS, Kim YS (2021) A graphene–based enzymatic biosensor using a common–gate field–effect transistor for L–Lactic acid detection in blood plasma samples. *Sensors* 21(5):1852.
142. Iijima S (2002) Carbon nanotubes: past, present, and future. *Physica B: Condensed Matter* 323(1–4):1–5.
143. Iijima S, Ichihashi T (1993) Single–shell carbon nanotubes of 1–nm diameter. *Nature* 363(6430):603–605.
144. Prasek J, Drbohlavova J, Chomoucka J, Hubalek J, Jasek O, Adam V, Kizek R (2011) Methods for carbon nanotubes synthesis. *J Mater Chem* 21(40):15872–15884.
145. Andreoni W ed., (2000) *The physics of fullerene–based and fullerene–related materials* (Vol. 23). Springer Science & Business Media.
146. Georgakilas V, Perman JA, Tucek J, Zboril R (2015) Broad family of carbon nanoallotropes: classification, chemistry, and applications of fullerenes, carbon dots, nanotubes, graphene, nanodiamonds, and combined superstructures. *Chem Rev* 115(11):4744–4822.
147. Ebbesen TW, Ajayan PM (1992) Large–scale synthesis of carbon nanotubes. *Nature* 358(6383):220–222.
148. Iijima S (1993) Growth of carbon nanotubes. *Mater Sci Eng B* 19(1–2):172–180.
149. Arora N, Sharma NN (2014) Arc discharge synthesis of carbon nanotubes: Comprehensive review. *Diam Relat Mater* 50:135–150.
150. Gupta P, Tsai K, Ruhunage CK, Gupta VK, Rahm CE, Jiang D, Alvarez NT (2020) True picomolar neurotransmitter sensor based on open–ended carbon nanotubes. *Anal Chem* 92(12):8536–8545.

151. Chen P, Wang Z, Zong S, Chen H, Zhu D, Zhong Y, Cui Y (2014) A wide range optical pH sensor for living cells using Au@Ag nanoparticles functionalized carbon nanotubes based on SERS signals. *Anal Bioanal Chem* 406(25): 6337–6346.
152. Saravanan RK, Naqvi TK, Patil S, Dwivedi PK, Verma S (2020) Purine-blended nanofiber woven flexible nanomats for SERS-based analyte detection. *Chem Commun* 56(43):5795–5798.
153. Zhang H, Nie P, Xia Z, Feng X, Liu X, He Y (2020) Rapid quantitative detection of Deltamethrin in Corydalis yanhusuo by SERS coupled with Multi-Walled carbon nanotubes. *Molecules* 25(18):4081.
154. Yeh YT, Gulino K, Zhang Y, Sabestien A, Chou TW, Zhou B, Lin, Z, Albert I, Lu H, Swaminathan V, Ghedin E (2020) A rapid and label-free platform for virus capture and identification from clinical samples. *Proc Natl Acad Sci* 117(2):895–901.
155. Liang X, Li N, Zhang R, Yin P, Zhang C, Yang N, Liang K, Kong B (2021) Carbon-based SERS biosensor: from substrate design to sensing and bioapplication. *NPG Asia Mater* 13(1):1–36.
156. Hu C, Hu S (2009) Carbon nanotube-based electrochemical sensors: principles and applications in biomedical systems. *J Sens* 2009:1–40.
157. Vashist SK, Zheng D, Al-Rubeaan K, Luong JH, Sheu FS (2011) Advances in carbon nanotube based electrochemical sensors for bioanalytical applications. *Biotechnol Adv* 29(2):169–188.
158. Wang J (2005) Carbon-nanotube based electrochemical biosensors: A review. *Electroanalysis* 17(1):7–14.
159. Gupta S, Murthy CN, Prabha CR (2018) Recent advances in carbon nanotube based electrochemical biosensors. *Inter J Biol Macromol* 108:687–703.
160. Savk A, Ozdil B, Demirkan B, Nas MS, Calimli MH, Alma MH, Asiri AM, Şen F (2019) Multiwalled carbon nanotube-based nanosensor for ultrasensitive detection of uric acid, dopamine, and ascorbic acid. *Mater Sci Eng C* 99:248–254.
161. Anirudhan TS, Alexander S, Lilly A (2014) Surface modified multiwalled carbon nanotube based molecularly imprinted polymer for the sensing of dopamine in real samples using potentiometric method. *Polymer* 55(19):4820–4831.
162. Huang Q, Lin X, Tong L, Tong QX (2020) Graphene quantum dots/multiwalled carbon nanotubes composite-based electrochemical sensor for detecting dopamine release from living cells. *ACS Sustain Chem Eng* 8(3):1644–1650.
163. Kan X, Zhou H, Li C, Zhu A, Xing Z, Zhao Z (2012) Imprinted electrochemical sensor for dopamine recognition and determination based on a carbon nanotube/polypyrrole film. *Electrochim Acta* 63:69–75.
164. Mazloum-Ardakani M, Sheikh-Mohseni MA, Mirjalili BF (2014) Nanomolar detection limit for determination of norepinephrine in the presence of acetaminophen and tryptophan using carbon nanotube-based electrochemical sensor. *Ionics* 20(3):431–437.
165. Kumar N, Rosy S, Goyal RN (2017) Palladium nano particles decorated multiwalled carbon nanotubes modified sensor for the determination of 5-hydroxytry ptophan in biological fluids. *Sens Actuators B Chem* 239:1060–1068.
166. Yapasan E, Caliskan A, Karadeniz H, Erdem A (2010) Electrochemical investigation of biomolecular interactions between platinum derivatives and DNA by carbon nanotubes modified sensors. *Mater Sci Eng B* 169(1–3):169–173.

167. Li C, Shi G (2014) Carbon nanotube–based fluorescence sensors. *J Photochem Photobiol C Photochem Rev* 19:20–34.
168. Barone PW, Baik S, Heller DA, Strano MS (2005) Near–infrared optical sensors based on single–walled carbon nanotubes. *Nat Mater* 4(1):86–92.
169. Hu K, Huang Y, Zhao S (2014) A carbon nanotubes based fluorescent aptasensor for highly sensitive detection of adenosine deaminase activity and inhibitor screening in natural extracts. *J Pharm Biomed Anal* 95:164–168.
170. Niu SY, Li QY, Ren R, Hu KC (2010) DNA/single–walled carbon nanotubes based fluorescence detection of Hg^{2+}. *Anal Lett* 43(15):2432–2439.
171. Li Y, Sun L, Qian J, Long L, Li H, Liu Q, Cai J, Wang K (2017) Fluorescent "on–off–on" switching sensor based on CdTe quantum dots coupled with multiwalled carbon nanotubes@ graphene oxide nanoribbons for simultaneous monitoring of dual foreign DNAs in transgenic soybean. *Biosens Bioelectron* 92:26–32.
172. Qian ZS, Shan XY, Chai LJ, Ma JJ, Chen JR, Feng H (2014) DNA nanosensor based on biocompatible graphene quantum dots and carbon nanotubes. *Biosens Bioelectron* 60:64–70.
173. Cho ES, Hong SW, Jo WH (2008) A new pH sensor using the fluorescence quenching of carbon nanotubes. *Macromol Rapid Commun* 29(22):1798–1803.
174. Huang H, Zou M, Xu X, Liu F, Li N, Wang X (2011) Near–infrared fluorescence spectroscopy of single–walled carbon nanotubes and its applications. *TrAC Trend Anal Chem* 30(7):1109–1119.
175. Patolsky F, Timko BP, Zheng G, Lieber CM (2007) Nanowire–based nanoelectronic devices in the life sciences. *MRS Bulletin* 32(2):142–149.
176. Allen BL, Kichambare PD, Star A (2007) Carbon nanotube field–effect–transistor–based biosensors. *Adv Mater* 19(11):1439–1451.
177. Tran TT, Mulchandani A (2016) Carbon nanotubes and graphene nano field–effect transistor–based biosensors. *TrAC Trend Anal Chem* 79:222–232.
178. Tran TL, Nguyen TT, Tran TTH, Tran QT, Mai AT (2017) Detection of influenza A virus using carbon nanotubes field effect transistor based DNA sensor. *Physica E* 93:83–86.
179. Gui EL, Li LJ, Zhang K, Xu Y, Dong X, Ho X, Lee PS, Kasim J, Shen ZX, Rogers JA, Mhaisalkar SG (2007) DNA sensing by field–effect transistors based on networks of carbon nanotubes. *J Am Chem Soc* 129(46):14427–14432.
180. Liu S, Guo X (2012) Carbon nanomaterials field–effect–transistor–based biosensors. *NPG Asia Mater* 4(8):e23–e23.
181. Sorgenfrei S, Chiu CY, Gonzalez Jr RL, Yu YJ, Kim P, Nuckolls C, Shepard KL (2011) Label–free single–molecule detection of DNA–hybridization kinetics with a carbon nanotube field–effect transistor. *Nat Nanotechnol* 6(2):126–132.
182. Hu P, Zhang J, Li L, Wang Z, O'Neill W, Estrela P (2010) Carbon nanostructure–based field–effect transistors for label–free chemical/biological sensors. *Sensors* 10(5):5133–5159.
183. Dastagir T, Forzani ES, Zhang R, Amlani I, Nagahara LA, Tsui R, Tao N (2007) Electrical detection of hepatitis C virus RNA on single wall carbon nanotube-field effect transistors. *Analyst* 132(8):738–740.

Chapter 8

Intelligent flow sensor using artificial neural networks

Vaegae Naveen Kumar and
Komanapalli Venkata Lakshmi Narayana
Vellore Institute of Technology, Vellore, India

Komanapalli Gurumurthy
Vellore Institute of Technology, Amaravathi, India

CONTENTS

8.1 INTRODUCTION: BACKGROUND OF FLOW MEASUREMENT AND TAXONOMY

In process-control industries, measurement and control of fluid flow rate is an important task. In a typical industrial control loop, the output quantity of a flow sensor corresponding to input flow measured is compared with the set-point. According to the error signal of the comparator, the controller commands the actuator (mainly the flow-control valve) to force the actual flow close to the set point. Generally, low-cost flow sensors such as venture-meters and turbine flow meters are employed to measure the volumetric fluid flow in unit time [1]. But these flow sensors cause pressure drop in the flow of fluids and results in the consumption of more energy for pumping the fluids. This problem can be overcome by using non-contact flow sensors, such as electromagnetic flow sensors that don't obstruct the flow of fluids. The non-contact flow sensors have been designed and extensively used in

DOI: 10.1201/9781003288633-8

instrumentation and control industries, as there is no necessity for installation in the pipeline. Also introduction to the differential pressure across pipelines is not needed. The high cost of non-contact flow sensors restricts its use more than that of its conventional sensors.

In this chapter, an ANN based virtual flow sensor is proposed to conduct the flow measurement of fluids for flow-control valves. The main intention is to eliminate the pressure drop and the associated power loss that occur in basic flow sensors. The virtual flow sensor is designed as an alternative to high-cost non-contact flow sensors. The principle approach is to determine the relation between the flow rate and physical properties of the control valve for flow measurements. In this approach, the physical characteristics and properties of the control valve are embedded in the MLP-ANN model to determine the flow rate of fluids, instead of real measurement using a conventional flow sensor.

Neural networks have been extensively used to add intelligence to sensors. Linearization [2], self-calibration [3] or extension of the range of the sensor is also performed using ANNs in various studies [4]. The universal approximation properties of feed forward neural networks allow us to create virtual sensors where output through the neural network model would supply the output as if it were the physical sensor without the parametrical losses and investment of the said sensor [3]. A neural network model is preferred over statistical methods on account of outperforming, including capacity to learn from exemplars, plasticity of architecture, parallel computation, fault tolerant, robustness to noise, and ease of implementation on hardware environments such as the FPGA.

As with most hardware implementations, digital or analog paradigms are considered to realize ANNs on a hardware device. The digital implementation is more prevalent in practical adoption due to high accuracy, high repeatability, better noise immunity, testability, advanced reliability, and flexibility with all types of processing systems. Conversely, analog implementation is more problematic in design and can only be viable for specific domains and applications involving mass production. The ANN hardware implementations can be performed by application specific integrated chips (ASIC), digital signal processing modules (DSP) and FPGA units [5]. The ASIC implementation of ANN models does not offer re-configurability by the user. The DSP modules do not preserve the parallel topology of the neurons in a layer, as DSP implementation is sequential. FPGA units preserve the parallel topology and architecture of the neurons in any ANN model and hence are considered to be the most suitable hardware implementation components. Therefore, it is computationally viable and can be reconfigured by the user. FPGA implementations of feed forward networks have been implemented successfully in various studies [6]. A thorough activation function approximation has been made, and the multilayer perceptron is implanted on the Spartan-3E FPGA starter unit to obtain the proposed objective. The organization of the chapter is as follows. The mathematics of the control valve is described in section 8.2. The steps involved in the development of MLP-ANN using a BP algorithm are elaborated in section 8.3. The FPGA implementation is explained in section 8.4. The FPGA implementation of log-sigmoid activation function and FPGA implementation of ANN-MLP using the BP algorithm are illustrated in section 8.5 and section 8.6 respectively. The discussion of the proposed work and summary of the chapter are presented in section 8.7 and section 8.8 respectively.

8.2 THEORY OF THE CONTROL VALVE

The process control and instrumentation systems are immense domains. Mainly the process control applications range from simple domestic cooker control to a whole production unit or process. Various physical quantities, such as temperature, pressure, humidity, flow, density, and level, need to be continuously controlled and monitored [7]. One of the most important aspects of such control is typically the regulation of flow of fluid and liquids using control valves. In this work, an equal percentage valve is employed. The flow through the valve is given by the equation

$$Q = C_v f(\theta)\sqrt{\frac{\Delta P}{G}} \tag{8.1}$$

where Q represents the flow rate, C_v indicates the valve coefficient, G represents the specific gravity of the fluid, and the pressure drop across the flow-control valve is indicated by $\Delta P . f(\theta)$ is defined by

Linear valve:

$$f(\theta) = \theta \tag{8.2}$$

Quick opening valve:

$$f(\theta) = \sqrt{\theta} \tag{8.3}$$

Equal percentage valve:

$$f(\theta) = R^{(\theta-1)} \tag{8.4}$$

where, $f(\theta) \in (0, 1)$ and θ is the lift of the valve. It is the relative position of the valve globe with reference to the valve seat. R is the valve design parameter $R \in (20, 50) \in (20, 50)$which depends on the geometry of the valve employed. According to Equation (8.1), the flow rate through the control valve is a function of the differential pressure and lift. It can be rephrased mathematically without loss of generalization that,

$$Q = f(\Delta P, \%) \tag{8.5}$$

Where, ΔP is the differential pressure and % is the percentage opening of the valve which also indicates the stem position.

8.3 THE MULTILAYER PERCEPTRON NEURAL NETWORK

In the MLP-ANN, the input signals propagate in a forward direction, from input layer to hidden layer and to output layer respectively. The MLP-ANN has been used in sensor linearization and compensation applications with the back-propagation algorithm [8].

For the network, shown in Figure 8.1, the feed-forward output is given as,

$$y = g\left(\Sigma\left(f\left(\Sigma I_i \cdot v_{ij} + b_j\right) \cdot w_k\right) + b_{k'}\right) \tag{8.6}$$

where b_k', w_k and b_j, v_{ij} are the bias' and weights to the output and hidden layer respectively and $g(.)$ and $f(.)$ are the activations functions in the output and hidden neurons respectively.

The weights and bias' are updated as per the back-propagation algorithm, as illustrated by the flow chart shown in Figure 8.2.

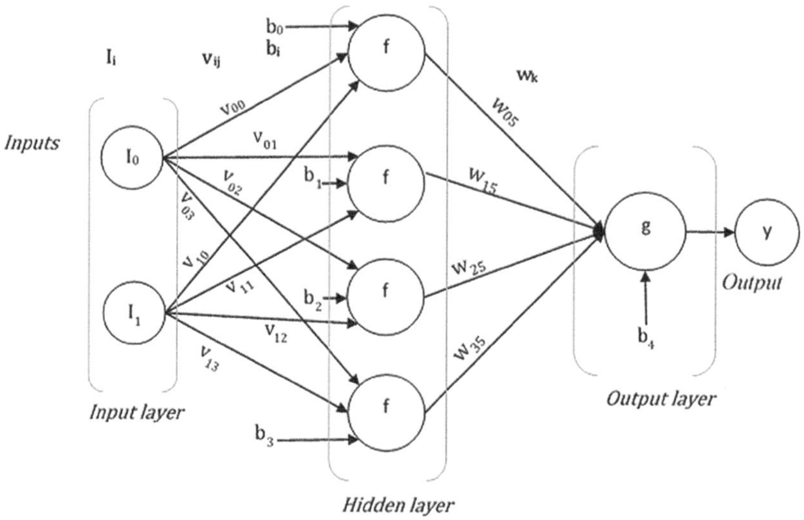

Figure 8.1 Multilayer perceptron feed-forward neural network.

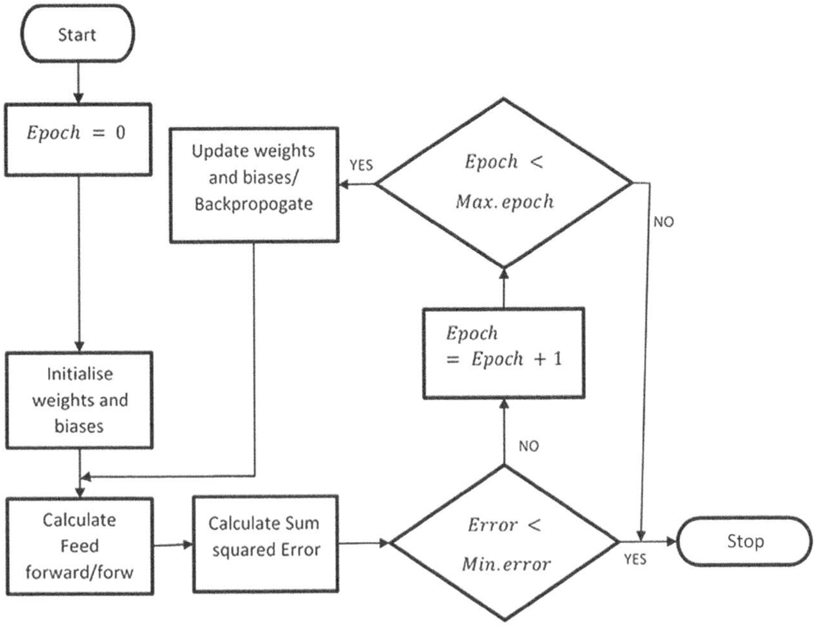

Figure 8.2 Flow chart of back propagation algorithm.

The back propagation algorithm as discussed consists of two distinct steps:

Forward pass: The network shown in Figure 8.1 gives the output as according to Equation (8.6).

Back propagation: In this step, there is the apparent computation in the backwards direction. The target data is given as y', hence the error is

$$e = y' - y \tag{8.7}$$

For the output layer, the error is

$$\delta_0 = e.\left[y(1-y)\right] \tag{8.8}$$

For the hidden layer, the error is

$$\delta_h = \left(\Sigma\delta_0.w\right)f' \tag{8.9}$$

where f' are the actual outputs in the hidden layer. This is followed by the updation of weights:

For the output layer:

$$w_k(new) = w_k(old) + \alpha\delta_0 f \tag{8.10}$$

$$b_{k'}(new) = b_{k'}(old) + \alpha\delta_0 \tag{8.11}$$

For the hidden layer:

$$v_{ij}(new) = v_{ij}(old) + \alpha\delta_0 f \tag{8.12}$$

$$b_j(new) = b_j(old) + \alpha\delta_0 \tag{8.13}$$

where α is the learning rate and f is hidden layer output.

Using these tools the neural network is constructed to emulate a flow sensor. As illustrated in Equation (8.5), the proposed artificial neural network requires 2 inputs namely the differential pressure and stem position of valve and one output, the flow. A dataset of these parameters are carefully obtained experimentally.

The back-propagation (BP) algorithm with a learning rate parameter of 0.3 is implemented for training the MLP. The stem position of the valve and the differential pressure are normalized and are considered as inputs to the MLP-ANN. Similarly, the flow through control valve is normalized and is considered as the target for the ANN. All the inputs and target are applied and the BP algorithm updates the weights and biases of MLP-ANN. The final updated weights are shown in Table 8.1 and the performance of the network in terms of error in each epoch over all epochs is plotted in Figure 8.3

The optimum weights and biases of the MLP-ANN are stored for testing, validation and further processing. During the testing and validation process, the stored weights are applied into the MLP-ANN. The test inputs are fed to this MLP-ANN to validate the output. The output (i.e., the predicted flow) is compared with the expected flow rate and is illustrated in Figure 8.4.

Table 8.1 ANN parameters

Neural network	*Multilayer Perceptron*	
Number of hidden neurons	Four	
Activation function	Hidden neurons – log-sigmoid, output neuron – linear	
Training algorithm	Back-propagation	
Epochs	1000	
Input weights to hidden neurons	V_{00} = 1.0522 V_{01} = 3.0286 V_{02} = −0.9959 V_{03} = 2.0069	V_{10} = 3.6616 V_{11} = 2.0187 V_{12} = −0.0884 V_{13} = −4.2753
Input biases to hidden neurons	b_0 = −0.3798, b_1 = 0.3076, b_2 = 0.3099, b_3 = −0.3674	
Weights to output neuron	w_{05} = 5.1865, w_{15} = −3.0431, w_{25} = −3.9924, w_{35} = 2.3473	
Bias to output neuron	b_4 = −0.1034	

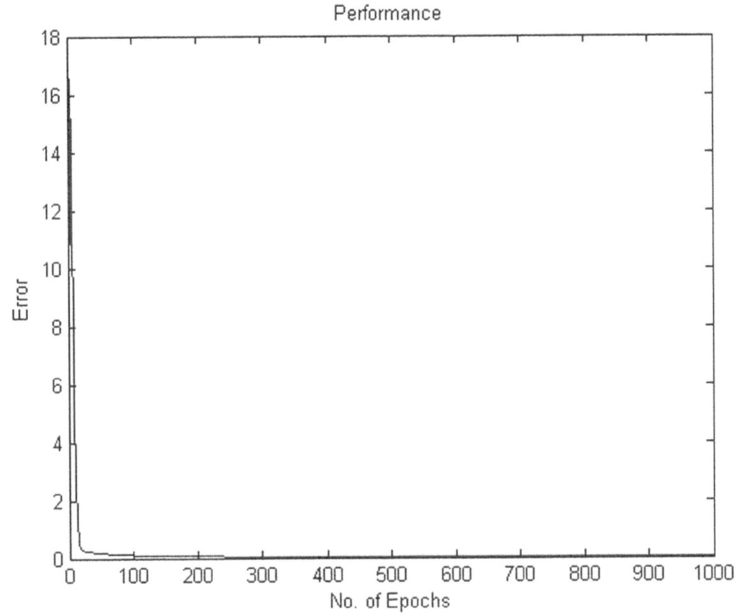

Figure 8.3 Performance of MLP-ANN in terms of error over subsequent epochs for all epochs.

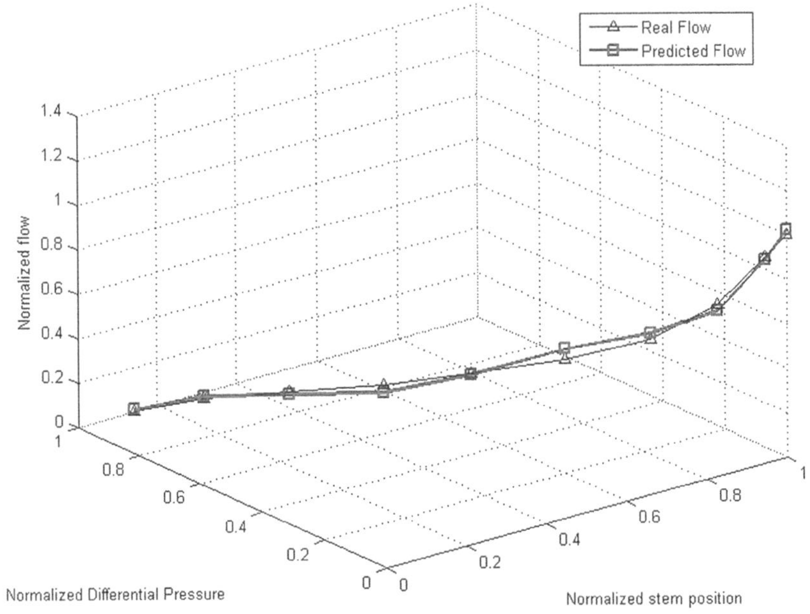

Figure 8.4 Valve characteristics: Expected & predicted from MLP-BPNN direct modeling.

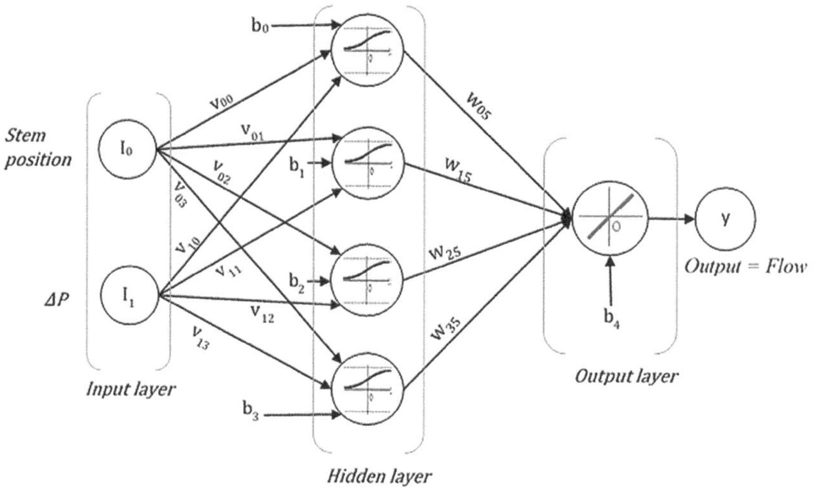

Figure 8.5 MLP network to approximate flow rate.

For optimum training of the neural network, these values are normalized. The MLP network with four hidden neurons, with log-sigmoid function and linear activation functions respectively, as shown in Figure 8.5 is coded using MATLAB.

8.4 FPGA IMPLEMENTATION

FPGAs are reprogrammable digital ICs that were developed in the mid 1980's. A personal computer can be used to design a digital circuit which is then compiled into a special programming file; will realize the circuit once it is downloaded to the FPGA. FPGA, available from the manufactures, currently consists of thousands of reconfigurable logic elements. Each logic element typically contains some number of D-flip-flop, look-up tables, NAND gates, and NOR gates. The wiring that connects these components, as well as the wiring that connects neighboring logic elements, is electrically reconfigurable. Therefore, components can be wired together to create one design. Then it can be erased and wired together differently to create a new one.

In this study, the FPGA design is accomplished with the help of the Xilinx System Generator toolbox [9]. It consists of various blocks of specific functions that are required and interlinked to compose algorithms. These blocks do not only serve to simulate but can also generate VHDL or Verilog code. In addition, the library provided by Xilinx to Simulink, contain main blocks that are basic for the ANN, except a few functions such as sigmoid function. We perform both a purely software simulation on Simulink as well as a hardware/software co-simulation through the JTAG emulator generated by the Xilinx System Generator toolbox.

8.5 FPGA IMPLEMENTATION OF THE SIGMOID FUNCTION

The log-sigmoid function consists of an infinite exponential series and cannot be implemented directly [10]. The simplified version of log- sigmoid function is considered for computational implementation. There are several methods to approximate the said function: Look-up tables, piecewise linear approximation among others [5]. Another study employs the Taylor series approximation [6] truncated to the second order terms. The log-sigmoid activation function is:

$$f(x)=\frac{1}{\left(1+e^{-x}\right)} \tag{8.14}$$

The Taylor series expansion of any function $F(x)$ in its defined domain can be given as:

$$F(x)=k_0+k_1x+k_2x^2+k_3x^3+k_4x^4+\ldots \tag{8.15}$$

The log-sigmoid function is asymptotic between the lines $y = 0$ and $y = 1$. The asymptotic distance between the function and the asymptotes becomes negligible in the range $(-\infty, -4) \cup (4, \infty)$. Thus the function $f(x)$ is linear in the range $(-\infty, -4) \cup (4, \infty)$ with the value equal to that of the asymptotes. The Taylor series can be truncated to the second-degree term to reduce complexity by minimizing the required number of multipliers to implement the function in FPGA. So now we have the Taylor series as:

$$F(x)= a_0+ a_1x+ a_2x^2 \tag{8.16}$$

The constants a_0, a_1, a_2 can be easily calculated and the log-sigmoidal function is expressed as:

$$f(x)=\begin{cases} 0 \ \forall x\in(-\infty,-4] \\ 0.5+0.2543x+0.0332x^2 \ \forall x\in(-4,4) \\ 1 \ \forall x\in[4,\infty) \end{cases} \tag{8.17}$$

Referring to the Equation (8.17), the log-sigmoid block is constructed using the Xilinx sysgen library in Simulink as shown in Figure 8.6.

8.6 FPGA IMPLEMENTATION OF THE MLP NEURAL NETWORK

FPGA implementation requires a series of interconnected multipliers and adders to construct so as to resemble the MLP network. This has been

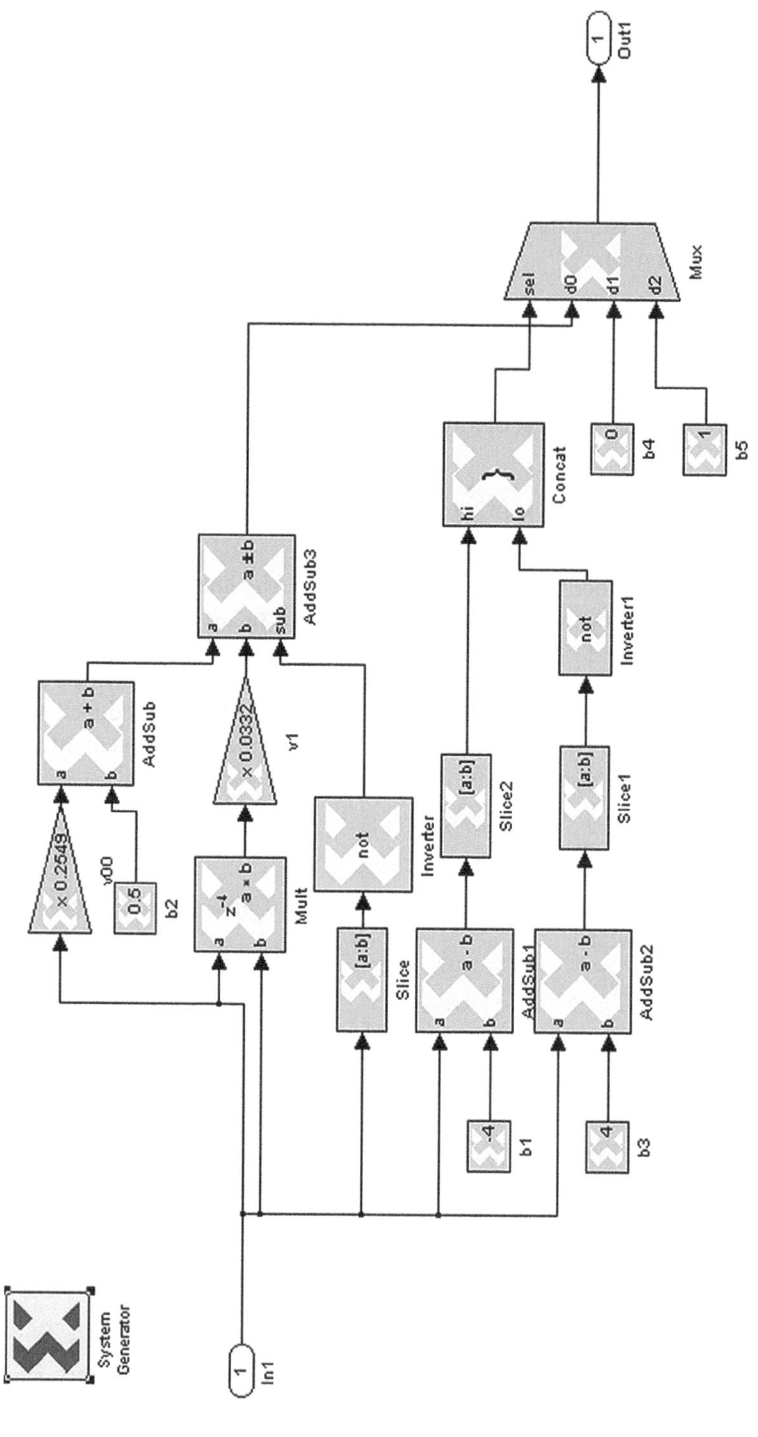

Figure 8.6 Simulink model - Taylor series approximation of the log-sigmoidal function using Xilinx sysgen blocks.

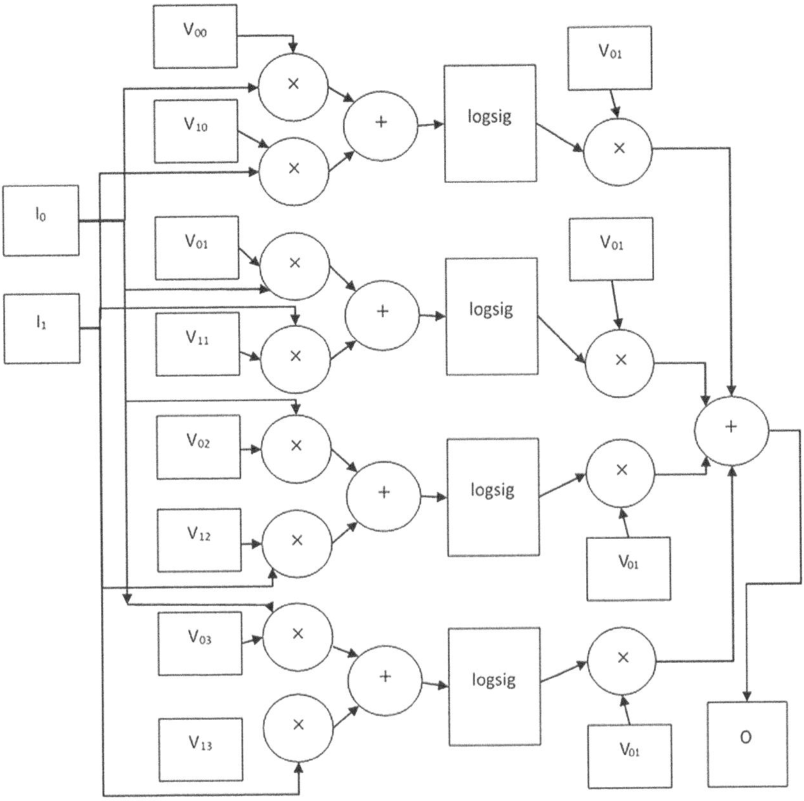

Figure 8.7 Block diagram of the MLP network made up entirely of adders and multipliers and the activation function.

illustrated in Figure 8.7. The activation function, log-sigmoid, has to be constructed as derived and exhibited earlier.

The system general library block is placed as seen in Figure 8.8 and the necessary sampling rates are applied for the system. Care has to be taken while assigning sampling rates, as rate should satisfy the Nyquist sampling theorem and the FPGA board maximum frequency. The Gateway in blocks samples, quantizes, and discretizes the input data and sends it to the other blocks, which now act as components of an FPGA environment, subject to necessary constrains. Gateway out block again converts data to be readable by MATLAB; in this system manually input the test data one after the other, recording the respective output. This simulation gives the tentative outputs to expect after dumping this model on the FPGA board.

Now this model is dumped onto an FPGA board. Sysgen allows converting this block into Verilog and/or VHDL code ready for implementation in FPGA. Next step is to perform a hardware co-simulation of the model via

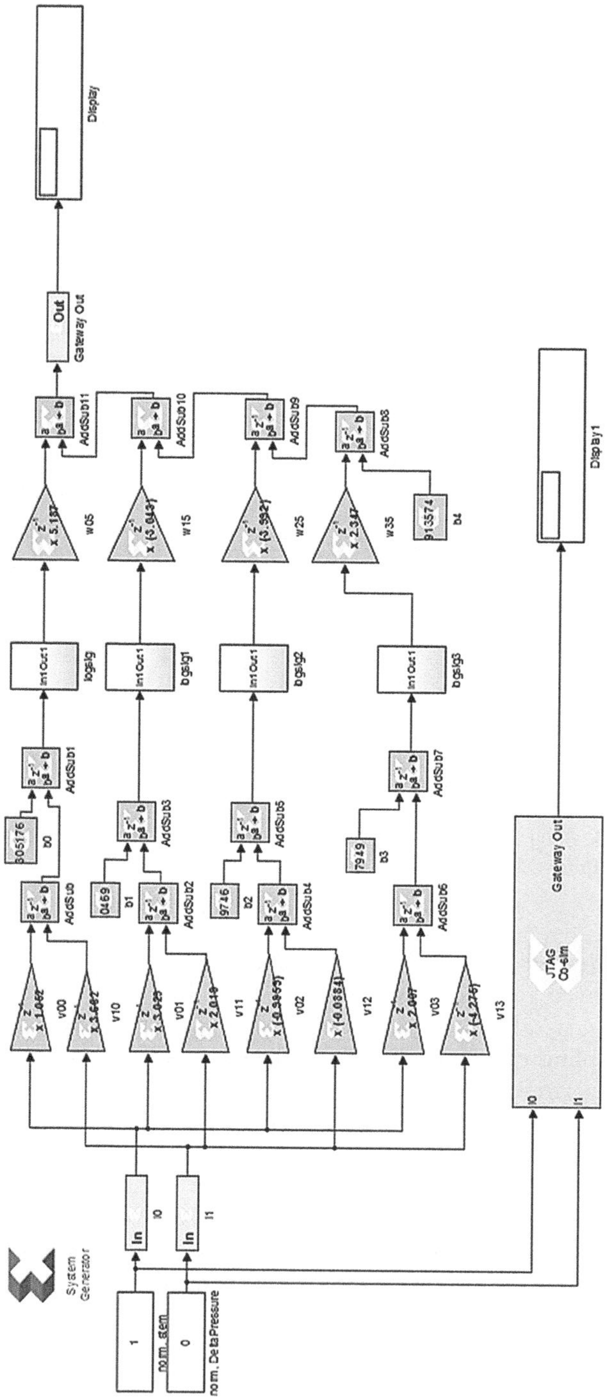

Figure 8.8 Xilinx Sysgen model of the trained MLP network along with the JTAG – emulator block for hardware co-simulation.

compiling a JTAG – emulator block. The sysgen block set, after compiling the VHDL code of the block set in Figure 8.8, occupies just 48.9% of the space in the FPGA, constructs the necessary bitstream file for communication between the computer and the FPGA board. The Spartan 3E starter kit is configured as the target co-simulation device, and the block is created. The output is then recorded.

8.7 DISCUSSION

The developed work has two distinct sections, namely, the neural network training and simulation and the FPGA implementation of the MLP-ANN with BP algorithm. The BP algorithm for training of MLP-ANN is coded in MATLAB, and the updated weights are used to test the network. The output from the training data after 1000 epochs has the sum of squared error of 0.0092 and the output from the test data offered a mean squared error of 4.306×10^{-4}. The FPGA implementation is accomplished through the Xilinx system generator block-set. The Simulink model of the neural network using the Xilinx System Generator is set and the simulation on Simulink is set up; the Simulink output and the FPGA hardware board co-simulation output has the mean square error with respect to the expected flow as 0.0057 and 48.9% board space utilization. The study can further include the live interfacing of the sensors from the control valve with the FPGA board. Moreover, the back-propagation algorithm itself can be implemented on the FPGA with the device to be used to its peak flexibility. This study provides yet another experimental validation of the universal approximation theorem pertaining to neural networks and can thus be used for several other systems, by substituting expensive sensors.

8.8 SUMMARY

In this chapter, a virtual flow sensor using artificial neural networks (ANN) is proposed to improve the efficiency of an industrial flow control loops. In a conventional flow-control loop, flow meters used for sensing flow rate in the feedback path cause pressure drop in the flow. This may increase the energy usage for propelling the fluid. The functional relation between the flow rate and the physical properties of the flow through the final control element (such as control valve) is known and the said properties (namely pressure drop, temperature, and valve position) are yielded from an experimental set-up. These properties are used as training data for ANN models to yield the fluid flow rate through the control valve. Here, the ANN acts as a virtual flow sensor. The feasibility of the proposed methodology is validated by using real measurement of flow and used them to model virtual flow sensor using the multi-layer perceptron artificial neural networks (MLP-ANN) with back propagation (BP) algorithm. Moreover, its practical proof of concept is demonstrated by implementing the trained MLP-ANN on a Spartan-3E-starter Field Programmable Gate Array (FPGA) unit through a hardware co-simulation.

BIBLIOGRAPHY

1. P. Pallás-Areny and J. G. Webster, *Sensors and Signal Conditioning*, 2nd ed. New York: Wiley, 2013.
2. V. N. Kumar and K. V. Lakshmi Narayana, "Development of an intelligent pressure sensor with temperature compensation," *Journal of Engineering Science and Technology*, vol. 12, no. 7, pp. 1723–1739, 2017.
3. J. C. Patra, A. C. Kot, and G. Panda, "An intelligent pressure sensor using neural networks," *IEEE Transactions on Instrumentation and Measurement*, vol. 49, no. 4, pp. 829–834, Aug. 2000.
4. V. Naveen Kumar and K. V. Lakshmi Narayana, "Development of thermistor signal conditioning circuit using artificial neural networks," *IET Science, Measurement & Technology*, vol. 9, no. 8, pp. 955–961, Nov. 2015.
5. S. Himavathi, D. Anitha, and A. Muthuramalingam, "Feedforward neural network implementation in FPGA using layer multiplexing for effective resource utilization," *IEEE Transactions on Neural Networks*, vol. 18, no. 3, pp. 880–888, 2007.
6. S. A. Al-Kazzaz and R. A. Khalil, "FPGA implementation of artificial neurons: Comparison study," *3rd International Conference on Information and Communication Tchnologies: from Theory to Applications*, 7–11 Apr. 2008.
7. D. E. Seborg, T. F. Edgar, and D. A. Mellichamp, *Process Dynamics and Control*, 3rd ed. New York: John Wiley & Sons, 2011.
8. B. Bavarian, "Introduction to neural netwroks and control," Special Section on Neural Networks for Systems and Control, Apr. 1988.
9. Xilinx Inc., "Xilinx, "Spartan™-3E Platform FPGAs complete data sheet," Datasheet Sheet DS312," 2013.
10. D. E. Khodja, A. Kheldoun, and L. Refoufi, "Sigmoid function approximation for ANN implementation in FPGA devices," CSECS '10: Proceedings of the 9th WSEAS international conference on Circuits, systems, electronics, control & signal processing. pp. 112–116, 2010.

Chapter 9

Smart sensor systems for military and aerospace applications

Parul Raturi

Omkaranand Sarswati Government Degree College, Devpryag, India

Bijit Choudhuri

National Institute of Technology Silchar, Silchar, India

P. Chinnamuthu

National Institute of Technology Nagaland, Nagaland, India

CONTENTS

9.1 INTRODUCTION

Smart sensors play a significant role in detecting any chemical or physical change and automating the device after processing the collected data. The effectiveness of a smart sensor system is decided by its ability to integrate various types of sensor components and establish communication between different components without requiring human intervention [1]. There are a variety of smart sensors which can sense the changes temperature, proximity, chemical environment, humidity, pressure, motion, etc. Various sensors are used for multiple applications to enable a smart environment. It picks input signals from the surrounding environment and performs the predefined functions after detecting specific inputs using inbuilt resources. The evolution of smart sensing tools emerged as a result of transformative

DOI: 10.1201/9781003288633-9

advancement in sensor technology. These sensor systems are made up of very basic sensing elements, but with a high level of embedded intelligence.

In the standard combat zone situation of the military engagement, there are well-defined, well-known enemies on the land, in the air, and at sea. Moreover, it also involves worldwide operations, urban environment operations, and operations except for the war, such as disaster relief and peacemaking missions. The signals which are essential to be sensed for military applications include sound, electromagnetic waves, pressure, and light resulting from the explosions and gunfires [2]. The performance of the sensor used for the military application depends not only on the signal that is sensed, but also on the capacity of the sensor used. Irrespective of the purpose, wireless sensors are the most widely used sensors to provide a cost-effective way to gather information about the surroundings in aerospace and defense applications.

By using the smart sensor systems, advantages over ordinary sensor systems can be achieved. This advantage includes superior signal-to-noise ratio, improvement in performance, execution of logical functions and commands, swift signal conditioning, standardization of the output signal, compatibility with the future data highway, compactness, and high sensor system reliability [3, 4]. Moreover, smart sensor systems exhibit auto-calibration, self-testing, and prevention of failure. One of the most challenging issues among sensor designers is the packaging. The package must be cheap, small, and durable for successful operation. Also, the packaging must protect the electronic interface from the envoirnment. Proper protection and packaging are crucial for sensors in a destructive and harsh environment, and the packaging cost for this type of sensors becomes one important factor. The cost of packaging for such sensors might take a significant portion of the overall cost of the sensor [5]. Therefore it is essential to take care of the cost of smart sensors to achieve improved performance and reliability of such sensors. Sensor technology has progressed immensely, but still, there are many scopes to solve challenging issues that must be overcome. The foremost challenge is developing the sensor circuit fabrication method to maintain a high yield without affecting the fabrication cost. Further, without compromising the performance, low fabrication cost of the sensing elements is favorable. The term smart sensor was introduced around the mid 1980s, and became common to develop sensors with smart features. These smart devices acquire their intelligence by using advanced functioning and smart components such as digital signal processors, microcontroller units, and application-specific integrated circuits. Various semiconductor manufacturers are progressively working on developing smarter semiconductor devices for the input and output part of the control system.

Before microelectronics came into the picture, the sensosrs were like a meter used to progressively sense physical quantity such as pressure, temperature, flow, etc. System corrections were carried out to modify the sensor system. However, to achieve unattended measurement and control, sensors must have an electrical output that can be easily interfaced. Furthermore, they also require amplification of the analog signal and conversion into the digital before processing further. A typical 5 V power supply is used in ssmart sensor units such as micro controller units and analog to digital converters. However, it is challenging to reduce this required power supply to the 3 V or even less for less power consumption, making its operation convenient and less complex. The use of separate integrated systems offers a solution to the issue, but by introducing complexity to the sensor system.

9.2 REQUIREMENT FOR AN IDEAL DESIGN OF SMART SENSOR SYSTEMS

For high-performance sensors, performance is preferred over the cost. In this case, the performance level of the smart sensor is much more important than the cost of the smart sensor unit. However, for large-scale sensor applications, low cost and high performance are desirable. Another challenging issue for sensor technology is the packaging. The aerospace industry requires various smart sensors for multiple applications in this field. The primary concern associated with smart sensors is the weight of associated cables and data integrity. By self-monitoring, features of the sensor can significantly improve safety. British designers for the sensor systems first deployed radar technology to give early warning to the Nazi bombers for crossing the channels to attack the towns and cities of England. Initially, Radar could only provide rudimentary information about the direction and contact speed.

Modern radar systems can yield digitized signals very quickly. Moreover, radar systems have imaging capabilities. They can be networked together to serve several functions so that the total system's net effective performance must be superior to the performance of its components. These modern radar systems can simultaneously perform tasks like target tracking, weather monitoring, fire control, and broad area search. However, separate systems were required in the case of old generation radar systems to fulfill the same jobs. The speed and ease associated with analog to digital conversion of current radar information open up various options for signal processing. Moreover, it also makes radar information available for real-time applications on the digital battlefield. Initially, radar systems worked by bouncing the radio waves and detecting the return signal. The preliminary radar systems were tube-based, mechanically steered, and included radio frequency (R.F.) transmitters, receivers, video display, and signal processors. A later improvement was carried out in the sensitivity of these radar systems, where they can virtually monitor and detect ocean waves and insects in the air. However increased sensitivity resulted in unwanted reflected signals from the objects which were of no use.

The solid-state device-based technology explored the first horizon in radar technology. Solid-state radars and radio frequency (RF) chips were introduced, enabling the electronically active steered arrays. Solid-state technology resulted in digital signal processing, high-speed digital networking, efficient processing, and broadened new perspectives. The use of new semiconducting materials such as GaN and GaAs helps improve system efficiency and reduce the size of the systems. For example, the Raytheon project with Next Generation Transmit/Receive Integrated Microwave Module (NGT) uses GaN and GaAs technology to upgrade major radar systems. One of the most critical developments in radar systems is multifunctionality, which provides a system that can simultaneously perform various applications in a single system. Multifunctionality of the radar systems can be helpful to save space, power, and weight by assembling various stand-alone radars into a single system. It might be proven beneficial in reducing the need for human resources by combining techniques and functions.

9.3 HEMT FOR RADAR APPLICATION

The recent advancement in fabrication technology is driving the development of various potential nano/microstructures of new and novel materials. A wide range of heterojunctions with tunable bandgaps and high mobility semiconductors can be easily prepared using metal-organic chemical vapour deposition (MOCVD)/metal-organic vapour phase epiaxy (MOVPE) and molecular beam epitaxy (MBE) techniques. Due to its high thermal conductivity, GaN finds its extensive use in the fabrication of high electron mobility transistors (HEMT) among the III-V semiconductors. Due to its high power compatibility and large bandwidth, GaN HEMT devices are preferred in RADAR power amplifiers over the conventional traveling wave tube (TWT) devices, laterally diffused metal oxide semiconductor (LDMOS), and GaAs technology. With these superior performances, the GaN-based RF device industry is estimated to reach $2 billion by 2024 [6].

Under a high doping concentration level, impurity scattering degrades the carrier mobility in a semiconductor significantly. In HEMT devices, the conduction carriers are confined within an undoped semiconductor heterojunction's triangular potential well. Due to the high mobility of carriers, HEMT is used in very high-frequency applications, such as microwave and millimeter-wave communication, military and commercial RADAR, astronomy, etc. Kikuchi and co-workers [7] fabricated GaN HEMT for X-band RADAR applications using MOVPE and i-line optical lithography. The breakdown voltage for fabricated HEMT was 290 V. A 10 dB power gain in the 8.5-10.0 GHz frequency range was observed in device characteristics. A maximum of 333W was observed as the highest saturated output power [7]. Del Prato et al. [8] reported the enhancement in pulse to pulse stability in a GaN HEMT power amplifier. The amplitude pulse to pulse stabilities were improved by 10 dB, while satisfying the critical stability limit of –55 dB. Before each RF pulse, a systematic gate bias pulse was applied to achieve this improvement [8].

9.4 PHOTODETECTORS IN MILITARY APPLICATIONS

The military relies greatly on optical sensing for mapping, sensing, and identifying the enemies over a significant distance. Photonics provides highly efficient communication devices and high-quality sensing. Multispectral imaging is a photonics-based technology to extract essential information about surroundings. It can be used for tasks such as uncovering enemies' movements, locating explosives, and pinpointing the depth of hidden bunkers. Spectrometers and holographic imagers based on photonics are used in the military. Spectrometers can be used for the detection of explosives in solids and liquids. A holographic imager produces three-dimensional visualization of mountainous and urban terrain. One of the advantages of photonic devices is that these devices are small, lightweight, and easily portable for on-field military operations. Following are a few examples of the photonics technologies used in the military.

9.4.1 Photoresistors

These light-dependent resistors are sensitive to light, and the resistance depends on light intensity. Resistance of these sensors decreases on the increase in light intensity and increases on the decrease in intensity. Thus these photoresistors can be utilized in the light-sensitive detector circuits.

9.4.2 Photodiodes

These devices produce an electric current after consuming light energy. Photodiodes are referred to as photosensors or photodetectors. General applications of the photodiode include detection of color, light, intensity, and position. A combination of photodiodes and light-emitting diodes is commonly used in various optical sensor systems.

9.4.3 Infrared sensors

These sensors are generally used to sense and detect objects and distance from the target. It emits an infrared pulse from the emitter. This emitted infrared light is reflected from the object present and hits on the detector. On the basis of the angle of return, the distance of the object can be determined. Infrared sensors are most widely used for distance measurements, terrain sensing, detection, and avoidance of targets.

9.5 SOLAR BLIND PHOTODETECTORS FOR MILITARY APPLICATIONS

Solar blind photodetectors are optoelectronic devices with paramount importance due to their applications in an extensive range of critical and sensitive domains such as military, defense, astronomical observations, flame detection, chemical and biological threat detection, power line, etc. Thus, achieving control on growth parameters is essential to improve their crystallographic properties and photoconversion efficiency. Previous reports demonstrate that dislocations associated with strain between subsequent layers become very important parameters in degrading the performance of the device. The doping concentrations, the thickness of the film, etc., significantly impact the device's performance. With the design of a sophisticated fabrication facility, it is now possible to control the growth even in atomic layer scales.

Recently, the government of India has taken firm initiatives to develop indigenous technologies for applications in various domains, such as defense technology, space research, new and renewable energy sectors, etc. Among them, the "Make in India" program is calculated to have saved Rs. 1 trillion in the defense sector in the timeframe of just three years [9]. Various wide-bandgap semiconductors, such as AlN, GaN, $Al_xGa_{1-x}N$, diamond, BN, $LaAlO_3$, and $In_2Ge_2O_7$, are used to fabricate solar-blind photodetectors. These detectors don't require Wood's filters to eliminate the longer wavelength and heavy cooling systems compared to conventional sensors [10]. Among these materials, GaN enjoys a very high bandgap, high electron mobility, fast response, radiation hardness, etc. [11]. Consequently, the U.S. Missile Defense Agencies are upgrading their AN/TPY-2 ballistic missile defense radar with GaN semiconductor technology [12]. More recently, the UV bandpass cut-off was tuned in $Al_xGa_{1-x}N$ structures with a higher concentration of aluminium and thus a solar-blindness with a high rejection ratio of near UV and visible spectra was achieved [13]. To reduce the detector's dark current, reducing the defect and dislocation density in the grown layers is essential. MBE and MOCVD are two most dominant growth techniques used to grow defect-free nanostructures [14, 15]. Among these two techniques, the drawbacks associated with MOCVD include using toxic precursor and by-product materials, frequent maintenance of scrubbers, hydrogen-related burn-in issue, etc. [16]. The Figure 9.1 display the cross-sectional schematic (Figure 9.1.a) and SEM image (Figure 1.b) of solar blind photodetectors.

Mendoza et al. reported solar blind photodetection by sulfur doped diamond films grown using hot filament CVD (HFCVD). In planar configuration, the diamond surface hosts the electrode directly, and the bias electrode is kept hanging above the diamond surface in electron field emission configuration. Under 220 nm UV illumination, the field emission configuration exhibited a maximum of 10 mA/W responsivity comparable to the

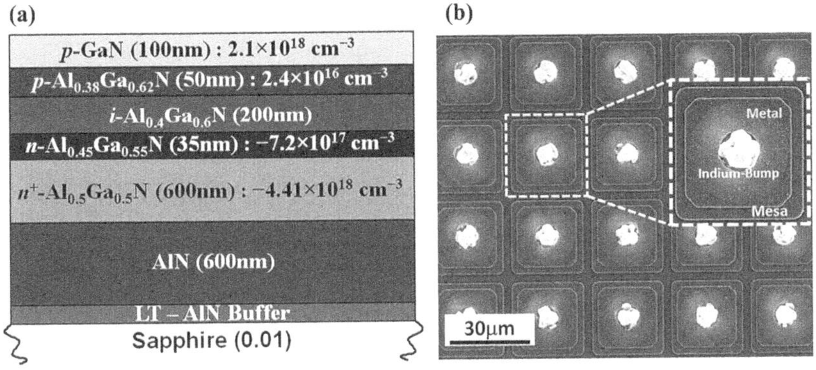

Figure 9.1 (a) Cross-sectional shematic of photodetector. (b) Micrograph image of the processed photodetector array with inset showing a single photodetector. [Figure reproduced from E. Cicek, R. McClintock, C.Y. Cho, B. Rahnema, and M. Razeghi, "AlxGa1-xN-based back-illuminated solar-blind photodetectors with external quantum efficiency of 89%" Applied Physics Letters 103, 191108 (2013) [13]].

commercial photodiode. The field emission configuration showed ~40% superior photodetection compared with the planar configuration [17]. Chen and co-workers synthesized 180 nm α-Ga_2O_3 on a-plane ZnO substrate using laser MBE for solar-blind photodetection. The detector showed a dark current of less than 1 pA. At 230 nm, the device showed zero bias responsivity and a specific detectivity of 3.42 mA/W and 9.66 × 10^{12} cm $Hz^{1/2}$ W^{-1}, respectively [18]. Qian et al. developed a Ga_2O_3 metal-semiconductor-metal (MSM) photodetector on a c-plane sapphire substrate using radio frequency (R.F.) magnetron sputtering. The dark current and the responsivity was 338.6 pA at 10 V bias and 70.26 A/W, respectively. Detectivity was 1.26 × 10^{14} cm $Hz^{1/2}$ W^{-1} [19]. Solar-blind ultraviolet photodetector was developed using TiO_2 and $ZnTiO_3$ nanowire (NW) heterojunction on SiO_2/Si substrate using the electrospinning and pyrolysis method. The detector exhibited a very low dark current of 0.42 μA. The device exhibited very high responsivity and external quantum efficiency of 1.1 × 10^6 and 4.3 × 10^8%, respectively [20]. Bao et al. designed Ni contact/Al nanoparticlesAlGaN on (0001) sapphire substrate MSM, using MOCVD and e-beam evaporation. The device exhibited a dark current and responsivity of 10^{-10} A and 0.288 A/W, respectively [21]. M. Razeghi and co-workers developed a p-i-n Solar Blind ultraviolet photodetector using Ni/Au/p-GaN/i-AlxGa1-xN/n-AlGaN/n-GaN/Ti/Au/sapphire substrates structure deposited using horizontal flow low-pressure MOCVD. The detector exhibited responsivity and external quantum efficiency of 0.11 A/W and 59 %. The internal quantum efficiency was found to be > 90 % [22]. The same research group also prepared Ni/Au contact/p-GaN/p-Al 0.38Ga0.62N/i-$Al_{0.4}Ga_{0.6}N$/n-$Al_{0.45}Ga_{0.55}N$/n^{+} -$Al_{0.5}Ga_{0.5}N$/Ti/Al/AlN/LT ALN on silicon substrates and reported dark current density and responsivity to be 1.6 10^{-8} A/cm2 at 10 V reverse bias and 18.3 mA/W at 290 nm. The external and internal quantum efficiencies were 7% and 10%, respectively [23]. By replacing the n^+ layer of $Al_{0.5}Ga_{0.5}N$ with $Al_{0.4}Ga_{0.6}N$ and silicon substrate with a sapphire substrate, the same research group reported enhancement in performance. The dark current density was reduced to 2 × 10^{-9} A/cm^2 at 10 V, and responsivity, external and internal quantum efficiencies were increased to 176 mA/W at 275 nm, 98%, and 80%, respectively [13]. They have also developed a solar-blind UV detector by MOCVD growth technique. $Al_xGa1\text{-}_xN$ based 320 × 256 focal plane arrays exhibited unbiased peak external quantum efficiency and responsivity of 49% and 109 mA/W, respectively, and a dark current density less than 7 × 10^{-9} A/cm^2 [24]. Chen et al. deposited β-Ga_2O_3 NW on sapphire substrates using a partial thermal oxidation process of Ga and gold Schottky junction. The detector exhibited dark current and unbiased responsivity of 10 pA at −30 V and 0.01 mA/W, respectively [25]. $Mg_{0.46}Zn_{0.54}O$ film was deposited on quartz substrate by the R.F. magnetron sputtering system. Au was deposited to form the Schottky junction. Under 265 nm wavelength illumination, the unbiased responsivity for the detector was 3.4 A/W [26]. Wei and co-workers developed a *β*-Ga_2O_3 active layer/indium zinc oxide (IZO) transparent electrode MSM detector on a sapphire substrate using the MOCVD technique. The detector exhibited a dark current and responsivity of ~1 nA at 200 V at 10 V and 0.32 mA/W, respectively. The overall quantum efficiency of 0.2% was observed. The detectivity and noise equivalent power were observed to be ~2.8 × 10^{10} cm Hz$^{1/2}$ W^{-1} and3.53 × 10 $^{-11}$ W, respectively[27]. Ozbay et al. reported the preparation of Si_3N_4/N-AlGaN/ N$^+$-AlGaN/N$^+$ GaN/undoped GaN/sapphire substrate-based solar-blind photodetector using MOCVD, reactive ion etching (RIE), and plasma-enhanced chemical vapor deposition (PECVD) system. The detector's dark current density and responsivity were 1.8 nA/cm^2 at 25 V and 89 mA/W under 267 nm wavelength illumination. The detectivity and external quantum efficiency were 5.6 × 10^{12} cm Hz$^{1/2}$ W^{-1} and 42% at 250 nm, respectively [28]. Hu et al. developed Au-Ga_2O_3- Au MSM photodetector on a sapphire substrate using the MOCVD technique. At 20 V bias, the dark current and responsivity were 6.2 × 10^{10} A and 17 A/W, respectively. The device showed quantum efficiency and detectivity of 8228% and 7.0 × 10^{12} cm Hz$^{1/2}$ W^{-1}, respectively[29]. Oh, and co-workers deposited Ti/Au electrode/ β-Ga_2O_3 film MSM detector on Al_2O_3 substrate using the MOCVD process. Under 254 nm wavelength illumination, the detector showed a responsivity of 1.45 A/W [30]. Shi et al. demonstrated the synthesis of aluminum nanoparticle arrays using microsphere lithography on PECVD grown diamond film. At 5 V bias, the device showed a dark current and responsivity of 10^{-12} A and 28 m A/W at 225 nm [31]. Patil-Chaudhari and co-workers developed aluminum contact - Nanotextured *β*-Ga_2O_3 Films on GaAs substrate by oxidation of GaAs in Argon atmosphere. The dark current, responsivity and quantum efficiencies of the detector were ~10 nA, 291.9 mA/W and 1.34%, respectively [32].

Xing et al. reported the preparation of Solar-blind deep-ultraviolet photodetectors by growing gold electrodes on a $LaAlO_3$ single crystal using electron-gun evaporation. The detector exhibited dark current, responsivity, and quantum efficiency of 25 pA at 20 V bias, 71.8 mA/W at 200 nm, and 44.6% [33]. Zhao and co-workers fabricated ZnO $-Ga_2O_3$ core-shell microwires on a glass substrate. This solar-blind avalanche photodetector showed a dark current and responsivity of 10^{-11} A and 1.3×10^3 A/W under −6 V bias. The device showed quantum efficiency and detectivity of 2.53×10^6 % and 9.91×10^{14} cm $Hz^{1/2}$ W^{-1}, respectively [34].

9.6 INFRARED PHOTODETECTORS AS NIGHT VISION DEVICES

A night vision device (e.g., camera, goggles, thermal imaging device, etc.) is now an integral part of any defense system due to their precise tracking ability of objects even in a dark environment. The infrared photodetectors are widely used in the implementation of night vision devices. Infrared detectors can detect radiation with its wavelength of 700 nm – 3 mm. Recently, researchers have been trying to improve the infrared photodetectors' working temperature, limitations in cut-off wavelength adjustability, and reduce their production cost. Although the III-V semiconductors and mercury cadmium telluride (HgCdTe) (MCT) based detectors are popular for commercial purposes, nanostructures of two-dimensional materials, graphene, metal oxides are also investigated to solve the mentioned problems.

An infrared photodetector consists of light-absorbing active material. Under photon illumination, this material absorbs the energy, and electrons move to the conduction band and get swept to the electrode, thus resulting in a current pulse in the detector. According to Planck's law, the wavelength of the light absorbed is determined by the material's bandgap. For example, MCT based detectors are used in long-wavelength infrared (LWIR) detection (λ = 7–12 μm); InGaAs/InAsSb strained-layer superlattices are preferred in mid wavelength infrared (MWIR) detection (λ = 3–7 μm), and Si, InGaAs based detectors are used in short-wavelength infrared (SWIR) detection (λ = 0.7–2.5 μm). Although the investigation of several decades has evolved the MCT and III-V based detectors to be the most mature technology, the complexity of fabrication steps associated with MCT growth due to the very high mercury vapor pressure makes the technology an expensive process [35]. Also, the use of expensive CdZnTe substrate in MCT devices and toxicity associated with mercury restrict the MCT-based detectors to be used only for commercial purposes [36]. As a result, the attention has now been focussed on the characteristics of a quantum dot in well-infrared photodetectors (Q-DWell-IRPD), quantum well-infrared photodetectors (QWIRPD), and quantum dot infrared photodetectors (QDIRPD). The III-V semiconductor nanostructures are widely used in the active material of the quantum devices due to their superior mobility, high absorption cross-section, an adjustment in bandgap as per requirement, etc. To improve the working temperature of the photodetector, the following technologies are preferred to realize IRPD 1) Type-II superlattice (T2SL) barrier infrared detector (BIRD) and 2) High dynamic range 3D readout integrated circuit (ROIC). The BIRD technology employs the XBn as the active material where X is n or p-type contact material, B is the barrier layer with wide bandgap, and n is the narrow bandgap absorber material.

Savich and co-workers fabricated short-wave infrared photodetectors (SWIRPD) using III-V semiconductor material. MBE was used to grow the $In_xGa_{1-x}As$ absorber layer on an InP substrate. AlInAs was used as step graded buffer to reduce the lattice mismatch originated defects at the interface. At 300K, this InGaAs IRPD reduced the dark current by 400 times compared to conventional photodetector dark current. In the same work, they have used a lattice-matched $In_xGa_{1-xAsySb1-y}$ absorber layer and obtained ~ 1000 times improvement in dark current compared to the traditional photodetector. The quantum efficiency was 26% at 1.55 µm wavelength [37]. Kim et al. [38] reported the application of gold-coated III-V NW for photodetection. The gold coating exhibited surface plasmon resonance, thus resulting in very high absorption of infrared radiation. The growth of III-V NW usually takes place in <111> direction, which the selected area that the epitaxy process can achieve. Finite-difference time-domain (FDTD) simulation showed an enhancement in optical absorption of III-V NW-gold hybrid nanostructure due to their plasmonic-photonic hybrid modes [38]. Miao et al. reported the application of vertically stacked graphene – InAs NW heterojunction in infrared photodetection. The photoresponsivity was 500 mAW^{-1} and I_{photo}/I_{dark} was 500. This performance improvement was attributed to integrating multiple functional materials in the device's active region [39]. Wang et al. [40] synthesized p-$GaAs1\text{-}_xSb_x$ – n-InAs core-shell NW-based heterojunction and investigated its infrared photodetection properties. The detector efficiently detected the illumination in the 488-1800nm range under 0.3V biasing. The dark current, switching delay, and responsivity were 32pA, 0.45ms, and $0.12AW^{-1}$[40].

The excellent stability of SiO_2 as the gate dielectric in field-effect transistors led to the paramount dominance of silicon in the electronics industry. The complementary metal-oxide semiconductors (CMOS) have become an integral part of every electronic device due to their low leakage current and superior integration compatibility. The complete realization of infrared photodetector (IRPD) device fabrication and their integration with CMOS devices can be illustrated in Figure 9.2.

Krishna et al. [42] demonstrated a 320×256 two-color focal plane array using InAs/InGaAs DWELLl nanostructure. Using MBE, 15 DWELL layers were inserted between two n^+-GaAs contact layers. The specific detectivity of the 320 × 256 two-color focal plane array detector was 7.1×10^{10} cm $Hz^{0.5W-1}$ under 1V biasing at 78K temperature. Under this condition, a dark

[The six steps of digital IR imaging]

	Digital imaging step	Technologies	
1	Get light into the detector	Anti-reflection coating Substrate removal	Infrared detector material
2	Charge generation	Detector material growth	
3	Charge collection	p-n junctions	
4	Charge-to-voltage conversion	Amplifiers optimized for flux, speed and noise ▸ Source follower ▸ Capacitive Transimpedance Amplifier ▸ Direct Injection	CMOS integrated circuit
5	Signal transfer	Multiplexer Scanner	
6	Digitization	Analog-to-digital converters	

Figure 9.2 Six steps of digital infra-red imaging. [Figure reproduced from Thomas Sprafke and James W. Beletic, "High-Performance Infrared Focal Plane Arrays for Space Applications" Optics and Photonics News Vol. 19, Issue 6, pp. 22-27 (2008) [41]].

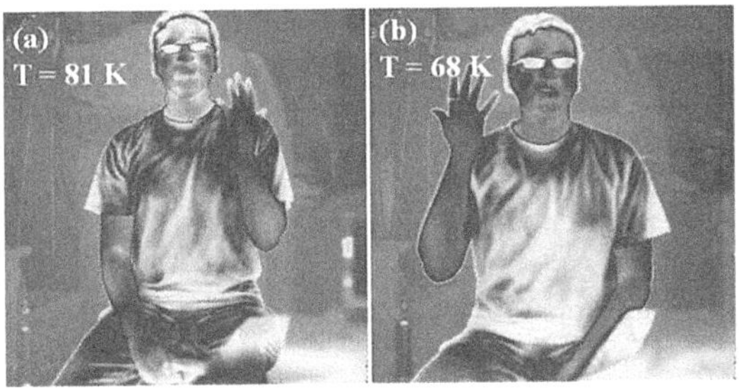

Figure 9.3 Pictures of a student taken with the 1k × 1k FPA at (a) 81 K and (b) 68 K. [Figure reproduced from Manurkar, et al. "High performance long wavelength infrared mega-pixel focal plane array based on type-II superlattices" Applied Physics Letters 97, 193505 (2010) [43]].

current density of 8.6×10^{-7}A/cm^2 was recorded [42]. Manurkar et al. [43] developed long-wavelength mega-pixel IRPD using type-II superlattice photodiodes. n-M-π-p-type superlattice of InAsSb/InAs/GaSb/AlSb/GaSb structure was prepared by MBE. The specific detectivity of the detector was 6×10^{11}cm Hz$^{0.5}$W^{-1} under 10μm wavelength light illumination at 77K temperature. The dark current density and noise equivalent differential temperature (NEDT) were 1×10^{-4} A/cm^2 and 23.6mK at 81K temperature, respectively. The pictures taken by the focal plane array is shown in Figure 9.3 [43].

Goma Kumari and co-workers [44] have also developed a 320 × 256 IR focal plane array using InAs/InGaAs/GaAs DWELL nanostructure. The NEDT at 70K was 197mK. The thermal images revealed that 95% of pixels were functional. The photoresponse exhibited a peak of 9.3 μm due to the intersubband transition [44].

9.7 CONCLUSION

This chapter discusses the progress of research on sensors deployed on the battlefield. The chapter presents the working principle of various semiconductor device-based sensors such as HEMT, U.V. detectors, I.R. sensors, etc. The fabrication methods necessary to prepare the sensors and the sensor's important characteristics were discussed thoroughly. The researchers are now trying to develop an autonomous weapon system for future battlegrounds. Overall, it is expected that global leaders will use all these development for the protection of our civilization rather than aggression.

REFERENCES

1. Mohd Javaid, Abid Haleem, Ravi Pratap Singh, Shanay Rab, and Rajiv Suman, "Significance of sensors for industry 4.0: Roles, capabilities, and applications," *Sensors International*, 2, 100110, 2021.
2. Mohammad Masoud, Yousef Jaradat, Ahmad Manasrah, and Ismael Jannoud, "Sensors of smart devices in the internet of everything (IoE) era: Big opportunities and massive doubts," *Journal of Sensors*, 2019, 1–26, 2019. https://doi.org/10.1155/2019/6514520.
3. Eugene Y. Song, Gerald J. FitzPatrick, and Kang B. Lee, "Smart sensors and standard-based interoperability in smart grids," *IEEE Sensors Journal*, 17, 7723–7730, 2017.
4. Newark element 14, "Smart sensor technology for the IoT" https://www.techbriefs.com/component/content/article/tb/pub/features/articles/33212
5. Guillermo Fuertes, Ismael Soto, Raúl Carrasco, Manuel Vargas, Jorge Sabattin, and Carolina Lagos, "Intelligent packaging systems: Sensors and nanosensors to monitor food quality and safety," *Journal of Sensors*, 2016, 1–8, 2016. https://doi.org/10.1155/2016/4046061
6. Aerospace & Defense, "GaN HEMTs for Pulsed Radar Applications" Wolfspeed, June 1, 2020, https://www.wolfspeed.com/knowledge-center/article/gan-hemts-for-pulsed-radar-applications
7. Ken Kikuchi, Makoto Nishihara, Hiroshi Yamamoto, Takashi Yamamoto, Shinya Mizuno, Fumikazu Yamaki, and Seigo Sano, "An 8.5-10.0 GHz 310 W GaN HEMT for radar applications," *IEEE MTT-S International Microwave Symposium (IMS2014)*, 2014.
8. Julien Delprato, Arnaud Delias, Pierre Medrel, Denis Barataud, Michel Campovecchio, et al. "Pulsed gate bias control of GaN HEMTs to improve pulse-to-pulse stability in radar applications," *Electronics Letters, IET*, 51(13), 1023–1025, 2015.
9. Ajit Kumar Dubey, "Modi's make in India just saved Rs 1 Lakh crore in defence sector. Here's how," Indiatoday, New Delhi, December 3, 2017.

10. Dung-Sheng Tsai, Wei-Cheng Lien, Der-Hsien Lien, Kuan-Ming Chen, Meng-Lin Tsai, Debbie G. Senesky, Yueh-Chung Yu, Albert P. Pisano, and Jr-Hau He, "Solar-blind photodetectors for harsh electronics," *Scientific Reports*, 3, 2628, 2013.
11. Zhongxin Wang, Guodong Wang, Xintong Liu, Shouzhi Wang, Tailin Wang, Shiying Zhang, Jiaoxian Yu, Gang Zhao, and Lei Zhang, "Two-dimensional wide bandgap nitride semiconductor GaN and AlN materials: properties, fabrication and applications," *Journal of Materials Chemistry C*, 9, 17201–17232, 2021.
12. Raytheon Co., "World's most capable X-band ballistic missile defense radar upgrades to GaN," *Microwave Journal*, April 6, 2017.
13. E. Cicek, R. McClintock, C. Y. Cho, B. Rahnema, and M. Razeghi, "AlxGa1-xN-based back-illuminated solar-blind photodetectors with external quantum efficiency of 89%," *Applied Physics Letters*, 103, 191108, 2013.
14. M. A. di Forte-Poisson, M. Magis, M. Tordjmann, and J. di Persio, "Chapter 5 – MOCVD growth of group III nitrides for high-power, high-frequency applications," in *Optoelectronic Devices: III Nitrides*, Editor: M. Razeghi and M. Henini, Elsevier Science, 69–94, 2005.
15. Ichiro Shibasaki and Naohiro Kuze, "Chapter 31 – Mass Production of sensors grown by molecular beam epitaxy," in *Molecular Beam Epitaxy From Research to Mass Production*, Editor: M. Henini, Elsevier, 2nd Edition, 693–719, 2018.
16. Rodney Pelzel, "A comparison of MOVPE and MBE growth technologies for III-V epitaxial structures," *CS MANTECH Conference*, May 13–16, New Orleans, LA, 105–109, 2013.
17. F. Mendoza, V. Makarov, B. R. Weiner, and G. Morell, "Solar-blind field-emission diamond ultraviolet detector," *Applied Physics Letters*, 107, 201605, 2015.
18. X. Chen, Y. Xu, D. Zhou, S. Yang, F.-F. Ren, H. Lu, K. Tang, S. Gu, R. Zhang, Y. Zheng, and J. Ye, "Solar-blind photodetector with high avalanche gains and bias-tunable detecting functionality based on metastable phase α-Ga_2O_3/ZnO isotype heterostructures," *ACS Applied Materials & Interfaces*, 9(42), 36997–37005, 2017.
19. L.-X. Qian, Z.-H. Wu, Y.-Y. Zhang, P. T. Lai, X.-Z. Liu, and Y.-R. Li, "Ultrahigh-responsivity, rapid-recovery, solar-blind photodetector based on highly nonstoichiometric amorphous gallium oxide," *ACS Photonics*, 4(9), 2203–2211, 2017.
20. H. Chong, G. Wei, H. Hou, H. Yang, M. Shang, F. Gao, W. Yang, and G. Shen, "High-performance solar-blind ultraviolet photodetector based on electrospun TiO_2-$ZnTiO_3$ heterojunction nanowires," *Nano Research*, 8, 2822–2832, 2015.
21. G. Bao, D. Li, X. Sun, M. Jiang, Z. Li, H. Song, H. Jiang, Y. Chen, G. Miao, and Z. Zhang, "Enhanced spectral response of an AlGaN-based solar-blind ultraviolet photodetector with Al nanoparticles," *Optics Express*, 22, 24286–24293, 2014.
22. R. McClintock, P. Sandvik, K. Mi, F. Shahedipour, A. Yasan, C. Jelen, P. Kung, and M. Razeghi, "$Al_xGa_{1-x}N$ materials and device technology for solar blind ultraviolet photodetector applications," *Proceedings, Photodetectors: Materials and Devices* VI, 4288, 2001.
23. E. Cicek, R. McClintock, C. Y. Cho, B. Rahnema, and M. Razeghi, "Al_xGa_{1-xN}-based solar-blind ultraviolet photodetector based on lateral epitaxial overgrowth of AlN on Si substrate," *Applied Physics Letters*, 103, 181113, 2013.

24. E. Cicek, R. McClintock, A. Haddadi, W. A. G. Rojas, and M. Razeghi, "High performance solar-blind ultraviolet 320 × 256 focal plane arrays based on $Al_xGa_{1-x}N$," *IEEE Journal of Quantum Electronics*, 50, 593–597, 2014.
25. X. Chen, K. Liu, Z. Zhang, C. Wang, B. Li, H. Zhao, D. Zhao, and D. Shen, "A self-powered solar-blind photodetector with fast response based on Au/β–Ga_2O_3 nanowires array film schottky junction," *ACS Applied Materials & Interfaces*, *8*, 4185–4191, 2016.
26. Q. Zheng, F. Huang, J. Huang, Q. Hu, D. Chen, and K. Ding, "High-responsivity solar-blind photodetector based on $Mg_{0.46}Zn_{0.54}O$ thin film," *IEEE Electron Device Letters*, 33, 7, 2012.
27. T.-C. Wei, D.-S. Tsai, P. Ravadgar, J.-J. Ke, M.-L. Tsai, D.-H. Lien, C.-Y. Huang, and R.-H. Horng, "See-through Ga_2O_3 solar-blind photodetectors for Use in harsh environments," *IEEE Journal of Selected Topics in Quantum Electronics*, 20, 3802006, 2014.
28. E. Ozbay, N. Biyikli, I. Kimukin, T. Kartaloglu, T. Tut, and O. Aytür, "High-performance solar-blind photodetectors based on alxga1-xn heterostructures," *IEEE Journal of Selected Topics in Quantum Electronics*, 10, 742–751, 2004.
29. G. C. Hu, C. X. Shan, Nan Zhang, M. M. Jiang, S. P. Wang, and D. Z. Shen, "High gain Ga_2O_3 solar-blind photodetectors realized via a carrier multiplication process," *Optics Express*, 23, 13554–13561, 2015.
30. S. Oh, Y. Jung, M. A. Mastro, J. K. Hite, C. R. Eddy Jr., and J. Kim, "Development of solar-blind photodetectors based on Si-implanted β-Ga_2O_3, 2015," *Optics Express*, 23, 28300–28305, 2015.
31. X. Shi, Z. Yang, S. Yin, and H. Zeng, "Al plasmon-enhanced diamond solar-blind U.V. photodetector by coupling of plasmon and excitons," *Materials and Technology*, 31, 544–547, 2016.
32. D. Patil-Chaudhari, M. Ombaba, J. Y. Oh, H. Mao, K. H. Montgomery, A. Lange, S. Mahajan, J. M. Woodall, and M. S. Islam, "Solar blind photodetectors enabled by nanotextured β-Ga_2O_3 films grown via oxidation of GaAs substrates," *IEEE Photonics Journal*, 9, 1–10, 2017.
33. J. Xing, E. Guo, K.-J. Jin, H. Lu, J. Wen, and G. Yang, "Solar-blind deep-ultraviolet photodetectors based on an $LaAlO_3$ single crystal," *Optics Letters*, 34, 1675–1677, 2009.
34. B. Zhao, F. Wang, H. Chen, Y. Wang, M. Jiang, X. Fang, and D. Zhao, "Solar-blind avalanche photodetector based on single ZnO–Ga_2O_3 core–shell microwire," *Nano Letters*, 15, 3988–3993, 2015.
35. Chee Leong Tan and Hooman Mohseni, "Emerging technologies for high performance infrared detectors," *Nanophotonics*, 7(1), 169–197, 2018.
36. R. R. LaPierre, M. Robson, K. M. Azizur-Rahman, and P. Kuyanov, "A review of III–V nanowire infrared photodetectors and sensors," *Journal of Physics D: Applied Physics*, 50, 123001, 2017.
37. Gregory R. Savich, Daniel E. Sidor, Xiaoyu Du, and Gary W. Wicks, Mukul C. Debnath, Tetsuya D. Mishima, and Michael B. Santos, Terry D. Golding and Manish Jain, Adam P. Craig, and Andrew R. J. Marshall, "III-V semiconductor extended short-wave infrared detectors," *Journal of Vacuum Science and Technology*, 35(2), 02B105, 2017.
38. Hyunseok Kim, Haneui Bae, Ting-Yuan Chang, and Diana L. Huffaker, "III–V nanowires on silicon (100) as plasmonic-photonic hybrid meta-absorber," *Scientific Reports*, 11, 13813, 2021.

39. Jinshui Miao, Weida Hu, Nan Guo, Zhenyu Lu, Xingqiang Liu, Lei Liao, Pingping Chen, Tao Jiang, Shiwei Wu, Johnny C. Ho, Lin Wang, Xiaoshuang Chen, and Wei Lu, "High-responsivity graphene/inas nanowire heterojunction near-infrared photodetectors with distinct photocurrent on/off ratios," *Small*, 11, 936–942, 2015.
40. Xinzhe Wang, Dong Pan, Yuxiang Han, Mei Sun, Jianhua Zhao, and Qing Chen, "Vis–IR wide-spectrum photodetector at room temperature based on p–n junction-type $GaAs_{1-x}$Sbx/InAs core–shell nanowire," *ACS Applied Materials & Interfaces*, 11, 38973–38981, 2019.
41. Thomas Sprafke and James W. Beletic, "High-performance infrared focal plane arrays for space applications," *Optics and Photonics News*, 19, 22–27, 2008.
42. Sanjay Krishna, Darren Forman, Senthil Annamalai, Philip Dowd, Petros Varangis, Tom Tumolillo Jr, Allen Gray, John Zilko, Kathy Sun, Mingguo Liu, Joe Campbell, and Daniel Carothers, "Demonstration of a 320 × 256 two-color focal plane array using InAs/InGaAs quantum dots in well detectors," *Applied Physics Letters*, 86, 193501, 2005.
43. Paritosh Manurkar, Shaban Ramezani-Darvish, Binh-Minh Nguyen, Manijeh Razeghi, and John Hubbs, "High performance long wavelength infrared mega-pixel focal plane array based on type-II superlattices," *Applied Physics Letters*, 97, 193505, 2010.
44. K. C. Goma Kumari, H. Ghadi, D. R. M. Samudraiah, and S. Chakrabarti, "Indigenous development of 320 × 256 focal-plane array using InAs/InGaAs/GaAs quantum dots-in-a-well infrared detectors for thermal imaging," *Current Science*, 112, 1568–1573, 2017.

Chapter 10

Magnetic biosensors

Need and progress

Meruga Udaya and Pratap Kollu
University of Hyderabad, Hyderabad, India

CONTENTS

DOI: 10.1201/9781003288633-10

10.1 BASICS OF BIOSENSORS

Biosensing is the process of detecting biomolecules and biological phenomena. Essentially biomolecules are chemicals that are linked to life processes. A biosensor is an instrument for the identification of a particular biological analyte that includes a biological component along with a physicochemical transducer and detector. A nano biosensor is essentially a biosensor at the nanoscale size. Molecular recognition is crucial to biosensing. Typically a nano biosensor consists of a biological element that uses a nanomaterial property or nanoscale phenomenon to detect a biological phenomenon in the analyte and convert it into a form that is suitable for chemical or physical transduction [1].

A biosensor consists of biorecognition, transduction, and interface elements. A biological recognition element or bio-receptor is a molecule such as an enzyme, an aptamer, antigen (Ag), an antibody (Ab), a cell, a tissue or a DNA sequence. It could either be biological or synthetic in nature. This recognition element is the entity that latches onto the target molecule that we intend to detect, so it must be highly specific, measured by a parameter called selectivity. It must be stable under storage conditions, since biomolecules have a tendency to degrade. Since it is a molecule binding onto the transducer surface, it should possess a high affinity to the transducer surface, a process called immobilization. The physicochemical transducer is an interface that measures the changes in the physical property of the bioreceptor during the reaction and converts that energy into a quantifiable output, which is then amplified and analyzed. The signal from the transducer is converted to appropriate units and displayed.

There are numerous types of nanobiosensors, based on their transduction mechanism, such as electrochemical, mechanical, field-effect, optical, magnetic, and piezo types (Figure 10.1).

10.2 EXISTING TECHNOLOGIES OF BIOSENSING

Immunoassays are the most widely used technique for the detection of diseases. The immunoassay test is based on the utilization of an antibody that binds selectively to the target antigen. These techniques are traditionally used in diagnostic medical laboratories.

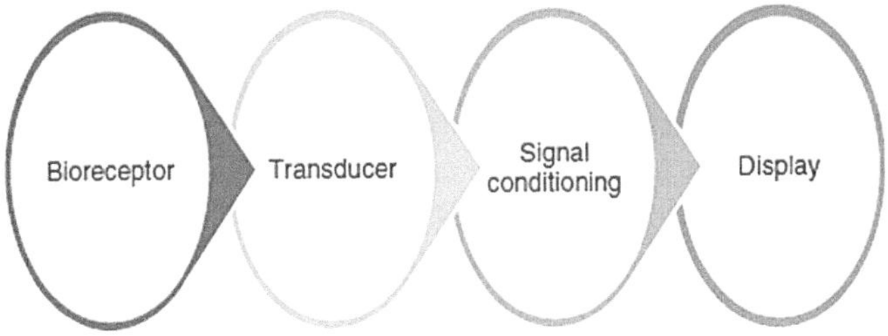

Figure 10.1 Schematic of a biosensor [3].

10.2.1 Hemagglutination Assay (HA)

The hemagglutination assay is a well-known method for detecting viruses. It is effective against viruses whose envelopes can connect to red blood cell surface molecules. This test depends on the fact that, under gravity, red cells sink to the bottom and form a button at the very lowest point of a round-bottomed jar. In the presence of sufficient amounts of the target virus, however, red blood cells form a shield. This occurs because the glycoproteins of the virus are capable of connecting to silac acids present on the outer membrane of the red blood cells, thereby cross-linking them, so that as the red blood cells fall, they form a lattice that spreads out at the bottom (Figure 10.2).

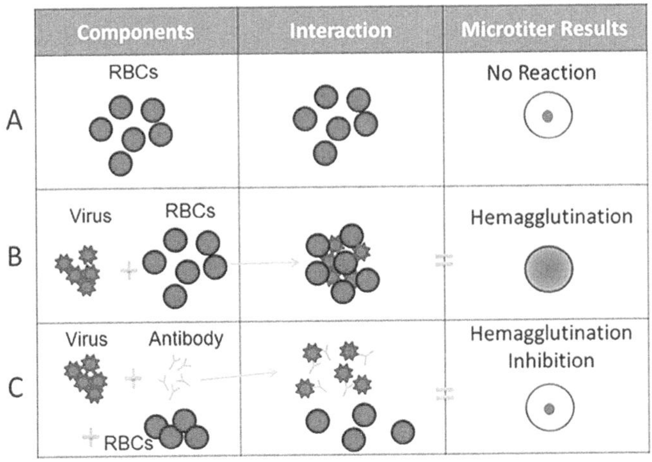

Figure 10.2 Interpreting results of hemagglutination (A) uninfected blood sample (B) sample infected with the virus but without antibodies (C) sample infected with the virus and has antibodies.

10.2.2 Enzyme-linked immunosorbent assay (ELISA)

ELISA is the method by which antibodies specific to proteins or an infection in solutions are identified. This method can also be used to determine its concentration of it. Specialized high protein-binding plates are coated with an excess of antibodies specific to the target protein; these are termed capture antibodies. When the analyte of interest is added, it binds to the capture antibody. It is immobilized on capture antibodies, while another protein-specific antibody, known as the detection antibody, is added. A third antibody binds to the detection antibody, which is tagged with an enzyme such as horseradish peroxidase (HRP). The presence of the analytes is then detected by the addition of enzyme substrates like tetramethyl benzidine, which, when exposed to HRP, changes from a colorless liquid to a blue-colored liquid and the intensity of the blue color will be directly proportional to the concentration of HRP, thus quantifying the analyte of interest (Figure 10.3).

10.2.3 Polymerase Chain Reaction (PCR)

PCR is a core technique in detecting several viruses, including HIV and COVID. Its operating principle lies in amplifying a specific region of DNA by using two primers that are complementary to the target of interest. This process

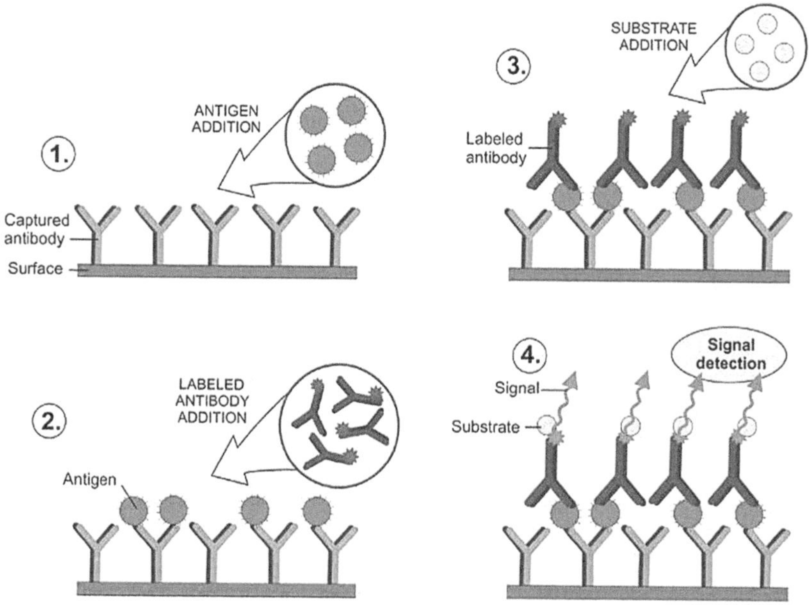

Figure 10.3 Schematic of sandwich ELISA (1) surface immobilized with capture antibodies (2) after addition of antibodies, they bind to capture antibodies. (3) protein specific detection antibodies are tagged with label bind with antigen (4) the labels are detected by addition of HRP enzyme.

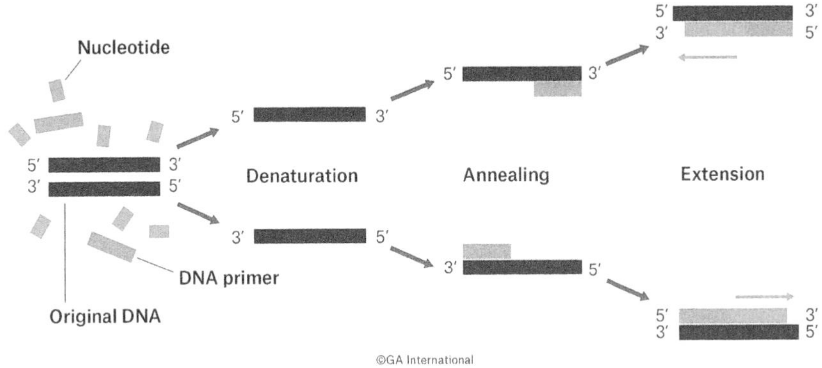

Figure 10.4 Steps for replicating DNA in PCR (a) Solution containing original DNA, DNA primer and nucleotide (b) Denaturation: splitting double-stranded DNA into two strands. (c) Annealing: DNA primer binding to denatured DNA. (d) Primer extending on both sides using dNTPs.

involves mixing the target DNA with a gene-specific primer or polymerase enzyme and the building blocks of DNA called dNTPs (Deoxynucleoside triphosphate), then cycling this mixture several times in stages through different temperatures. The mixture is heated to about 95°C to denature the double-stranded DNA into its constituent strands, then it is cooled to allow the annealing of the primers to the DNA template, followed by incubation to allow the polymerase enzyme to extend the primers on both strands, using the added dNTPs to create two new pieces of theater stranded DNA or amplicons. (An amplicon is a segment of chromosomal DNA that undergoes amplification and contains replicated genetic material) where before there was only one. Then, this process is repeated. Cyber green is added to the reaction that binds to DNA non-specifically. It becomes fluorescent only when it is bound to the DNA; therefore, the more amplicons in the sample, the more fluorescent it becomes. The fluorescence is quantified against a standard scale to determine the presence and concentration of the target (Figure 10.4).

10.3 ADVANTAGES AND LIMITATIONS OF THE EXISTING METHODS

Immunoassays can provide a rapid, simple, and cost-effective method of detection with equivalent or, in some situations, superior sensitivity and specificity than conventional methods. They are efficient in the detection of a wide variety of proteins and antibodies. However, they require expensive laboratory equipment with biosafety standards and well-trained professionals to handle and perform the procedures. Some procedures require a longer time to complete. For example, HA takes up to one hour, while PCR may take up to 24 hours to complete. Also, These techniques are ineffective while detecting at concentrations below 1 nMol/l. Detection at a fMol/l is essential for cancer detection, as early diagnosis can save lives. These limitations have prompted the investigation of other alternative methods, such as magnetic biosensors.

10.3.1 Characteristics of a biosensor

The intended application of a sensor dictates the design specification of a biosensor. However, other crucial aspects remain important to influencing the performance of these biosensors [2].

10.3.2 Selectivity

Selectivity is a biosensor's capacity to detect the desired target analyte from a sample comprising a mixture of undesired impurities. The selectivity of complex biomolecules is due to complementary three dimensionality, which is analogous to a lock and key system. Proteins, enzymes, and oligonucleotides all have a unique structure, owing to electronegativities of each functional group and the fact that a single C—C is free to rotate along the axis of the bond.

10.3.3 Reproducibility

Reproducibility refers to a biosensor's capacity to produce the same results every time an experiment is conducted. This is predominantly governed by the accuracy and precision of the transduction and electronic devices. The reproducibility of biosensor devices is crucial to the dependability of biosensor output.

10.3.4 Stability

Stability is the capacity of biosensors to remain unaffected by the environmental disturbances that may affect the expected output response during measurement. Despite the fact that accuracy and precision govern the reproducibility of biosensors, biosensor performance can also be compromised by stability. This is especially important in the design of biosensors, which demand a long working time or continual monitoring. Several variables, including temperature, membrane fouling, bioreceptor affinity, and aging can affect stability.

10.3.5 Sensitivity

Sensitivity refers to the minimum analyte concentration detectable by a biosensor. Sensitivity is a function of the geometry of the sensor and the transduction mechanism. It may range from nanomolar to femtomolar.

10.3.6 Linearity

Linearity is the precision of the output such that the concentration of the analyte is directly proportional to the output within a working range. In the fideal case, the linearity of a biosensor must be a straight line throughout the working range.

10.4 PRINCIPLES OF MAGNETIC BIOSENSING

The detection of a specific protein by a biosensor involves the formation of a three layered structure wherein two antibodies form a sandwich around the protein of interest. Two antibodies are called recognition molecules. They can be antibodies, aptamers, or nucleic acids, which are complementary to the target protein. Due to the complementary three-dimensional structure of antibody and antigen, which is analogous to a lock and key system, the analyte only binds with a specific antibody, with high selectivity. This allows the sensor to selectively capture and probe a single specific protein out of a large number of molecules in the biological solutions.

To construct a sensor, a capture antibody is directly immobilized onto the sensor surface. When the solution containing the analyte is exposed to the sensor, the capture antibody selectively immobilizes a target protein directly onto the surface. The second antibody, also known as the detection antibody, which has a tag attached, is then added to the sensor. The tag is an element that can be recognized by virtue of its fluorescent or magnetic or optical properties. When the detection antibody having the tag is delivered into the solution, it binds to the second epitope on the captured target protein. The tag, which is now immobilized over the protein, will alter either magnetic or electrical or optical properties of the sensor, depending on the type of tag. This change in the physical property can be detected by appropriate sensing elements to determine the presence and concentration of the target protein. A magnetic sense element such as a GMR sensor detects the field created by the magnetic NPs. Therefore, the higher the protein concentration that is in the sample, the more the detection antibodies bind to the target protein, and the more the magnetic tag accumulates on the sensor. As the concentration of magnetic tags increases on the sensor, the magnetoresistance (MR) of the transducing GMR sensor changes. The change in MR can be measured and interpreted. Thus, proteins can be quantitatively detected by biosensors (Figure 10.5).

10.4.1 Magnetic nanoparticles

Recently, much work has been done towards the development of MNPs based biosensing applications. MNPs have several advantages, including reduced production costs. They can be synthesized in a variety of sizes.

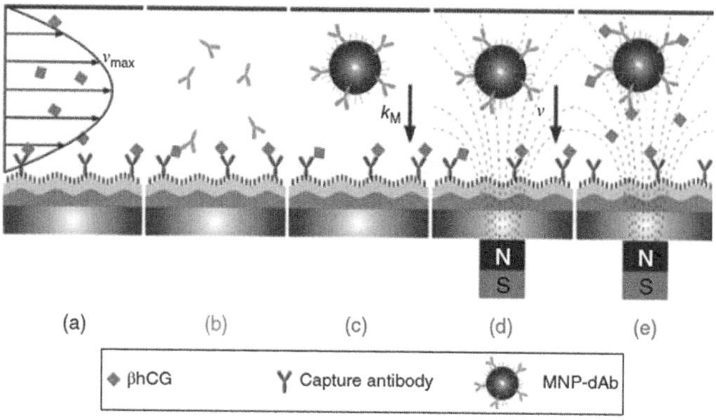

Figure 10.5 Types of magnetic biosensing (a) direct detection where the target β-human chorionic gonadotropin antigen (βhCG), which binds onto capture antibody (b) sandwich assays where the antigen binds with the capture antibody on one side and detection antibody on the other (c) MNP functionalized with antibody (MNP-dAb) is used without applied magnetic field (d) MNP functionalized with antibody is used as a magnetic field (e) Detection using pre-incubated MNP-dAb with βhCG followed by sandwich assay placed in magnetic field [2].

The 10–20 nm MNPs appear to garner the most interest because of their superparamagnetic properties. There are several uses for MNP-based biosensors, including medical applications, food research, and environmental monitoring. However, biomedical research necessitates additional concerns. They are as follows:

- MNP size should be between 10 and 50 nm in order to preserve their efficiency and provide a wider surface area for colloidal stability and to avoid aggregation. At this scale, it is stable in water at pH 7.7, and it is also feasible to prevent MNPs from dissolving in the water. It also prevents precipitation due to the gravitational influence.
- MNPs should be nontoxic, biocompatible, and nonhazardous.
- The saturation magnetization of MNPs should be high so that a moderate external magnetic field can control the path of MNP in blood and bring it into proximity to the desired tissue.

Uncoated MNPs often agglomerate rapidly and must thus be coated. Several coatings, including surfactants, polymers, silica, gold, and rare earth metals have been utilized. Recent interest in their application in the biological and biosensing domains is largely attributable to their small size and ease of conjugation with biological molecules.

Rather than relying solely on diffusion, the position of MNPs can be controlled by electromagnets. This technique also proves efficient for drug delivery. MNPs tagged with antibodies can be attached to their antigen in serum, and saliva media. A first electromagnet is used to attract MNPs toward a sensing surface like a surface plasma resonance chip, carrying a secondary antibody, and a second electromagnet is then used to wash away the unattached MNPs. This combination of MNPs with optical detection approaches are known to be effective.

10.4.2 Classification of MNPs based on their magnetic nature

These nanostructures may be broken down into five distinct categories: (1) ferromagnetic, (2) antiferromagnetic, (3) diamagnetic, (4) ferrimagnetic, and (5) paramagnetic. Ferromagnetic materials, which may be obtained from Fe, Co, or Ni, are employed the most often. These MNPs have a net magnetic moment because of the unpaired electrons, and they are made up of domains, each of which has several atoms with parallel magnetic moments, which results in a net moment in the domain. The net moments of the domains are arranged in a random manner. However, in the event that the ferromagnetic particle is positioned within a magnetic field, the alignment of the magnetic moments of each domain will be in the same direction as the external magnetic field, which will result in a significant increase in the total net magnetic moment. However, magnetic domains do not occur in paramagnetic materials. This is because the atoms in these materials have a net magnetic moment owing to the presence of unpaired electrons. In the event of exposing the paramagnetic materials to a magnetic field, alignment of the magnetic moments of the atoms occurs such that it is parallel to a magnetic field, which results in a weak net magnetic moment. Diamagnetic materials like copper(Cu) and silver(Ag) are another class of MNPs. Materials are said to be diamagnetic if they may be magnetized freely when they are placed in a magnetic field. However, the magnetization is in the direction that is opposite to the direction that the magnetic field is pointing. Diamagnetic materials do not have any unpaired electrons in their atoms, they have zero net magnetic moments. As a result, diamagnetic MNPs show a very weak response towards the external magnetic field because the electron orbits realign under the applied magnetic field. In the absence of the magnetic field, they do not exhibit any signs of the magnetic moment. Ferrimagnetic materials also consist of distinct atoms like Fe_3O_4 and γ-Fe_2O_3 which have different lattice locations in their atoms, leading to antiparallel magnetic moments. However, the magnetic moments do not cancel each other out, which results in a net spontaneous magnetic moment. The antiferromagnetic materials, such as MnO, NiO, $CuCl_2$, and CoO, occupy distinct locations in the crystal lattice, and both of these atoms have magnetic moments. Though magnetic moments are equal, they are opposite in direction, which results in a net magnetic moment that is zero. In a magnetic field, the behavior of ferrimagnetic and antiferromagnetic compounds is analogous to that of ferromagnetic substances. Magnetism is measured with the use of a magnetometer, which monitors the magnetization of a substance with respect to an applied magnetic field. The characterization of MNPs is usually done with techniques such as SEM, TEM, AFM, NSOM.

10.4.3 Need for magnetic nanosensors

Although immunoassay tests are widely used due to their effectiveness, they require biosafety laboratories and expensive instruments. The equipment used are often stationary, and patient samples must be sent to the lab for analysis. Trained professionals are required to perform the procedures. This prohibits the performance of quantitative immunoassay testing outside of a laboratory setting. On the other hand, magnetic nanosensors are sized in the nanoscale which enables thousands of sensors to be integrated into a single chip. These implications point to reliable lab-of-chip devices for point-of-care applications.

MNP-based Magnetic nanosensors have many advantages, like low cost, variety, flexibility, and simplicity. They can be synthesized in a variety of sizes and are easily functionalized. Due to their size, MNPs are biocompatible. They can easily bind to molecules such as proteins, antibodies, and enzymes. Functionalization of nanoparticles is particularly crucial to prevent the agglomeration of nanoparticles and tailor their physical and optical properties [3]. MNPs can easily be functionalized with various molecules and functional groups to achieve desired properties for different applications. Most sensors use MNPs having a size of 10–20 nm, which are superparamagnetic, i.e. they quickly respond to external magnetic fields, enabling Magnetic nanosensors to be employed for real-time detection due to their quick response time and high sensitivity.

While other sensing methods rely on the particle to randomly diffuse through the medium until it reaches the sensor surface, magnetic biosensors are able to accelerate the binding of the target molecule onto the receptor by pulling the MNPs in an external magnetic field, allowing the particles to be moved to the sensor surface rapidly, thus reducing the response time. Also, MNP based sensing offers a high signal to noise ratio [4–6].

10.5 TYPES OF BIORECOGNITION ELEMENTS

10.5.1 Enzymatic biosensors

Leyland Clark et al. [7] developed one of the early biosensors by coating an oxygen electrode with a thin film of the glucose oxidase enzyme and a dialysis membrane. Enzymes are ideal for use in biosensors due to their high selectivities. Glucose oxidase interacts exclusively with glucose and has no effect on other sugars. Enzymes have rapid substrate turnovers due to their high catalytic activity. It is an essential characteristic because, otherwise, they would quickly get saturated and would no longer be able to create enough active species for detection. However, they have downsides, such as the possibility that there is no suitable enzyme for the target of interest. Some enzymes are also difficult and costly to extract in significant amounts, unstable, fast degrading, and become ineffective after some time. They are also susceptible to poisoning by numerous species [8, 9].

10.5.2 Biosensors using DNA and RNA

DNA is the blueprint for protein synthesis and can be seen as a molecular data storage device. It also serves as a messenger between DNA and the ribosomes that make proteins. RNA has a vast array of biological functions. A single oligonucleotide chain is typically attached to an appropriate transducing element, such as an electrode, SPR (surface plasmon resonance) chip, QCM (quartz crystal microbalance), etc., when exposed to a solution consisting of an oligonucleotide strand of interest. The surface bound oligonucleotide is chosen such that it is complementary to the oligonucleotide of interest, and the recognition event involves sequence specific hybridization between the binding and solution strands [10, 11].

10.5.3 Biosensors using antibody

Antibodies are naturally occurring Y-shaped proteins present in living organisms to protect it from invading germs, viruses, and bacteria. They attach to specific antigens with an extraordinarily high specificity via a combination of hydrogen bonds and other noncovalent interactions. An advantage of antibodies is that they can be harvested from laboratory animals. The animal's natural defense response is to produce antibodies against the specific antigen. They can be extracted from animals. Once the antibody has been created, it can be immobilized onto a transducer surface to create a biosensor. The limitation when using antibodies is that when they attach to their antigens to create a complex, byproducts such as electrons and redox active species that are difficult to quantify are formed [12–14].

10.5.4 Aptasensors

Aptamers are a type of oligonucleotides that can attach to a variety of substrates, including peptides, proteins, medicines, and cells. Typically, when an aptamer binds to its targets, the aptamer undergoes conformational changes. Aptamers are ideally suited for sensing applications, since such structural changes are simple to detect. They have additional benefits compared to other recognition elements because, unlike enzymes, they can be synthesized in vitro, without an animal host, and with a high degree of specificity and selectivity towards almost any target, whether a tiny molecule or peptides [15–18]. Once the appropriate aptamer for a specific target has been identified, it may be commercially manufactured in its purest form and frequently exhibits more stability than other biomolecules, so is named "chemical antibody" (Figure 10.6).

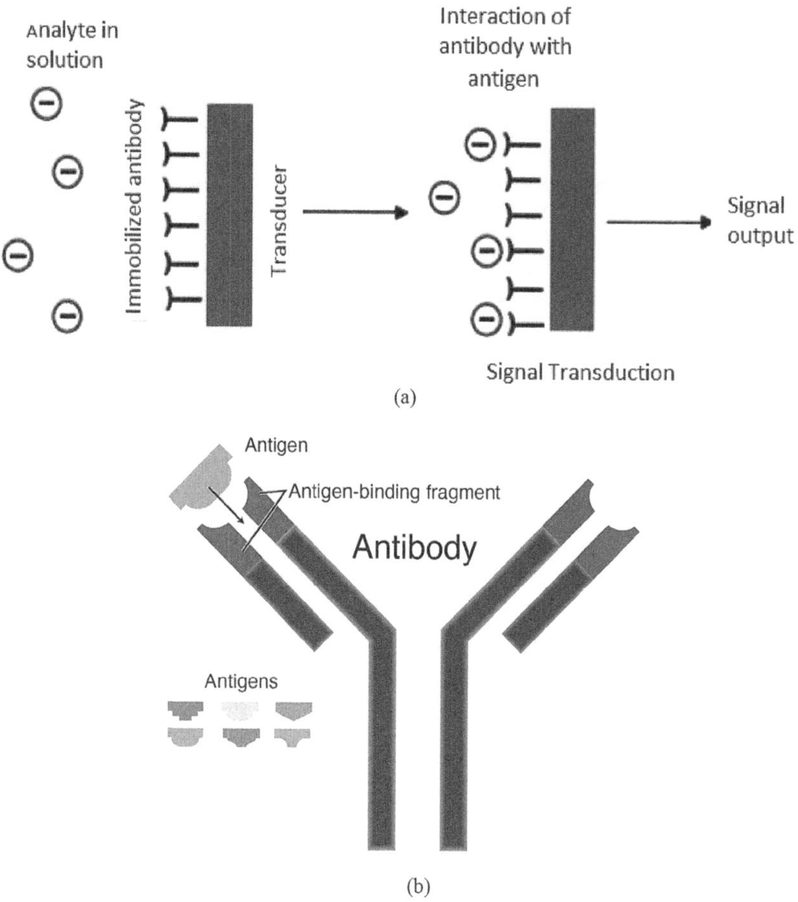

Figure 10.6 (a) Transduction mechanism (b) structure of antibody.

10.5.5 Peptide based molecular sensors

Peptides are polymers that can occur naturally or be synthesized from amino acids and share the same structural components as proteins. Like proteins, peptides with the proper amino acid sequence can bind to targets with high selectivity and specificity. Short peptides provide a variety of advantages over long proteins, including greater conformational and chemical stability and a significantly lower propensity for denaturation. In addition, they may be synthesized with precise sequences by solid phase synthesis procedures, and labeling groups can be easily changed without impacting their activity. Peptide are tagged on one or both ends with fluorescent groups for several applications [19].

10.6 PRINCIPLES OF MAGNETIC NANOSENSORS

10.6.1 Magnetoresistance

Magnetoresistance is the property of a material that refers to the occurrence of a change in resistance when kept in a magnetic field, measured as magnetoresistance ratio. It is expressed in terms of maximum resistance and minimum resistance. MR ratio can exceed 100%

$$\text{Magne-toresistance ratio} = MR\% = \frac{R_{max} - R_{min}}{R_{min}} \times 100$$

10.6.2 The Hall effect sensors

The Hall effect is the phenomenon in which a voltage difference occurs across a current carrying electrical conductor that is exposed to a orthogonal magnetic field. The induced electrical potential is perpendicular to both the current flow and the magnetic field. Consider a conductive plate kept in a magnetic field B (B = μH), perpendicular to its surface in Z-direction and current 'I' flowing along its length in X-direction, then according to Lorentz law, an electron traveling in X-direction with a certain velocity and placed in a perpendicular magnetic field experiences a force and the direction perpendicular to both velocity and magnetic field. This forces the path of electrons to curve in negative Y-direction according to Fleming's left hand rule. This causes a voltage V_H to build along Y-direction (Figure 10.7).

Hall voltage is given by $V_H = \frac{IB}{nte}$, while Hall coefficient is given by $R_H = \frac{E}{JB}$.

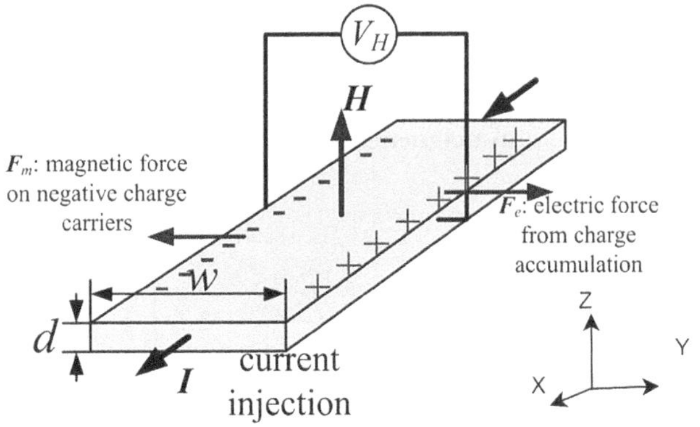

Figure 10.7 Description of Hall effect.

Where V_H is the voltage induced due to Hall effect, "I" is current flowing through the conductor, $B = \mu H$ is the magnetic field, "n" is the carrier density in the conductor, "t" is the thickness of conductor and "e" is the charge of the electron. Hall coefficient R_H is related to electric field "E" and current density "J".

10.6.3 Anisotropic Magnetoresistance Sensors (AMR)

A magnetoresistive effect called anisotropic magnetoresistance (AMR) is seen in ferromagnetic materials [20]. Resistance of the ferromagnetic material changes as a current is passing through a conductor that is placed in magnetic field. This change depends on the angle between the current flow and the magnetization of the current carrying conductor. The direction of the magnetization is controlled by the applied magnetic field. AMR occurs in all metallic magnetic materials. The material exhibits maximum resistivity for current flowing along the direction of magnetization and minimum resistivity for current flowing orthogonally.

A material deposited as a thin film has only one easy axis of magnetization in the plane of the film. So, in a stripe made from this film, the current flows in a direction parallel to the magnetization, resulting in maximum resistance. But when a magnetic field is applied at right angles to the easy axis, the vector of magnetization rotates around its original position at some angle. Since the probability of electron scattering is maximum along the direction of magnetization and the magnetization vector has shifted from its parallel alignment with the current flow direction, therefore a decrease in resistance of the film is observed. Hence the film can be used as a sensor for a perpendicularly applied magnetic field. The maximum resistance is observed when the direction of magnetization is parallel to that of the current as represented in Figure 10.8.

10.6.4 Giant Magnetoresistance Sensors (GMR)

GMR is another magnetoresistance that is observed in a layered structure consisting of two or more ferromagnetic metal (like Fe, CoFe) layers which are separated by thin, non-magnetic metal layers, like Cr, Ru, called spacer

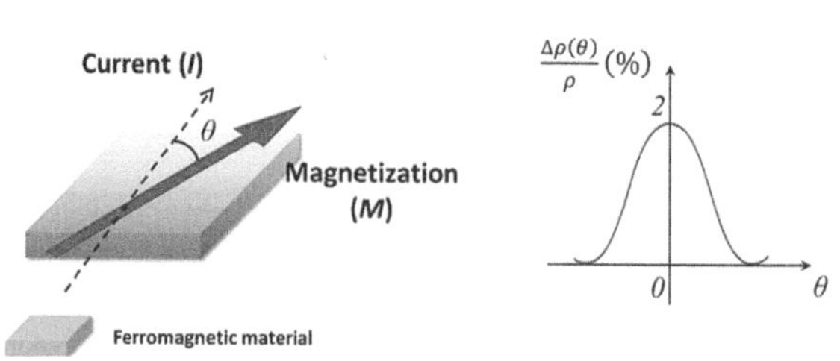

Figure 10.8 Phenomenon of Anisotropic magnetoresistance.

layers. The spacer layers must be less than a micrometer in thickness [21]. When a magnetic field is applied to a GMR structure, the electric current is greatly influenced by the orientation of the magnetization of each magnetic layer. This results in a huge reduction in the resistance of the multilayered structure, which can reach up to 50% at room temperature.

This effect was discovered to be significantly greater than all other magnetoresistive effects, and was so named "giant magnetoresistance."

At low temperatures, GMR can exceed 100% in Fe/Cr and Co/Cu layers. The latest GMR structures can attain MR values up to 200% at room temperature. The two fundamental GMR effect configurations are: current-in-plane (CIP) where the current is parallel to ferromagnetic layers and current-perpendicular-to-plane (CPP) where the current is perpendicular to ferromagnetic layers.

GMR can be understood qualitatively through the Mott model, which describes the conductivity of metals in terms of two largely separate conducting channels, related to spin-up and spin-down electrons. Regardless of the nature of the scattering centers, the scattering rates of spin-up and spin-down electrons are significantly distinct.

The number of spin up and spin down electrons in all energy bands is equal in nonmagnetic conductors but different in ferromagnetic conductors. According to quantum physics the probability of an electron scattering while passing through a ferromagnetic conductor relies on the nature of the electron spin. Electrons with a positive spin move freely and cover greater distances without being scattered.

In one of the layers, spin-up and spin-down electrons are substantially dispersed because of their antiparallel spin. Therefore, overall resistance of the multilayer is high in this instance. Consequently, both spin up electrons and spin down electrons encounter high resistance in one layer and low resistance in the other layer, Therefore the overall resistance is

$$R_{\text{Antiparallel}} = \left(\frac{1}{2}\right)\left(R_{\uparrow} + R_{\downarrow}\right)$$

In an anti-parallelly aligned multilayer, spin up and spin down electrons are strongly scattered in one of the ferromagnetic layers because of antiparallel magnetization direction in one of the layers (Figure 10.9).

For parallelly aligned magnetic layers, spin-up electrons starting from one layer traverse the structure with minimal scattering, as its spin is parallel to the magnetization of the layers. Low resistance is experienced by spin-up electrons, while high resistance is experienced by spin-down electrons. The total resistance is given as

$$R_{\text{parallel}} = \frac{2R_{\uparrow}R_{\downarrow}}{R_{\uparrow} + R_{\downarrow}}$$

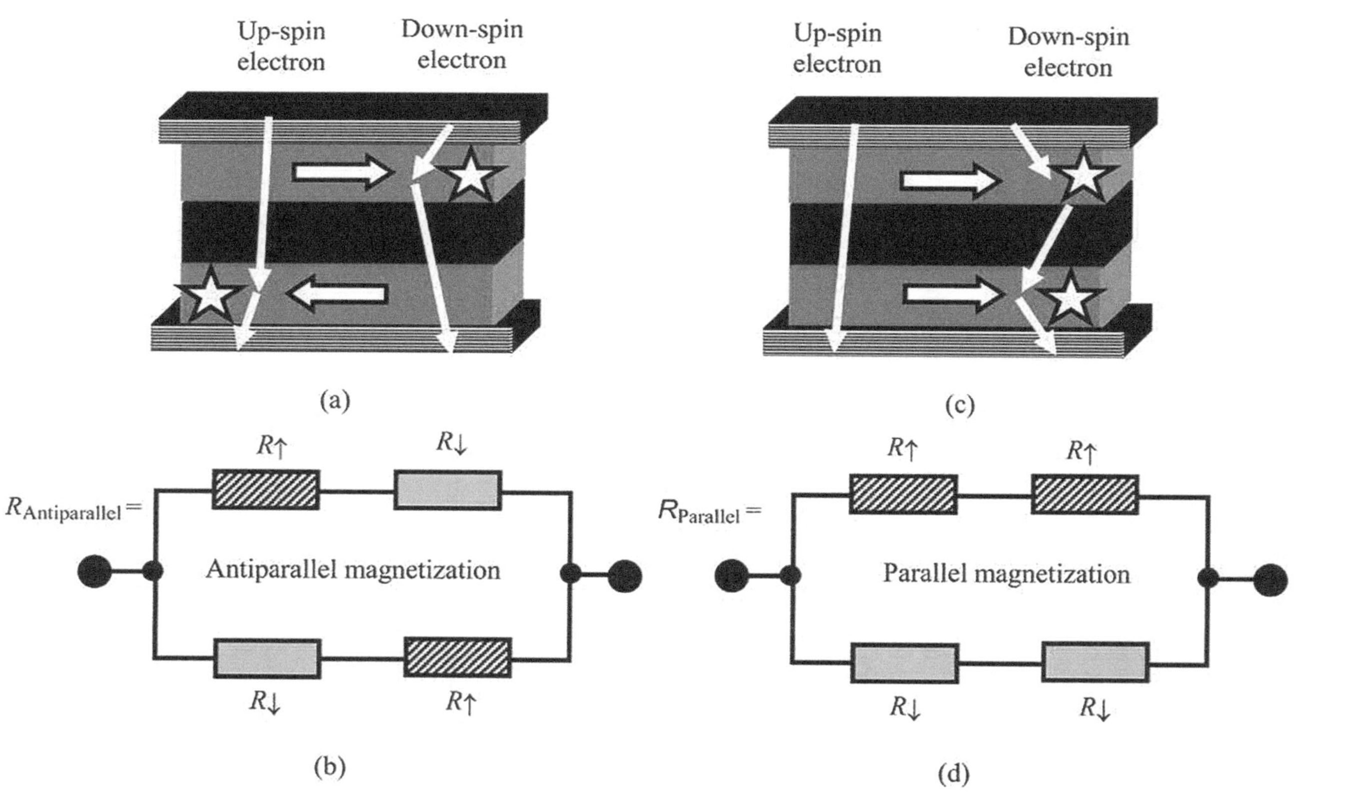

Figure 10.9 Giant magnetoresistance in antiparallel alignment and parallel alignment [1].

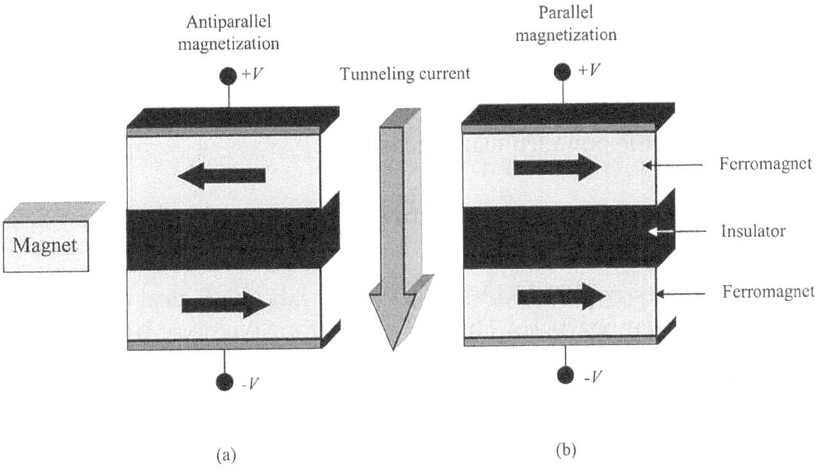

Figure 10.10 TMR in parallel and antiparallel alignment [1].

10.6.5 Tunneling Magneto Resistance (TMR)

TMR devices have structure as that of GMR devices except that they use a very thin insulating layer instead of a conductor to separate two magnetic layers. This layer's thickness is relatively thin so that electron tunneling can occur. Using an Al_2O_3 barrier approximately 1 nm thick, sensor prototypes are manufactured [1]. Both electrodes are composed of a ferromagnetic metal. Quantum-mechanical tunneling allows electrons to move from one layer to the next through an insulator. TMR devices utilize the multilayer configuration of a magnetic layer that is pinned and a magnetic layer that is free.

Figure 10.10 depicts the configuration of TMR. A very thin insulating barrier separates a fixed ferromagnetic plate and a free ferromagnetic plate. The angle between the magnetization of the top and bottom magnetic layers modulates the probability that an electron will tunnel between them. The two alternative configurations are designated as parallel and antiparallel. When the magnetization of the layers is parallel, the bottom layer contains many states that can accommodate the spin-polarized electrons that tunnel through from the top layer. However, if the directions of magnetization are antiparallel, electrons cannot tunnel because they lack the proper orientation to enter the bottom layer. This type of tunneling is known as spin-dependent tunneling (SDT).

10.6.6 Magnetic nanosensors for biomedical applications

Magnetic biosensors have been around since 1996. Baslet et al. describe a bead array counter method for detecting and characterizing the interactions between biomolecules, such as, antibody–antigen or DNA–DNA interactions [22]. The method is based on AFM, where magnetic beads are used instead of AFM. The target DNA or antigen bridges both the substrate and the magnetic bead. Magnetic beads settle on the substrate bonding to it by a single target molecule. Then, an external magnetic field is used to remove unbounded and non-specifically bound beads. A magnetoresistance sensor is used to sense the bounded magnetic beads.

10.6.7 Magnetic Relaxation Immunoassays (MARIA) using Fluxgate Sensor

Zitchoff et al. designed MARIA sensors which label biological targets using supermagnetic nanoparticles. Fluxgate magnetorelaxometry is used for the measurement of reaction kinetics. Binding assays with the streptavidin–biotin (StAv) is used to detect two types of analytes. It was known that the use of 10 ng StAv–MNPs enables a lower detection limit of biotin agarose. The amount of bound StAv–MNPs is directly proportional to biotin agarose. The limit of detection of 14–71.3 mmol/L was obtained.

10.6.8 Planar Hall Magnetoresistive (PHR) aptasensor for thrombin detection

B. Sinha et al. proposed a sensitive aptamer based sandwich type sensor for detecting human thrombin employing a PHR sensor with superparamagnetic labeling [23]. It can function as a high-resolution biosensor due to its high SNR, linear response, and low offset voltage. Prior to the binding of the isolated DNA primary aptamer, the sensor in this device has an active area of 50 m × 50 m with a 10-nm gold layer placed on the sensor surface. A 600-m-diameter and 1-mm-tall polydimethylsiloxane well was fabricated around the sensor's surface to retain the specific area and volume for each sensor. By adding streptavidin-functionalized magnetic labels to the sensor, the response of the sensor could be monitored in real time. The detection limit of 86 pM was reached. The proposed aptasensor has significant diagnostic application potential.

10.6.9 Anisotropic Magnetoresistance (AMR) biosensors

Zing et al. describe a lab-on-a-chip biosensing device consisting of AMR sensors designed to detect single magnetic beads and electrodes for manipulating and sorting the beads, all of which are integrated into a microfluidic channel [24]. Beads with various magnetic moments are sorted into different microfluidic channels by applying a magnetic field gradient.

10.6.10 Eigen Diagnosis Platform (EDP) using giant magnetoresistance biosensors

J Choi et al. describe the EDP device, which consists of three primary components: a disposable cartridge, the reader station, and smartphone interfacing. The GMR biosensor chip contains an array of 8 × 8 GMR sensors kept inside the cartridge. When the cartridge is inserted into the reader station, the reader station detects the magnetic sensors on the cartridge and records the electrical resistance in each sensor in real-time from the multiplexed assay of human immunoglobulin IgG and IgM antibodies. EDP was demonstrated to exhibit sensitivities as low as 0.33 nanomolar [25].

10.6.11 Tunneling Magnetoresistance (TMR) biosensors

Wu et al. described magnetic biosensors replacing the typical "sandwich" design with an electromagnetic trap. The device demonstrates the detection of *E. coli* [26]. A current-carrying microwire forms the trap by attracting magnetic beads into a detecting area of a TMR sensor. A signal that is proportional to the size of beads in the sensing space is produced. Since biological targets increase the size of the beads, it facilitates their detection. Individual *E. coli* bacteria within the sensing space were shown to be detectable using superparamagnetic beads with a diameter of 2.8 µm. The electromagnetic trap reduces the detection process to two steps. First, bacteria are mixed with magnetic beads. Secondly, a sample solution is applied to the sensor for measurement, which can be completed in approximately 30 minutes for a µl of the sample. This biosensor can be easily cleaned and reused.

10.7 CONCLUSION

MNPs' unique and smart properties make them useful in biological research, preclinical, and clinical settings. Surface modification chemicals offer exciting prospects for biosensor-based illness diagnosis, cancer screening, and real-time biomonitoring applications. Their usage in biosensors offers sensitive, reproducible, and reliable detection of cancer, neurological illnesses, and cardiac issues.

Among all magnetic sensing technologies, the Hall sensor exhibits high linearity. Although GMR and TMR sensors have high sensitivity, they are nonlinear, and this drawback limits their use as a replacement of Hall sensors for a large number of applications where sensitivity is not crucial. Hysteresis is also a degrading factor for the sensitivity of GMR sensors. Since effective biosensing requires operating at extremely small analyte concentrations, sensors must be designed to overcome this problem.

Also, the reduction of the magnetic noise is of utmost priority while designing GMR and TMR sensors because at low frequencies, the flicker noise is more dominant compared to the thermal noise. However, compared to the Hall effect and AMR sensors, GMR sensors have many advantages. They allow for large magnetic field-dependent changes in resistance, a larger output signal, and the ability to have a larger air gap (i.e., distance from the sensor to the target analyte). GMR devices are not susceptible to damage from large magnetic fields.

TMR magnetic sensors have a number of attractive properties, such as a pronounced spin-tunneling effect; it gives sensitivity in the range of pico tesla. Compared to AMR sensors using similar materials, TMR magnetic sensors offer higher operational temperature and frequency ranges. Extremely small TMR devices, with high resistance, can be fabricated using photolithography, allowing very dense packing of magnetic sensors. This could allow for multiple analytes to be sensed within a smaller chip area for point-of-care devices.

REFERENCES

1. Vinod Kumar Khanna, Nanosensors: Physical, chemical, and biological', *Contemp. Phys.*, vol. 53, no. 4, pp. 207–215, 2012. doi: 10.1080/00107514.2012.689351.
2. Zeynep Altintas, *Biosensors and nanotechnology: Applications in health care diagnostics*, 2017, p. 3, 226. John Wiley & Sons: Hoboken, NJ.
3. D. Saini, 'Synthesis and functionalization of graphene and application in electrochemical biosensing', *Nanotechnol. Rev.*, vol. 5, no. 4, pp. 393–416, Aug. 2016, doi: 10.1515/ntrev-2015-0059.
4. V. Urbanova, M. Magro, A. Gedanken, D. Baratella, F. Vianello, and R. Zboril, 'Nanocrystalline iron oxides, composites, and related materials as a platform for electrochemical, magnetic, and chemical biosensors', *Chem. Mater.*, vol. 26, no. 23, pp. 6653–6673, 2014.
5. M. Holzinger, A. Le Goff, and S. Cosnier, 'Nanomaterials for biosensing applications: a review', *Front. Chem.*, vol. 2, 2014, https://www.frontiersin.org/article/10.3389/fchem.2014.00063 (accessed: Jun. 02, 2022).
6. M. Martín, P. Salazar, R. Villalonga, S. Campuzano, J. M. Pingarrón, and J. L. González-Mora, 'Preparation of core–shell Fe3O4@poly(dopamine) magnetic nanoparticles for biosensor construction', *J. Mater. Chem. B*, vol. 2, no. 6, pp. 739–746, Jan. 2014, doi: 10.1039/C3TB21171A.
7. L. C. Clark Jr. and C. Lyons, 'Electrode systems for continuous monitoring in cardiovascular surgery', *Ann. N. Y. Acad. Sci.*, vol. 102, no. 1, pp. 29–45, 1962, doi: 10.1111/j.1749-6632.1962.tb13623.x.
8. I. García, J. Gallo, N. Genicio, D. Padro, and S. Penadés, 'Magnetic glyconanoparticles as a versatile platform for selective immunolabeling and imaging of cells', *Bioconjug. Chem.*, vol. 22, pp. 264–273, Feb. 2011, doi: 10.1021/bc1003923.
9. A Razgulin, 'Strategies for in vivo imaging of enzyme activity: an overview and recent advances', *Chem. Soc. Rev.* (RSC Publishing). https://pubs.rsc.org/en/content/articlelanding/2011/cs/c1cs15035a (accessed Jun. 02, 2022).
10. T. Tanaka, R. Sakai, R. Kobayashi, K. Hatakeyama, and T. Matsunaga, 'Contributions of phosphate to DNA adsorption/desorption behaviors on aminosilane-modified magnetic nanoparticles', *Langmuir ACS J. Surf. Colloids*, vol. 25, no. 5, pp. 2956–2961, Mar. 2009, doi: 10.1021/la8032397.
11. M. Uhlen, 'Magnetic separation of DNA', *Nature*, vol. 340, no. 6236, Art. no. 6236, Aug. 1989, doi: 10.1038/340733a0.

12. J. Lee et al., 'Multifunctional magnetic gold nanocomposites: Human epithelial cancer detection via magnetic resonance imaging and localized synchronous therapy', *Adv. Funct. Mater.*, vol. 18, no. 2, pp. 258–264, Jan. 2008, doi: 10.1002/adfm.200700482.
13. S. Ryan et al., 'Single-domain antibody-nanoparticles: promising architectures for increased *staphylococcus aureus* detection specificity and sensitivity', *Bioconjug. Chem.*, vol. 20, no. 10, pp. 1966–1974, Oct. 2009, doi: 10.1021/bc900332r.
14. S. Puertas et al., 'Taking advantage of unspecific interactions to produce highly active magnetic nanoparticle–Antibody conjugates', *ACS Nano*, vol. 5, no. 6, pp. 4521–4528, Jun. 2011, doi: 10.1021/nn200019s.
15. 'Aptamer–nanoparticle complexes as powerful diagnostic and therapeutic tools'. *Exp. Mol. Med.*, https://www.nature.com/articles/emm201644 (accessed Jun. 02, 2022).
16. K. Saha, S. S. Agasti, C. Kim, X. Li, and V. M. Rotello, 'Gold nanoparticles in chemical and biological sensing', *Chem. Rev.*, vol. 112, no. 5, pp. 2739–2779, May 2012, doi: 10.1021/cr2001178.
17. M. K. Yu, D. Kim, I.-H. Lee, J.-S. So, Y. Y. Jeong, and S. Jon, 'Image-guided prostate cancer therapy using aptamer-functionalized thermally cross-linked superparamagnetic iron oxide nanoparticles', *Small Weinh. Bergstr. Ger.*, vol. 7, no. 15, pp. 2241–2249, Aug. 2011, doi: 10.1002/smll.201100472.
18. T. Chen et al., 'Smart multifunctional nanostructure for targeted cancer chemotherapy and magnetic resonance imaging', *ACS Nano*, vol. 5, no. 10, pp. 7866–7873, Oct. 2011, doi: 10.1021/nn202073m.
19. D. Sosnovik and R. Weissleder, 'Magnetic resonance and fluorescence based molecular imaging technologies', in *Imaging in Drug Discovery and Early Clinical Trials*, P. L. Herrling, A. Matter, and M. Rudin, eds. Basel: Birkhäuser, 2005, pp. 83–115. doi: 10.1007/3-7643-7426-8_3.
20. D. R. Baselt, 'A biosensor based on magnetoresistance technology'. 1998. https://pubmed.ncbi.nlm.nih.gov/9828367/
21. I. Ennen, D. Kappe, T. Rempel, C. Glenske, and A. Hutten, 'Giant magnetoresistance: basic concepts, microstructure, magnetic interactions and applications', *Sensors*, vol. 16, p. 904, Jun. 2016, doi: 10.3390/s16060904.
22. E. Heim, F. Ludwig, and M. Schilling, 'Binding assays with streptavidin-functionalized superparamagnetic nanoparticles and biotinylated analytes using fluxgate magnetorelaxometry', *J. Magn. Magn. Mater.*, vol. 321, no. 10, pp. 1628–1631, May 2009, doi: 10.1016/j.jmmm.2009.02.101.
23. B. Sinha, T. Ramulu, K. Kim, R. Venu, J. Lee, and C. G. Kim, 'Planar Hall magnetoresistive aptasensor for thrombin detection', *Biosens. Bioelectron.*, vol. 59C, pp. 140–144, Mar. 2014, doi: 10.1016/j.bios.2014.03.021.
24. Z. Jiang, J. Llandro, T. Mitrelias, and J. A. C. Bland, 'An integrated microfluidic cell for detection, manipulation, and sorting of single micron-sized magnetic beads', *J. Appl. Phys.*, vol. 99, p. 08S105, Apr. 2006, doi: 10.1063/1.2176238.
25. J. Llandro, J. J. Palfreyman, A. Ionescu, and C. H. W. Barnes, 'Magnetic biosensor technologies for medical applications: a review', *Med. Biol. Eng. Comput.*, vol. 48, no. 10, pp. 977–998, Oct. 2010, doi: 10.1007/s11517-010-0649-3.
26. Y. Wu, Y. Liu, Q. Zhan, J. P. Liu, and R.-W. Li, 'Rapid detection of *Escherichia coli* O157:H7 using tunneling magnetoresistance biosensor', *AIP Adv.*, vol. 7, no. 5, p. 056658, May 2017, doi: 10.1063/1.4977017.

Index

Page numbers in *italics* refer figures and those in **bold** refer tables.